10.5.77

Textile Yarns

TEXTILE YARNS

Technology, Structure, and Applications

B. C. GOSWAMI

Textile Research Institute
Princeton, New Jersey

J. G. MARTINDALE

Galashiels College of Textiles
Galashiels, Scotland

F. L. SCARDINO

Philadelphia College of Textiles and Science
Philadelphia, Pennsylvania

A Wiley-Interscience Publication
JOHN WILEY & SONS
New York · London · Sydney · Toronto

Copyright © 1977 by John Wiley & Sons, Inc.

All rights reserved. Published simultaneously in Canada.

No part of this book may be reproduced by any means, nor transmitted, nor translated into a machine language without the written permission of the publisher.

Library of Congress Cataloging in Publication Data

Goswami, Bhuvenesh Chandra, 1937-
  Textile yarns.

  "A Wiley-Interscience publication."
  Includes index.
  1. Yarn. I. Martindale, J. G., joint author.
II. Scardino, F. L., joint author. III. Title.

TS1449.G66   677'.02862   77-398
ISBN 0-471-31900-7

Printed in the United States of America

10 9 8 7 6 5 4 3 2 1

Foreword

Textile technology has made great strides during the past two decades with many important processing, marketing, and product innovations. Particularly significant developments have taken place within the realm of textile product performance. For example, the consumer of apparel and household textiles has come to routinely expect such fabric characteristics as wash-and-wear, durable-press, dimensional stability, wash and light fastness, soil resistance, soil release, bright color patterning, and durability. At the same time, the producers of industrial textiles have provided other industries with a variety of efficient and effective products, many of which are specially designed and engineered to perform specific functions. Increasingly, these industrial textiles are able to provide high strength-to-weight ratios, dimensional integrity, thermal stability, and generally high levels of performance reliability. Indeed, the textile industry offers an extremely broad range of apparel, household, and industrial products. With the possible exception of the food industry, the textile industry offers a wider range of products to the consumer than any other industrial sector. Furthermore, these products are offered at a relatively modest cost; thus the consumer spends only a small fraction of his disposable income on textile and textile-related products. **1980274**

The virtual revolution that has taken place during the past two decades in textile technology and particularly in textile product quality and performance is due in large part to the introduction of man-made fibers. These materials have complemented the properties of the important natural fibers and have made possible many of the major advances in product performance. The worldwide production and consumption of textile fibers has increased dramatically in recent years, from approximately 20 billion pounds in 1950 to 59.1 billion pounds in 1974. The two principal naturally occurring fibers, cotton and wool, accounted for approximately 82% of this production in 1950, whereas, in 1974, the cellulosic and synthetic man-made fibers accounted for approximately 44% of worldwide textile fiber production. The increase in total fiber production reflects principally the growth in population, whereas the trend from naturally occurring fibers to man-made fibers reflects many factors, including the desirable physical properties of the man-made fibers, their uniformity, stability of supply, and in many cases an advantageous price structure. The man-made fibers become particularly effective and desirable when they are blended with either cotton or wool.

While important strides have been made in textile product quality and performance, equally important developments have taken place in the area of textile processing. For example, textile productivity in the United States, which is the per capita production of goods or services expressed as the fiber consumption per employee in the textile industry, has increased more than twofold since 1950. Obviously, the use of new technology is largely responsible for this improvement in textile productivity. Although there are many facets to any new technology, the increased productivity of the textile industry reflects principally the trend toward new, high-speed, efficient, and automated processing equipment that replaced the older machinery in the years following the Second World War. Textile yarns and fabrics are now being produced by new technologies that rely on the most modern concepts of heat and mass transfer. The textile industry has become increasingly technologically intensive, relying to an ever greater extent on science and technology and on the engineering approach to the development of new products and processes.

One of the most important developments that has taken place during the past two decades is what might be referred to as the yarn revolution. Although formed fabrics (nonwovens) are gaining in importance, woven and knitted textiles continue to dominate the textile scene. In these products, the elementary building unit is the textile yarn, which ultimately determines the quality and performance of the final fabric, while at the same time reflecting the chemical and physical properties of the component fibers. Spun and continuous-filament yarns are the basic building blocks that are used to construct the wide range of woven and knitted textile fabrics.

Since the textile yarn is indeed such an important building element in textile production, it is therefore quite surprising that no comprehensive text has been available to provide authoritative and quantitative information about the production, properties, and characteristics of textile yarns. Accordingly, the authors of the present book have served a very useful purpose in providing an up-to-date and comprehensive discussion of textile yarns, which should prove extremely useful to all students and practitioners of textile technology. This book serves as an excellent complement to the advanced treatise on yarn and fabric technology entitled "Structural Mechanics of Fibers, Yarns, and Fabrics," by Hearle, Grosberg, and Backer. Taken together with this treatise, the current book makes available a full spectrum of knowledge and understanding of the many diverse textile yarns that are used in the worldwide textile industry today.

Dr. Ludwig Rebenfeld
President
Textile Research
Princeton, New Jersey

Preface

The study of the production, processing, and properties of fibers, yarns, and fabrics is what forms the unified subject called textile technology. This book attempts to fulfill a long-felt need by presenting a systematic and comprehensive study of the textile yarns, their technology, structure, properties, and applications, with an additional chapter on fibers.

The material, designed primarily as a textbook, is based on our teaching experience and is intended for scientists, engineers, and students of textile technology in the industry, universities, and colleges. Although an understanding of some parts of the text requires some basic knowledge of physics and mathematics, most of the subject matter has been presented in a form that can be easily comprehended by those who have not studied mathematics beyond the high school level. The subject matter is appropriate for a student specializing in some other branch of textiles, such as weaving or knitting. It provides an adequate basis for a more advanced study of the structure and mechanics of textile yarns. It is hoped that the book will be useful to those engaged in textile design, to textile and science graduates entering the industry, and to a large number of persons already pursuing a career in the textile industry.

The book begins with the classification of yarn types such as staple, continuous-filament, novelty, and various high-bulk and stretch yarns presently used in the industry. This is followed by a brief account of the basic fiber types and an introduction to the idea of the physical properties desired in fibers that make them suitable for processing into yarns. A detailed and critical discussion of the role played by twist in modifying the structure and the physical and mechanical properties of yarns is given. The effect of physical properties of yarn such as hairiness, covering power, luster, and softness on the esthetic and tactile properties of fabrics is discussed briefly.

The fundamentals of the various systems of conversion of staple fibers into yarns, blending of various fiber types, and the importance of the uniformity characteristics of staple yarns are also presented. The last two chapters deal with the manufacturing technologies of continuous-filament yarns and the various texturing methods employed in the modification of the character and properties of these yarns.

Until recently, the student of textile technology has been required to consult separate books and publications for information regarding the processing and properties of staple and continuous-filament yarns. It is hoped that this comprehensive treatment in a single text will provide sufficient background of knowledge and understanding of the subject of textile yarns.

However, it must be realized that some areas of yarn technology such as the unconventional staple yarn spinning techniques and the texturing modifications are in a constant state of flux. And each additional development offers a new dimension in the study of textile yarns. It is expected that these developments will be incorporated in later editions of this text.

B. C. Goswami
J. G. Martindale
F. L. Scardino

Princeton, New Jersey
Galashiels, Scotland
Philadelphia, Pennsylvania
November 1976

Contents

Textile Yarns

1

Classification of Yarns

DEFINITION OF YARN

In general, yarn may be defined as a linear assemblage of fibers or filaments formed into a continuous strand, having textile-like characteristics. The textile-like characteristics referred to include good tensile strength and high flexibility. Many nontextile materials can be designed to have similar strength and pliability in continuous-strand form. To be considered a yarn, however, these strands must be processable on conventional textile equipment or must possess visual and tactile characteristics (aesthetics) that are usually associated with textile products.

As illustrated by the idealized models in Figure 1.1, yarn may be composed of one or more continuous filaments or of many noncontinuous and rather short fibers (staple). To overcome fiber slippage and to be formed into a functional yarn, staple fibers are ususaly given a great amount of twist or entanglement. Yarns made from staple fiber are often referred to as spun yarns. Two or more single yarns can be twisted together to form ply or plied yarns (Figure 1*d* and *e*). Plied yarns can be further twisted into various multiples. Combination yarns are plies of dissimilar components such as staple and continuous-filament yarns.

CLASSIFICATION OF YARNS

From the variety of yarns that are made commercially, it would appear that there is no limit to the number of functional and aesthetic design possibilities and to the number of distinctly different yarns. Natural, regenerated, and synthetic fibers are processed alone and in a multitude of blend combinations on

1

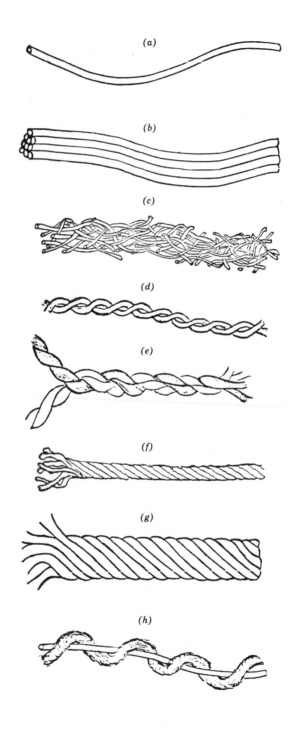

staple yarn systems. Several combinations of continuous-filament and staple fiber yarn blends are also made. Even when a yarn is made from a particular staple fiber or continuous filament, a great number of variations are possible. Through subsequent processing of a chemical or mechanical nature, basic staple or continuous-filament yarns can acquire substantially different structural features that can dramatically change the appearance and functional performance of the original yarns.

Notwithstanding the seemingly infinite variety, yarns may be conveniently classified according to their physical properties and performance characteristics. The physical properties and performance characteristics of yarns depend on the physical properties of the constituent fibers or filaments and on yarn structure. A classification of yarns, based on physical properties and performance characteristics, is given in Table 1.

STAPLE YARNS

There are four basic staple yarn manufacturing systems that have become fairly well standardized. These staple yarn systems are the carded cotton, the combed cotton, the woolen (carded woolen), and the worsted (combed woolen). The carded and combed cotton systems were developed to convert short (1 in.) and long (1.5-2-in.) cotton and cotton-like fibers into yarn. The woolen and worsted systems were developed to convert short (up to 2.5-in.) and long (3-9-in.) wool and wool-like fibers into yarn. Most other staple yarn manufacturing systems are adaptations of one of the four basic systems. Man-made fibers are usually tailored to a fiber length, diameter, and crimp resembling that of cotton or wool for processing on these systems. A yarn made on any one of these systems has a specific structural geometry (fiber contiguity) characteristic of the system regardless of fiber content. The differences in structural geometry of yarns produced on these staple systems are discussed in another chapter.

In a fabric, staple yarns categorically have excellent tactile qualities (hand, good covering power, and excellent comfort factor) and are aesthetically pleasing (a natural textured appearance). However, staple yarns as a group are not as strong or as uniform as continuous-filament yarns of equal linear density. Finally, because staple fibers are processed as a mass rather than individually, the

---

Figure 1.1. Idealized diagrams of various yarn structures. (*a*) Monofilament—solid, single strand of unlimited length. (*b*) Multifilament—many continuous filaments with some twist. (*c*) Staple—many short fibers twisted together tightly. (*d*) Two-ply yarn—two single yarns twisted together. (*e*) Multiply yarn—plied yarns twisted together. (*f*) Thread—hard, fine ply yarn. (*g*) Cord or cable—many plied yarns twisted into a course structure. (*h*) Combination— two dissimilar yarns plied together.

TABLE 1. Yarn Classification by Physical Properties and Performance Characteristics

Yarn Type	General Yarn Properties
Staple yarns	Excellent hand, covering power, comfort and textured appearance
Combed cotton	
Carded cotton	Fair strength and uniformity
Worsted	
Woolen	
Continuous-filament yarns	Excellent strength, uniformity, and possibility for fineness
Natural	
Man-made or synthetic	Fair hand and poor covering power
Novelty yarns	Excellent decorative features or characteristics
Fancy	
Metallic	
Special end-use or industrial yarns	Purely functional; designed to satisfy a specific set of conditions
Tire cord	
Rubber or elastic	
Core	
Multiply	
Coated	
High bulk yarns	Great covering power with little weight, good loftiness or fullness
Staple	
Continuous filament (Taslan)	
Stretch yarns	Stretchability and cling without great pressure, good hand and covering power
Twist-heat set-untwist	
(Helanca, Fluflon)	
Crimp-heat set	
(Ban-lon)	
Stress under tension	
(Agilon)	
Knit-deknit	
Gear crimp	

number of fibers per yarn cross section varies considerably along the yarn length. This condition limits the fineness of staple yarn that can be spun on a commercial basis.

CONTINUOUS-FILAMENT YARNS

Before the advent of man-made fibers, silk was the only continuous-filament yarn available. Briefly, a given number of the naturally occurring double filaments of a specific fineness are extruded and extended by unwrapping select cocoons. The desired frequencies and directions of twist are then added to form the singles and, subsequently, multiply yarns.

In the manufacture of man-made filaments, a solution is forced through very fine holes of a spinneret, at which point the solution solidifies by coagulation, evaporation, or cooling. Usually the number of holes in the spinneret determines the number of filaments in the yarn. Also, the size of each hole and the amount of drawing, if any, determine the diameter of each filament. As the individual filaments solidify, they are brought together with or without slight twist or entanglement to form a continuous-filament yarn.

If the filaments are to be processed on a staple yarn system, several thousand are brought together into a twistless linear assemblage known as tow, for subsequent crimping and cutting. One of the advantages of the man-made fibers is the control that it is possible to exercise over each step of the production process. Fibers can be tailored to fit a wide variety of end-uses that require physical or chemical properties not found in the parent fiber or in the natural fibers.

Continuous-filament yarns in fabric form usually have excellent strength and uniformity. As indicated by the fine monofilament and multifilament yarns that have found commercial acceptance, continuous-filament yarns can be made much finer in linear density and diameter than staple yarns. In an untextured form, however, continuous-filament yarns are not thought to possess a combination of good covering power, tactile qualities, comfort, and a pleasing appearance, except for limited apparel applications such as sheer hosiery and lingerie. In industrial and nonapparel applications, however, this combination of properties is usually not important, and the continuous-filament yarn excels quite often.

NOVELTY YARNS

Effect threads or novelty yarns are designed for decorative rather than functional purposes. Very seldom is a fabric composed entirely from novelty yarn, except possibly in drapery applications. Most novelty yarns are basically either of a fancy effect or metallic type. Combination yarns are used quite often to obtain the desired effect.

Fancy yarns are usually made by the irregular plying of staple or continuous-filament yarns and are characterized by abrupt, periodic effects. The periodicity of these effects may be random or uniform. Often quite large or noticeable, the novelty effect is brought about by programmed variation in twist frequency or rate of input in one or more components during the plying of the yarns. This usually results in differential bending or wrapping among the components or in segments of buckled yarn permanently entangled in the composite structure.

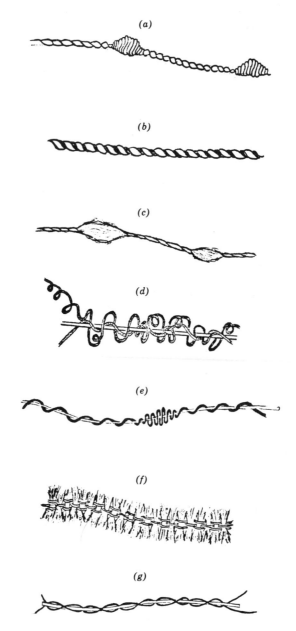

Figure 1.2. Examples of various fancy yarn structures. (*a*) Seed yarn. (*b*) Spiral yarn. (*c*) Slub yarn. (*d*) Bouclé yarn. (*e*) Knop bouclé yarn. (*f*) Chenille. (*g*) Diamond–metallic core.

Another variation is to entrap short segments of a novelty-effect material into regularly plied base yarns. Examples of fancy yarns are shown in Figure 1.2.

Metallic novelty yarns are characterized by a glittering appearance and a rectangular cross-sectional shape. Durability is added to the metallic yarn by protecting with a transparent film the aluminum foil or metallized material that produces the glittering effect. Metlon, acetate, or Mylar metallic yarns are examples of durable glossy yarns used for decorative designs.

SPECIAL END-USE YARNS

Industrial design requirements of special end-use yarns are purely functional in character. These yarns are engineered for predictable performance under specific conditions. Many industrial yarns do not have the visual and tactile qualities of yarns designed for apparel or home-furnishing applications. Examples of yarns that are engineered for a specific end-use are: tire cord, twine, sewing thread, rubber or elastic threads, asbestos and glass yarns, core-spun yarns, wire yarn, heavy monofilaments, and split-film yarns.

HIGH-BULK YARNS

A high-bulk yarn is a staple or continuous-filament yarn that has a normal extensibility but an unusually high degree of loftiness or fullness. These yarns retain their bulkiness under both relaxed and stressed conditions. Great covering power with little weight is possible in fabrics composed of high-bulk yarns.

Some high-bulk yarns are made on staple systems from thermoplastic fibers that shrink differentially in treated yarn form. The difference in shrinkage results in massive buckling or crimping of fiber segments between points of entanglement in the yarn structure. Continuous-filament yarns can acquire substantial bulkiness in the same manner or by the creation of nonlinearity and loop formation in the individual filaments and by entraping the loops with twist. This is commonly referred to as jet stream texturizing because a stream of air or steam is used to create nonlinearity in the filaments. Examples of high-bulk yarns are shown in Figure 1.3. Very fine, yet bulky structures can be made from the texturized continuous-filament yarns. Combination yarns can also be texturized with differential bulk effects.

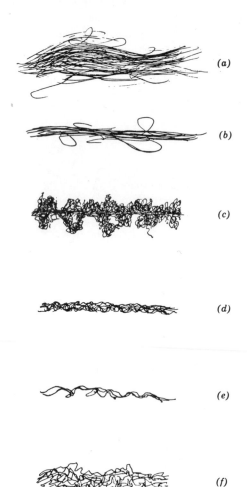

(a)

(b)

(c)

(d)

(e)

(f)

(g)

Figure 1.3. Diagrams of typical high-bulk and stretch textured yarn structures. (*a*) High-bulk stable yarn shrunk-preshrunk blend. (*b*) Air-jet filament high-bulk yarn. (*c*) Combination yarn differential bulking. (*d*) False-twist stretch filament yarn. (*e*) Knife-edge stretch filament yarn. (*f*) Stuffer-box stretch filament yarn. (*g*) Knit-de-knit stretch filament yarn. (*h*) Gear-crimped stretch filament yarn.

(h)

STRETCH YARNS

Textured yarns that have been programmed for extraordinary extensibility are known as stretch yarns. Most stretch yarns can be extended from one and one-half to twice their normal or relaxed length; some can be extended up to three or four multiples of their relaxed length. These structures are not only highly extensible but highly elastic as well. The term highly elastic is used here to suggest a quick and practically complete recovery toward original yarn geometry or configuration after substantial extension in the direction of the yarn axis.

In their relaxed state, stretch yarns resemble high-bulk yarns. On extension of the stretch yarns, however, the bulkiness is considerably reduced. At full extension, the stretch yarn resembles an ordinary continuous-filament yarn, combination or staple yarn, whichever may be the parent yarn or precursor.

Most stretch yarns are made by texturizing thermoplastic continuous-filament yarns. Texturizing brings about considerable nonlinearity or crimp in the individual filaments. This nonlinear configuration of the filaments is heat set for permanency but not entangled as in the case of the high-bulk yarns. Consequently, the stretch yarn extends and recovers as the filaments collectively straighten and recoil. This phenomenon is very similar to the action of a coiled spring. In addition to the aforementioned advantages, stretch yarns provide cling without great pressure and good covering power in fabrics. However, the tactile qualities are not as good as high-bulk or staple yarn. Also, some serious disadvantages such as filament snagging are associated with stretch yarns. Examples of various stretch yarn geometries are shown in Figure 1.3. The methods of inducing stretch and the degree and character of stretch vary considerably among these yarns. Stretch yarns should not be confused with rubber or elastomeric fiber yarns; they are normally used in power stretch fabrics where more than a subtle pressure is required. These heavy duty, truly elastomeric yarns are not as widely applicable as stretch yarns and therefore are considered to be special end-use yarns.

RELATIVE CONSUMPTION OF YARN

It is estimated that 65% of all yarns consumed annually in the United States are staple type yarns and 35% are continuous filament type yarns. For economic and technological reasons, the trend has been toward a larger share of the yarn market for continuous-filament yarns. Recent developments in the texturing of continuous-filament yarns have created somewhat of an acceleration in this trend. One-half of the consumption of filament yarn is for tires, carpet, and textured apparel; the other half is for knitted and woven fabrics.

Although the demand for staple yarns has been growing at a slower rate, a growth trend is still quite apparent. The supply of natural fiber with desirable processing characteristics has been fairly constant. At this time, it is estimated that one-half of the staple yarns consumed are composed of natural fibers, and one-half are composed of man-made fibers.

YARN DESCRIPTION

To properly describe a specific yarn for communicative purposes, a great deal of information is required. First, the fiber content must be identified generically, and in the case of a blend, by proportion of the total weight of the yarn. The physical properties of the constituent fibers (fiber length, fineness, crimp, cross-sectional shape, delusterant, etc.) should be described also. Second, the yarn constructional features (staple or continuous filament; singles, ply, or combination) should be indicated. In the case of a stretch or a bulky yarn, the technique for texturizing should be made clear. Third, the linear density of the yarn should be expressed. If the yarn is a ply or combination yarn, the linear densities of the individual components and of the resulting structure should be stated. Furthermore, twist direction and frequency should be identified in singles yarn and in the individual components in the case of ply yarn. Certain performance characteristics should also be given. Whereas indications of strength and breaking extension might be appropriate for some yarn, industrial and special end-use yarns would require much more information relative to mechanical and chemical properties. Staple yarns usually require an expression of the evenness and appearance of the structure. Finally, it should be realized that the yarns that are dyed or finished before conversion into fabric or textile products require considerably more stated specifications in a description than do unfinished (greige) yarns. All these basic elements of proper yarn identification and description are discussed at considerable length in the following chapters.

2

Raw Materials

INTRODUCTION

This chapter is concerned with a general description of the fibers that in most cases form the raw material from which yarns and fabrics are produced (yarns may also be made from ribbons, slit films, and by splitting of plastic films). Until recently, most textiles were produced from fibers of primarily natural origin. The important natural fibers presently in use for apparel applications are cotton, flax, wool, and silk. These fibers have certain inherent characteristics that make them suitable for conversion into yarns and ultimately into the fabrics that are most commonly used in day-to-day wear. In addition, there are many other natural fibers, such as jute, hemp, ramie, sisal, and kapok, that have been utilized for certain specific end-uses.

In recent years, this list has been supplemented by a variety of new fibers called "man-made" fibers (which include "regenerated" and "synthetic" types). Most of the developments in the field of regenerated and synthetic fibers have been directed toward simulating the properties of natural fibers and toward utilizing the production techniques already employed in the processing of natural fiber yarns. However, the introduction of these fibers has caused great changes in technology and processing techniques; some of these are quite different from those used in the manufacture of conventional spun yarns. The intention here, therefore, is to discuss the general classification of the textile fibers and the properties that make them suitable for conversion into yarns and fabrics. For more specific and detailed information on the chemical composition, structure, and properties, the reader is referred to Cook (1), Moncrieff (2), and Morton and Hearle (3).

CLASSIFICATION OF FIBERS

The textile fibers may be divided into two major groups, namely, (a) natural fibers and (b) man-made fibers. Table 2.1 describes the different fiber types classified under each major group.

The natural fibers have been further classified into six subgroups: (a) seed fibers, for example, cotton; (b) animal fibers, for example, wool and hair; (c) silk; (d) soft vegetable fibers; (e) hard vegetable fibers; and (f) mineral fibers, for example, asbestos.

The man-made fibers are divided into two broad categories, for example, (a) regenerated (natural polymer) and (b) synthetic fibers. The regenerated fibers are those in which the fiber-forming material is of natural origin; the second class of fibers is made by the chemical synthesis of simple polymer-forming materials. The regenerated fibers are further divided into the following four subgroups:

1. Cellulose fibers: rayon, such as viscose; polynosic (in which the fiber is either wholly or mainly cellulose).
2. Cellulose esters: acetate and triacetate.
3. Protein fibers: casein and zein.
4. Miscellaneous: alginate, rubbers, etc.

Synthetic fibers are classified according to their chemical structure. They fall into six broad groups. Because the synthetic fibers are often made from copolymers or from modifications of polymers, a fiber may belong to two or more of the chemical subgroups.

1. Polyamides: nylon 6, -7, 6-6, -10, -11;
2. Polyesters: Dacron®, Terylene®, Trevira®
3. Polyvinyl derivatives:
   a. Polyacrylonitrile
      i. acrylic, for example, Orlon®, Acrilan®, Creslan®, Zefran®, etc.
      ii. Modacrylic, for example, Verel®, Dynel®
   b. Polyvinylchloride (PVC) for example, Rhovyl®, Vinyon®, etc.
   c. Polyvinylidine chloride, for example, Saran®
   d. Polyvinyl alcohol (PVA) for example, Vinal® and Mewlon®
   e. Polytetrafluoroethylene for example, Teflon®
   f. Polyvinylidene dinitrile for example, Darvan®
   g. Polystyrene for example, Durabass®
   h. Miscellaneous polyvinyl derivatives

®Registered trademark

Table 2.1. Claffication of Textile Fibers

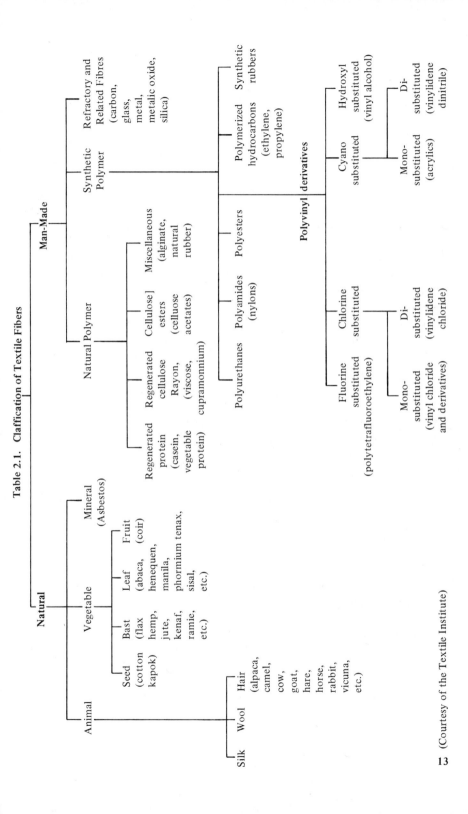

(Courtesy of the Textile Institute)

13

4. Polyolefins:
   a. polyethylene
   b. polypropylene for example, Herculon® and Marvess®
5. Polyurethanes: lycra or spandex
6. Miscellaneous synthetics: glass and metallic, carbon: PBI (polybenzimida-zole), aramid (Kevlar® and Nomex®); and novoloid (Kynol®).

During the past two decades, there has been a great surge of activity in the development and production of new synthetic fibers. This has given rise to considerable confusion about the nomenclature of synthetic textile fibers. Consequently, the United States Federal Trade Commission has established rules and regulations under the Textile Products Identification Act [Section 7(c)] for fiber identification, first promulgated on March 3, 1960. Following are the generic names and the definitions for the most commonly used textile man-made fibers, according to the Commission.

1. Acrylic—a manufactured fiber in which the fiber-forming substance is any long-chain synthetic polymer composed, at least 85% by weight, of acrylonitrile units.

$$(-CH_2-CH-)$$
$$|$$
$$CN$$

2. Modacrylic—a manufactured fiber in which the fiber-forming substance is any long-chain synthetic polymer composed, less than 85% but at least 35% by weight, of acrylonitrile units.

$$(-CH_2-CH-)$$
$$|$$
$$CN$$

[except fibers qualifying under subparagraph (b) of paragraph (10) of this section and fibers qualifying under paragraph (17) of this section.]

3. Polyester—a manufactured fiber in which the fiber-forming substance is any long-chain synthetic polymer composed, at least 85% by weight, of an ester of a substituted aromatic carboxylic acid, including but not restricted to substituted terephthalate units,

$$p(-R-O-\underset{\underset{O}{\|}}{C}-C_6H_4-\underset{\underset{O}{\|}}{C}-O-),$$

and parasubstituted hydroxybenzoate units,

$$p(-R-O-C_6H_4-\underset{\underset{O}{\|}}{C}-O-)$$

(As amended September 12, 1973.)

4. Rayon—a manufactured fiber composed of regenerated cellulose, as well as manufactured fibers composed of regenerated cellulose in which substituents have replaced not more than 15% of the hydrogens of the hydroxyl groups.
5. Acetate—a manufactured fiber in which the fiber-forming substance is cellulose acetate. When no less than 92% of the hydroxyl groups are acetylated, the term triacetate may be used as a generic description of the fiber.
6. Saran—a manufactured fiber in which the fiber-forming substance is any long-chain synthetic polymer composed, at least 80% by weight, of vinylidene chloride units

$$(-CH_2-CCl_2-)$$

7. Azlon—a manufactured fiber in which the fiber-forming substance is composed of any regenerated, naturally occurring proteins.
8. Nytril—a manufactured fiber containing at least 85% of a long-chain polymer of vinylidene dinitrile

$$(-CH_2-C(CN)_2-)$$

where the vinylidene dinitrile content is no less than every other unit in the polymer chain.
9. Nylon—a manufactured fiber in which the fiber-forming substance is a long-chain synthetic polyamide in which less than 85% of the amide

$$(-\underset{\underset{O}{\|}}{C}-NH-)$$

linkages are attached directly to two aromatic rings.
(As amended January 11, 1974.)
10. Rubber—a manufactured fiber in which the fiber-forming substance is comprised of natural or synthetic rubber, including the following categories:
a. a manufactured fiber in which the fiber-forming substance is a hydrocarbon such as natural rubber, polyisoprene, polybutadiene, copolymers of dienes and hydrocarbons, or amorphous (noncrystalline) polyolefins.
b. a manufactured fiber in which the fiber-forming substance is a copolymer of acrylonitrile and a diene (such as butadiene) composed, not more than 50% but at least 10% by weight, of acrylonitrile units.

$$(-\underset{\underset{CN}{|}}{CH_2-CH}-)$$

The term lastrile may be used as a generic description for fibers falling within this category,

c. a manufactured fiber in which the fiber-forming substance is a poly-chloroprene or a copolymer of chloroprene in which at least 35% by weight of the fiber-forming substance is composed of chloroprene units.

$$(-CH_2-\underset{\underset{Cl}{|}}{C}=CH-CH_2-)$$

11. Spandex—a manufactured fiber in which the fiber-forming substance is a long-chain synthetic polymer comprised, at least 85%, of a segmented polyurethane.

12. Vinal—a manufactured fiber in which the fiber-forming substance is any long-chain synthetic polymer composed, at least 50% by weight, of vinyl alcohol units

$$(-CH_2-CHOH-)$$

and in which the total of the vinyl alcohol units and any one or more of the various acetal units is at least 85% by weight of the fiber.

13. Olefin—a manufactured fiber in which the fiber-forming substance is any long-chain synthetic polymer composed, at least 85% by weight, of ethylene, propylene, or other olefin units, except amorphous (noncrystalline) polyolefins qualifying under category (1) of Paragraph (j) of Rule 7.*

14. Vinyon—a manufactured fiber in which the fiber-forming substance is any long chain synthetic polymer composed, at least 85% by weight, of vinyl chloride units.

$$(-CH_2-CHCl-)$$

15. Metallic—a manufactured fiber composed of metal, plastic-coated metal, metal-coated plastic, or a core completely covered by metal.

16. Glass—a manufactured fiber in which the fiber-forming substance is glass.

17. Anidex—a manufactured fiber in which the fiber-forming substance is any long-chain synthetic polymer composed, at least 50% by weight, of one or more esters of a monohydric alcohol and acrylic acid.

$$(CH_2=CH-COOH)$$

18. Novoloid—a manufactured fiber containing, at least 85% by weight, a cross-linked novolac.

19. Aramid—a manufactured fiber in which the fiber-forming substance is a long-chain synthetic polymide in which at least 85% of the amide

*Rule 7— Generic Names and Definitions for Manufactured Fibers (Federal Trade Commission, Textile Products Act).

$$(-\overset{\displaystyle }{\underset{\displaystyle O}{C}}-NH-)$$

linkages are attached directly to two aromatic rings.

REQUIRED PHYSICAL PROPERTIES OF FIBERS

Textile materials are generally soft to the touch, flexible, capable of being transformed into desired shapes without resistance, and durable over a reasonable period of wear. They derive these properties from fibers and yarns that form the building units arranged or interlaced in various forms. The yarn, in turn, is formed by twisting a bundle of fibers together. It is therefore obvious that the properties of the ultimate textile structure will depend very largely on the characteristics of the fibers from which they are made.

Fibers have been defined by the Textile Institute as "units of matter characterized by flexibility, fineness, and a high ratio of length to thickness." In individual textile fibers, the length/width ratio is at least 1000/1. Some additional characteristics, such as stability at high temperature and a certain minimum strength and extensibility, might be included if the fiber is to be used for textile end-uses. Table 2.2 gives examples of the length/diameter ratio of some natural

Table 2.2. Length/Diameter Ratio of Textile Fibers

Fiber	Typical Length	Typical Diameter	Length/Diameter
Cotton	25 mm	17μ	1,500
Wool	75 mm	25μ	3,000
Flax (ultimate)	25 mm	20μ	1,250
Jute (ultimate)	2.5 mm	15μ	170
Ramie	150 mm	50μ	3,000

fibers. These dimensional characteristics of fibers form the basis of their use as textile raw materials, and the importance of each of these characteristics is discussed in what follows. The following table illustrates some of the most essential and other desirable properties that must be considered in making a choice of a fiber for use as a textile material:

1. Dimensional and physical characteristics
    length
    fineness

cross-sectional shape
crimp
density
2. Mechanical properties
strength
elasticity
extensibility
rigidity (stiffness)
3. General
surface characteristics—frictional, softness,
environmental stability—resistance to sunlight,
thermal stability, resistance to chemical and organic-solvents
pliability
durability
abrasion resistance
dimensional stability
moisture absorption
resistance to bacteria, fungi, mildew, moths, etc.
static buildup
color
wetting characteristics

A detailed discussion of the fiber properties enumerated above is out of the scope of this book; hence the reader is referred elsewhere (1,2). However, it is appropriate to discuss the technical significance of some of the most important properties of fibers that have a profound influence on the processing behavior and the end-use characteristics of yarns.

Table 2.3. Range of Lengths of Textile Raw Materials

Cotton:	Bengals	$\frac{1}{2}$-$\frac{5}{8}$ in. staple length
	Surats	$\frac{3}{4}$-$1\frac{1}{16}$ in. staple length
	American Uplands	$\frac{3}{4}$-$1\frac{1}{16}$ in. staple length
	Egyptian Uppers	$1\frac{1}{16}$-$1\frac{1}{4}$ in. staple length
	Egyptian Karnak	$1\frac{1}{4}$-$1\frac{3}{8}$ in. staple length
	Sea Island	$1\frac{1}{2}$-$1\frac{3}{4}$ in. staple length
Wood:	80s Australian Merino	$2\frac{1}{2}$-3 in. staple length
	56s Southdown	3-$3\frac{1}{2}$ in. staple length
	48s Romney Marsh	5-6 in. staple length
	36s Lincoln	10-12 in. staple length
Flax:	strand length, approx.	12-36 in.
Hemp:	strand length, approx.	4-10 ft
Jute:	strand length, approx.	5-12 ft

1 in. = 25 mm; 1 ft = 304.8 mm
From Morton and Hearle (3).

Fiber Length

Like all other physical properties of natural textile fibers, fiber length varies considerably within any one sample. The variability in terms of the coefficient of variation may be as high as 40% for cotton and about 50% for wool. Table 2.3 shows the range of staple lengths for various natural fibers. Man-made staple fibers are, generally, much more uniform; however, they may have a coefficient of variation of as much as 10%. This variation is caused partly by faulty stapling machines and partly by fiber breakage that occurs during processing. Some idea of the variation in fiber length for cotton may be obtained from the fiber array in Figure 2.1*a*; 2.1*b* and *c* show a comparison of the fiber array for a lot of "Fibro" before and after processing. Man-made fibers may be produced in any length to meet the requirements of the spinner and to suit the existing machine design. Usually, however, man-made fibers are available in lengths of 1¼ in.,

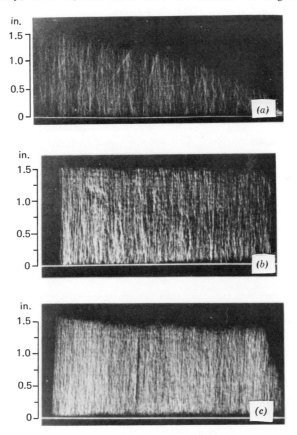

Figure 2.1. Fiber arrays. (*a*) Cotton, (*b*) raw fibro, (*c*) fibro from card sliver [from Morton and Hearle (3)].

1 $7/16$ in., and 2 in. for processing on cotton systems; 3 in., 4 in., 6 in., and 8 in. for processing on woolen and worsted machinery; and 10 in. and 15 in. for processing on flax and jute systems.

Staple yarns are made by drafting the fibers into a more or less parallel bundle form and then twisting the strand so that the fibers are gripped firmly. The process of drafting on most processing systems is achieved by means of roller-drafting. To produce a satisfactory yarn, it is essential that the drafting system have an effective fiber control capability. A roller-drafting system is generally designed to operate efficiently on a comparatively narrow range of staple lengths. Consequently, it is desirable that the raw material being processed have the lowest possible variation from the requirements set by the design of the processing unit. In the case of combed yarns, it is necessary to control not only the staple length but also the amount of short fiber present in the raw material. Too many short fibers in the bundle can result in uneven drafting and eventually in irregular yarns (see Chapter 7). Some idea of the relationship between the staple length of cotton and the count of yarn that can be spun commercially may be obtained from Table 2.4.

Table 2.4. Cotton Yarn Counts Usually Spun from Specific Staple Lengths

Staple Length (inches)	Carded Yarn Counts	
	Warp	Filling
Up to $7/8$	Up to 20s	Up to 20s
$7/8$ - 1 $1/32$	Up to 30s	Up to 30s
$1 1/32$ - $1 1/8$	30s - 50s	30s - 50s
	Combed Yarn Counts	
Up to $1 1/16$	Up to 20s	Up to 20s
$1 1/16$ - $1 1/8$	20s - 40s	20s - 40s
$1 1/8$ - $1 1/4$	40s - 60s	40s - 60s
$1 1/4$ - $1 3/8$	60s - 70s	60s - 70s

1 in. = 25.4 mm
From the American Cotton Hanbook, Volume I.

The purpose of twist in staple yarns is to generate lateral pressure and thus help in firmly gripping the fibers together. The greater the twist, the higher the lateral pressure and the more closely the fibers are held. If longer fibers are used in the production of yarns, they will have a larger surface area of contact (overlaps) over which the fibers can be made to cohere by means of twist. It must be pointed out that, in high-speed roller-drafting machines, the movement of the fibers in the drafting zones is controlled by installing constrictors. Thus, up to a reasonable limit, wider roll settings used for longer fibers will offer better accommodation for control devices, resulting in better quality yarns. Less twist is required to make yarns with essentially similar strength with longer fibers than is needed for short staple fiber yarns. Hence, from the standpoint of processing, yarn strength, softness (higher twist results in a harder yarn), and appearance, longer fibers are preferable. Longer fiber lengths and fewer short fibers also help in lowering the end breakage rate during spinning. However, in some special cases, where it is desired to produce fabrics with soft feel having a hairy or fuzzy appearance and warm handle, short fibers are preferred over long ones.

Fiber Fineness

Textile fibers, whether natural or synthetic, come in various forms and cross-sectional shapes. Some are circular in section (wool and synthetics), whereas others are of irregular cross-sectional shape (cotton, silk, and specially man-made synthetic fibers). The fineness or coarseness of a fiber has been sometimes defined in terms of its diameter. This term can be used advantageously where a fiber has a cylindrical cross section, but it has no useful meaning when characterizing other shapes such as elliptical, trilobal, etc. Following are some of the other parameters that are used in the industry, as well as in textile research, to characterize the transverse dimension of a fiber.

1. Width
   Because of its dependence on the fiber cross-sectional shape, the parameter 'mean fiber width' has limited application. One such example in which the maximum and the minimum widths of the twisted ribbon (Figure 2.9) vary considerably throughout the length of the fiber is cotton.
2. Area of cross section
   This quantity, for a given fiber type, is proportional to the limiting weight per unit length at a section. The area of cross section can be calculated from the knowledge of density (g/cm^3) and the linear density ($g/cm \times 10^{-8}$).
3. Wall thickness
   This parameter is used to describe hollow fibers such as cotton.
4. Linear density
   This is defined as the weight per unit length and is the most commonly used (at least for synthetic fibers) parameter to characterize the fineness of textile fibers. It is especially advantageous, since it corresponds to the parameter used in characterizing the yarn count (denier or tex).

The property of fineness or coarseness of textile fibers has been recognized as one of the most important of all the fiber characteristics affecting processing behavior and yarn properties. Finer and shorter wools are generally regarded as much more valuable to the worsted trade than long and coarse ones. On the other hand, in the cotton industry, the importance of fiber fineness is somewhat overshadowed by length considerations, mainly because of the fact that, generally, longer cottons are also finer ones. The following discussion considers some of the yarn and fabric properties that depend directly on the fineness of the fibers used in their production.

The stiffness, drape, and handle of fabrics are highly influenced by the bending behavior of fibers, whereas recovery from bending influences creasing. The resistance to bending or flexibility of a fiber depends on its shape, its tensile modulus, its density, and above all its fineness. In this relationship, fineness occurs as a squared term. The density of ordinary textile fibers varies from 0.91 for polypropylene to 1.52 for rayons, and the shape factor ranges from 0.59 for silk to 0.91 for nylon; obviously, these do not account for an important influence on bending. The modulus of textile fibers ranges from 1800 g wt/tex for bast fibers and highly drawn rayons to values as low as 300 g wt/tex for wool and acetate fibers (the values for glass are excluded). This would cover about a sixfold change in stiffness. However, the fineness of fibers varies from about 0.10 tex (1 μg/cm) for fine stretched rayons to approximately 1 tex (10 μg/cm) for coarse wools and even higher for man-made monofilaments. One can see that fineness thus becomes the most important factor in determining the bending rigidity of fibers. It can therefore be said that, for a given yarn linear density or for a given fabric weight made from a given fiber, the resistance to bending increases as the linear density (fineness) of the fiber increases. Consequently, fiber fineness is an important property in determining the stiffness of a yarn or fabric and eventually its softness, handle, and drapability qualities.

In the process of yarn formation, fibers are subjected to the twisting operation. The resistance to twisting of a fiber, its torsional rigidity, is dependent on its shape, density, shear modulus, and fineness (diameter). As in bending, the fiber fineness comes in as a squared term in the expression for torsional rigidity modulus. It can be shown that, as the fiber linear density (fineness) increases, other factors being equal, its resistance to torsion increases too, but at a faster rate than the fiber weight per unit length. It is obvious from the foregoing statement that the torque generated in a given yarn size by a given amount of twist increases as the fiber fineness decreases. In other words, for a given fiber, the amount of twist required to produce comparable internal stresses, which in turn control the cohesion characteristics and yarn strength, with all other things being equal, will be greater in a yarn composed of coarse fibers than in one composed of fine fibers. This factor is of profound importance when considered in terms of producing highly twisted yarns and sewing threads (snarls and kinks can have a detrimental effect on the quality of thread). Furthermore, the cohesiveness of fibers in a spun yarn depends on the interfiber friction (influenced by twist), which in turn is a function of the total area of fiber contact (fiber specific surface) and the fiber coefficient of friction. Thus fibers with larger specific surface (finer fibers), all other factors being equal, will require less twist compared to those having smaller specific surface (coarse fibers) to prevent slippage in a yarn.

In addition to fiber length, fiber fineness also plays an important role in determining the processing efficiency and the spinnability—the optimum yarn size to which a raw material can be spun—of a staple fiber stock. Moreover, the uniformity of a yarn is largely dependent on the average number of fibers in its cross section. Therefore it follows that, for a given yarn linear density, the finer the fibers (meaning larger numbers of fibers in a cross section), the more regular the yarn. It must be emphasized here that the uniformity characteristic has a considerable influence on such yarn properties as strength, extensibility, appearance, and end breakages in spinning, winding, warping, and weaving.

Fabric luster is yet another property that is affected by the fineness of fibers, since this determines the number of individual reflecting surfaces per unit area of the fabric. In other words, fine fibers will produce soft "sheen," whereas coarse fibers will generate hard "glitter." The rate of dye uptake by a fiber is also dependent on the total available surface for a given volume of the fiber material, that is, the specific surface of the fiber. It is evident, therefore, that finer fibers will require less time to exhaust the dye bath than coarse ones. This would eventually affect the overall appearance of dyed fabrics.

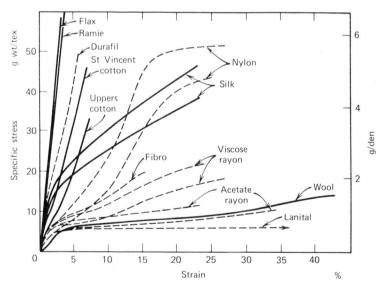

Figure 2.2(a). Stress-strain curves of various fibers broken at 65% r.h., 20°C, 90 gf tex^{-1} min^{-1} [after Meredith (4)]; (Durafil is the Lilienfeld rayon, from 1945; Fibro is staple viscose rayon; Lanital is a casein fiber).

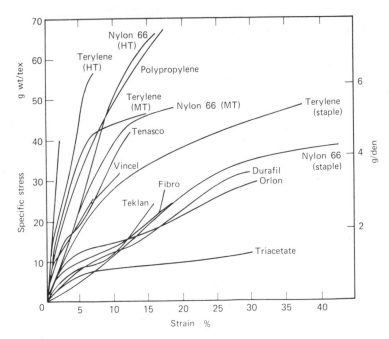

Figure 2.2(b). Stress-strain curves of various fibers [from Farrow (5,6) and Ford (7)]; [Viscose rayon variants are Fibro (regular-staple); Vincel (high-wet-modulus), and Tenasco (high-tenacity, industrial), Teklan is modacrylic; Terylene is polyester fibre; Orlon is acrylic fiber.)

Fiber Strength and Extensibility

The tensile properties of a staple yarn are partly governed by the strength of its constituent fibers and partly by the degree of cohesiveness achieved through twisting of the fiber assembly. Thus mechanical properties (particularly tensile behavior) are probably the most important fiber properties, since they contribute both to processing behavior and to the characteristics of the end product. Textile fibers are available in a very wide range of tensile properties. Generally speaking, fibers cover a range from high tenacity/low extension (such as flax, glass, and Fortisan fibers), to low tenacity/high extension (such as wool and acetate fibers). The stress-strain characteristics of some common textile fibers are shown in Figure 2.2a and b. However, a textile fiber must possess certain minimum strength and adequate extensibility characteristics if it is to be of significant use to the textile industry. It is therefore up to the user to select a fiber whose mechanical properties are best suited to a particular end use. Some of the end-use properties affected by the mechanical properties of textile fibers are the durability, low load deformations (wrinkle recovery, drape, etc.), resilience, stiffness, abrasion resistance, compressibility, and softness.

Miscellaneous Properties

In addition to the properties previously discussed, there are several other miscellaneous fiber properties that have a profound bearing on the overall performance of yarns and fabrics. Fiber density is one such property, because it is this property that controls the covering power of a fiber in a fabric. In a given weight of fabric, the lower the density, the greater the volume of fiber present. As a result, fabrics made from yarns of low-density fibers will have a fuller and bulkier appearance than those made from higher-density fibers. Cotton and viscose rayon fibers have a density of about 1.50, compared to 1.33 for wool and 1.14 for nylon; polyethylene and polypropylene fibers have a very low density of 0.92. Because of their low density, polyethylene and polypropylene float on water. Some other industrial fibers, such as polytetrafluoroethylene and glass, have very high densities, approximately 2.2 and 2.54, respectively.

Moisture absorption by fibers can be an asset insofar as the comfort and warmth characteristics of clothing is concerned. On the other hand, it may be a disadvantage and a nuisance when its effect on the changes in the dimensional stability of certain fibers is considered. It can add to the cost of drying the hygroscopic fibers, where it is necessary to remove absorbed moisture that is comparatively negligible in the nonhygroscopic synthetic fibers. Moisture absorption changes the properties of fibers. It causes the fibers to swell, eventually resulting in changes in fiber dimensions. Consequently, the size, shape stiffness, and permeability of yarns and fabrics are modified. The amount of absorbed moisture also affects the mechanical, frictional, and electrical properties (static) of fibers; all these changes influence the processing behavior and the end-use characteristics.

Normal changes in temperature conditions do not have any significant effect on the thermal stability of textile fibers. When subjected to high temperatures and then cooled, the newer synthetic fibers become "heat set." This means that the synthetic fibers can be usefully deformed by heat and pressure to make them aquire special textured effects (for a detailed discussion, see Chapter 12). This property of the thermoplastic fibers has made them quite attractive for use in blends with natural fibers where a reasonable permanence of folds or pleats is desired. All textile fibers are susceptible to degradation when exposed to very high temperature conditions. Heating causes decomposition, which results in the weakening of the structure. The decomposition characteristics of fibers are also very important when considered in the light of their flammability behavior. Natural fibers such as cotton, wool, and linen decompose into brown-colored products on overheating. On the other hand, the synthetic fibers and regenerated acetate soften and melt and can cause severe burn injuries. However, textile fibers can be made fireproof or flameproof by treatments with flame-retardant chemicals.

During ordinary wear, textile materials are generally exposed to all kinds of environmental conditions that can cause their discoloration and degradation.

Table 2.5. Physical Properties of Textile Fibers

Fiber Type	Name	Range of Diameter (μ)	Density (g/cm^3)	Initial Modulus (gf/tex)	Tenacity (gf/tex)	Breaking Extension (%)
Natural	Cotton	11-22	1.52	500	35	7
Vegetable	Flax	5-40[a]	1.52	1830	55	3
	Jute	8-30[a]	1.52	1750	50	2
	Sisal	8-40[a]	1.52	2500	40	2
Natural	Wool	18-44	1.31	250	12	40
Animal	Silk	10-15	1.34	750	40	23
Regenerated	Viscose rayon	12+	1.46-1.54	500	20	20
	High-tenacity rayon	12+	1.46-1.54	600	51	10
	Polynosic rayon	12+	1.49	800	30	8
	Fortisan	5+	1.49	1700	60	6
	Acetate	15+	1.32	350	13	24
	Triacetate	15+	1.32	300	12	30
	Casein	17+	1.30	350	10	60
Synthetic	*Nylon*					
	6	14+	1.14	250	32-65	30-55
	6.6	14+	1.14	250	32-65	16-66
	Qiana (du Pont)	10+	1.03		25	26-36
	Polyester					
	Dacron (du Pont)	12+	1.34	1000	25-54	12-55
	Kodel (Eastman)	12+	1.38	1000	40-50	35-45
	Acrylic					
	Orlon (duPont)	12+	1.16	650	20-30	20-28
	Acrilan (Monsanto)	12+	1.17	650	18-25	35-50
	Polyolefin					
	Polypropylene	–	0.91	800	60	20
	Polyethylene	–	0.95	–	30-60	10-45
	Aramid					
	Nomex (du Pont)	12+	1.38	–	36-50	22-32
	Novolid					
	Kynol (Carborandum)	–	1.25	–	16	35
	Spandex					
	Lycra (du Pont)	–	1.21	–	6-8	444-555
Inorganic						
	Glass	5+	2.54	3000	76	2-5
	Asbestos	0.01-0.30[a]	2.5	1300	–	–

[a] A cutal fibers—usual textile fibers are coarse bundles.
[b] Much lower with antistatic agents.

26

Work of Rupture (g/tex)	Elect. Resistance 65% r.h. (ohm-cm)	Moisture Regain 65% r.h. (%)	Melting Point (°C)	Strength Retentions 20 days 130°C-(%)	Attack by Chemicals Dissolved Degraded By
1.3	10^7	7	c	38	Strong acid,
0.8	10^7	7	c	24	strong alkalis,
0.5	10^7	12	c	–	mildew,
0.5	10^7	8	c	–	light
3	10^9	14	c	–	Strong alkalis,
6	10^{10}	10	c	–	acids, light
3	10^7	13	c	44	Acids,
4	10^7	13	c	–	strong alkalis, light, mildew
1	10^7	11	c	–	
2	10^7	11	c	28	
2	$10^{13\,b}$	6	230	–	Acids, alkalis, light,
2	$>10^{12\,b}$	4	230	–	acetone acids, and
4	10^9	14	c	–	alkalis, light
6-7	$>10^{12\,b}$	2.8-5	225	21	Strong acids,
6-7	$>10^{12\,b}$	2.8-5	250	21	oxydizing agents,
–	–	2.5	274	–	light
2-9	$>10^{12\,b}$	0.4	250	95	Strong acids Strong alkalis
9	$>10^{12\,b}$	0.4	250	95	Strong alkalis
5	$>10^{12\,b}$	1.5	Sticks at 235 d	–	Strong alkalis
5	$>10^{12\,b}$	1.5		91	Strong alkalis
8	$>10^{12\,b}$	0.1	165	–	Light
3	$>10^{12\,b}$	0	115	–	Very resistant
7.5	–	6.5	decomposes at 380°C	–	Resistant
5	–	6	Chars-Carbon 300-580	–	Strong Alkalis
18	–	1.3	230	–	Resistant
1	10^9	0	800	100	Very resistant
–	–	1	1500	–	Very resistant

cDecomposes first.
dDoes not melt.

Prolonged exposure to sunlight, high temperatures, moisture, and attack by microorganisms can cause severe damage to the strength of fibers and thus affect the durability of fabrics and garments. It is therefore important that the fibers have reasonable resistance to degradation caused by environmental conditions. Cotton fiber is prone to attack by bacteria and organisms when stored in damp conditions; dry fiber is not much affected. Ultraviolet rays can induce oxidation and thus cause severe damage to the fiber. Wool fibers are attacked by moths that lay eggs in the wool material; these eggs hatch out into grubs that eat the wool, forming holes in the fabric. Synthetic fibers are strongly resistant to mildew and fungi.

Although textile fibers are required to be strong and flexible, it is equally important that they should be resistant to attacks by chemicals. Most natural textile fibers are inert, possess a good resistance to mild alkalis and acids, and are practically insoluble in organic solvents and, of course, in water. However, some synthetic fibers have poor inertness characteristics and are susceptible to modifications by organic solvents. It is therefore important that inert solvents be used in dry-cleaning, otherwise materials like acetate, nylon, and rayon will swell or dissolve and lose their useful characteristics. Very strong acids and alkalis generally cause most textile fibers to degenerate.

The physical and mechanical properties of various textile fibers that concern the textile engineer and the consumer are summarized in Table 2.5.

NATURAL FIBERS

Vegetable Fibers

Vegetable fibers are divided into three major categories: (1) seed fibers, (2) bast fibers, and (3) leaf fibers. In the first group, the fibers are attached to the seeds and fruits of plants. The most important of the seed fibers is cotton, which has been put to innumerable textile end uses. It is grown in many parts of the tropical regions of the world.

Cotton

Cotton fiber is a seed-hair consisting of a single cell. It comes from a variety of plant species included in the *Gossypium* family. The cotton grows on the plant as long hairs attached to the seeds inside the boll.

There are numerous varieties of cotton grown all over the world. Such basic characteristics as length and fineness of the cotton fiber are dependent on the type of seed used. However, fiber properties are also sensitive to changes in environmental conditions during the growth period. All varieties of cotton inherently contain a small percentage of short and immature fibers, but any drastic changes in climatic conditions can result in the unbalance of the normal proportions. The proportions of short and immature fibers in a cotton are a major factor in determining its quality and are a source of nuisance during processing. Maturity of a cotton is characterized by the degree of the development of the cell wall. If a cotton has a well developed wall thickness, it is said to be mature; on the other hand, a cotton fiber with a thin and poorly developed cell wall is said to be immature. Maturity has been defined by Pierce (8) as the ratio θ of the cross-sectional area, A of the cell wall to the area A_o of a circle of the same perimeter. Hertel and Craven (9) have defined the maturity of cotton in terms of "immaturity ratio," I, which is equal to the reciprocal of maturity ratio. Geometrically:

$$\theta = \frac{1}{I} = \frac{A}{A_o} \ .$$

In most commercial cottons, the maturity count varies from 68 to 76%. Cottons with maturity counts below the level of 67 are considered to be immature and unfit for producing quality textiles. These immature fibers are the cause of "neppiness" in a yarn and of specky appearance in the fabric. The incidence of an abnormally high percentage of short fibers can cause excessive yarn irregularities and can make the yarn more fuzzy or hairy in appearance.

Cotton Fiber Staple Length

Cottons are generally classified (graded) according to their length. Technically the most important fiber length is called the "staple length." The staple length of a cotton has been defined as "a quantity estimated by personal judgement by which a sample of fibrous raw material is characterized as regards its technically most important—"fiber length." The importance of this quantity has been recognized since the earliest inception of the roller-drafting system in the spinning of cotton yarns, since there is a fairly close correspondence between staple length and roll setting. The measurement of staple length is carried out by classers who use hand stapling method, as shown in Figure 2.3. The staple length is then given by the distance between the well-defined edges, the well-defined edges are the areas in which the density of the tuft changes most rapidly.

Because of the reliance on the personal judgment of the classers, the staple length was never formally defined in terms of any statistic of length distribution; a true estimate of this quantity had a tendency to shift over a period of years. However, in order to minimize the fluctuations, the United States Department of Agriculture established physical reference standards for Upland cottons. The classer can check this standard of judgment against the reference periodically to project any shift in his level of judgment. Nevertheless, there is still a possibility of individuals differing by about 10% in their judgment.

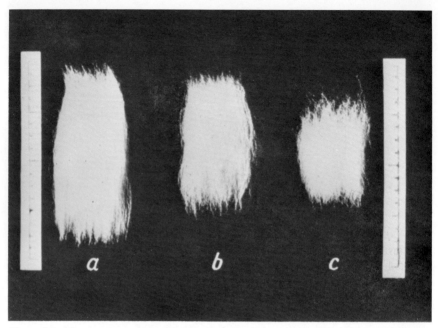

Figure 2.3. Hand staples. (*a*) Egyptain cotton of 1 7/16 in. staple, (*b*) American cotton of 1 1/8 in. staple, (*c*) Indian cotton of 7/8 in. staple. (*from Morton and Hearle*)

The first attempt to define the staple length more explicitly was made by Clegg (10). She started with a "Baer Sorter" diagram and geometrically derived a quantity called "effective length."

Lord (11) has reported a most exhaustive study of the measurement of staple length of cottons from all over the world both by judgment and by an instrument he designed to measure this quantity rapidly. He pointed out that, except for Egyptian cottons, the staple length is given by the modal, or most frequent (highest frequency), length of a numerical frequency distribution. The derivation and the construction of the diagram is shown in Figure 2.4. The effective length has been defined as follows:

The effective length may be defined statistically as the upper quartile of the fiber length distribution curtailed below the value equal to half the effective length (12).

In other words, effective length is the upper quartile of a numerical distribution from which some of the shortest fibers have been discarded by an arbitrary geometric construction. A detailed account of the frequency diagrams, and survivor and beard diagrams in relation to fiber length distribution is found elsewhere

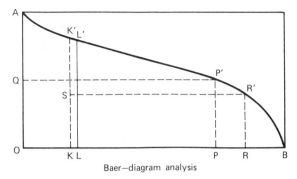

Baer—diagram analysis

Figure 2.4. Construction for estimating staple length.

(3). Lord and Underwood (12) have reported that "For American Upland cottons, from about ¾ in. to 1¾ in. staple and classed on the basis of American Staple Length Standards, a simple conversion formula is:

$$\text{Effective length} = 1.1 \times \text{American Staple."}$$

However, in case of Egyptian cottons where there are no universal standards in use, the staple length corresponds fairly closely to the effective length.

There are several methods that have been developed over a period of years for the measurement of fiber length. These include:

1. Individual fiber methods
   a. Oiled-plate method
   b. Semiautomatic single fiber testers (e.g., Wira Fiber Length machine)
2. Comb sorter methods (mechanical and semimechanical sorters)
   a. Comb sorters (e.g., Baer comb sorter; Suter-Webb comb sorter)
   b. Balls sledge sorter
3. Scanning Methods
   a. Thickness scanning (e.g., Uster Stapling Apparatus)
   b. Photoelectric scanning method (e.g., Fibrograph and Shirley P.E.M. Stapler)
   c. Capacitance methods (e.g., Wira Fiber Diagram Machine)
   d. Cutting and weighing methods (e.g., Chandler's; Ahmad and Nanjundayya's; and Muller's method).

One method, evidently the most reliable one, is to measure the length of each individual fiber under the microscope. This method requires placing a few fibers on a glass plate with a centimeter scale photographed or etched on it. The plate

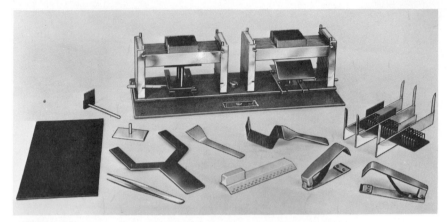

Figure 2.5(a).   Suter-Webb Comb Sorter. (courtesy Alfred Suter Co., Inc.)

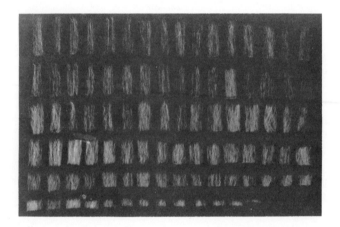

Figure 2.5(b).

is smeared with liquid paraffin to control the fibers. The fibers are straightened one by one over the slide with the tips of little fingers, and their lengths are measured. This is a very tedious and time-consuming method and is only used by research workers. For long man-made fibers and wool tops, Anderson and Palmer (13) have devised a semiautomatic Wira Fiber Length Machine to measure the single fiber length.

To avoid the time-consuming method of measurement of individual fibers and to accelerate testing of larger samples, a variety of mechanical and semimechanical sorters have been developed. In the sorter methods, a sample is either divided into a suitable number of groups, or the fibers are graded in the order of their lengths. Basically these methods consist of two steps: (1) the preparation of a tuft of fibers and (2) the withdrawal of the fibers from the tuft in order of either their increasing or their decreasing length. In the United Kingdom, the most commonly used apparatus for measuring the length of raw cotton is what is known as the "Baer Comb Sorter," which is used to prepare a Baer diagram (length distribution). The method used in the United States for measuring the length characteristics directly from a distribution by weight is that developed by ASTM (14) applied in conjunction with the Suter-Webb Sorter (15). A brief description of this method for determining fiber length is now given.

In this method, a standard weight of 75 ± 2 mg test specimen is used. A sorting apparatus of two banks of parallel combs is employed (as shown in Figure 2.5), and a three-stage process is followed to straighten and align the fibers. In the third and the final pull, the fibers are carefully withdrawn, and they are placed in order of their length on velvet-covered boards. The fiber groups are then formed with respect to length in increments of $\frac{1}{8}$ in. or 3 mm and the midpoint of the group range is taken to be its mean. These groups of fibers are then weighed; a reliable distribution by weight is obtained. From these weight-length data, the upper quartile length, mean length, and other characteristics of the material are computed. A detailed description of this and other comb sorter methods is given in (3).

Although the comb sorters are very useful in providing information on the length distribution and in determining the amount of short fibers present in the sample, the method is tedious and time consuming. To overcome these difficulties, a number of other devices have been developed in which a representative tuft of a material is scanned photoelectrically from end to end for a property linearly related to number. One of the most extensively used instruments in which the principle of photoelectric scanning is utilized is the Fibrograph (Figure 2.6) developed by Hertel (16) for the measurement of cotton fiber length. This method involves the presentation of the test specimen for scanning in the form of a pair of fiber tufts. The instrument is used mainly to obtain the average or mean length and the upper half mean length of fibers (approximately ¼ - 0.15 in. or 3.81 mm in the latest digital Model 230—and longer). The preparation of the fiber fringes is very important; consequently, the instrument makers suggest that the utmost care should be taken in the preparation of tufts. The fiber fringes are prepared by repeatedly working the test speciment between two combs until there is approximately the same amount of the material uniformly distributed over the full length of the comb. In the latest version, an accessory called a "Fibrosampler" is supplied to assist in charging the combs. The combs are then placed in the instrument, and the fringes are scanned slowly by a fluorescent light source. The light transmitted through the fringes is moni-

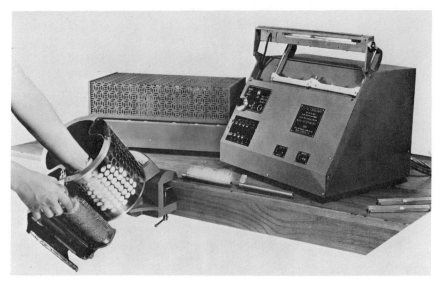

Figure 2.6. Fibrograph, digital modal. (courtesy Spinlab)

tored photoelectrically. The amount of light passing through the beard is taken to be proportional to the number of fibers that extend various distances from the comb.

In the manual and Servo-Fibrographs, a trace (curve) with the distance along the fibers as one axis and amount (number of fibers) as the other is recorded graphically. This is shown in Figure 2.7 and is called the Fibrogram. The Fibrogram is then analyzed to obtain various mean lengths, span lengths, and measures of length distribution, as shown in Figure 2.7. The most important quantities in the manual and Servo-Fibrograph are the mean length (M), and upper half mean length (UHM) and the uniformity ratio (M/UHM). In the digital and computerized Fibrograph, the 2.5% and 50% span length and 50/2.5 uniformity ratio are automatically recorded on digital counters during the scanning operation.

Following are the three major groups under which most of the commercial cottons fall:

1. Staple length 1 - 2½ in. Fibers in this group are very fine and are generally in the range of 10-15 μ (microns) in diameter. This group constitutes the soft and lustrous top-quality cottons such as Sea-Island, Egyptian, and some American varieties.
2. Staple length ½ - 1¼ in. These cottons have diameters ranging from 12 to 17 μ. Cottons in this group generally have medium strength and form the bulk of the world production. Most of the American, Peruvian and the American-Indian varieties fall into this group.

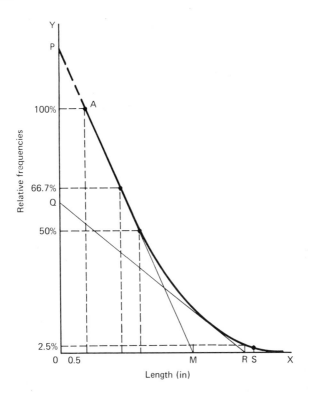

Figure 2.7. Fibrogram.

3. Staple length ⅜ - 1 in. Cottons in this particular length group are usually coarse, varying from 13 to 22 μ in diameter. Most of the Indian, Asiatic, and some varieties of Peruvian cottons belong to this category.

In a previous section, it is pointed out that the fineness of a fiber plays an important role in the processing and in the determination of the physical and the aesthetic properties of the end-use product. It is of even greater importance in the case of natural fibers (such as cotton and wool), where wide variations occur within a fiber as well as between varieties obtained from different sources. For example, some breeds of sheep invariably yield coarse wool, whereas others yield fine wool; the fineness of cotton also varies with the types and strains of plant. The importance of fiber fineness in relation to the evenness of yarns and the uniformity of blend composition are discussed in Chapters 7 and 10, respectively.

The linear densities of cottons range from about 340 mtex for a coarse Indian to about 100 mtex for a St. Vincent Sea Island, whereas the five Australian and

South African merino wools have a mean linear density in the range of about 450-600 mtex, compared to 1800-2000 mtex for the coarse Asiatic carpet wools.

Measurement of Fineness—Gravimetric Method

The linear density of a sample of fibers can be determined either by an individual fiber method or by the cut middles method. The former method can be combined with the individual fiber method used for measuring the fiber length, a known number of fibers whose length has been measured is weighed—the weight divided by the total length then gives the linear density of the fibers. The ASTM standard method for cotton* suggests the weighing of the groups of fibers of known length ranges (as obtained in the comb sorter). From each group, approximately 100 fibers are taken, and the weight and the count of the fibers are obtained (the two shortest groups and the groups that weigh less than 2 mg are ignored). From the weighings, the mean fiber linear density (micrograms per inch) is calculated.

In the latter method (cut-middles method), a known length (generally 1 cm) from the middle of tufts of parallelized fibers is sliced out by means of two razor blades set in a holder at a desired distance. Fibers in groups of 100 are counted from each of the tufts and weighed, care being taken to avoid fibers shorter than 1 cm. These methods are not suitable for highly crimped fibers because of the error introduced into the length measurement.

To speed up the process of measuring the fineness of fibers, some indirect methods of obtaining the desired results have been introduced. The most extensively researched and successfully employed methods include the porous plug or air-flow fineness testers, which are available commercially in a variety of forms. In these methods, the quantity measured is the specific surface, not the linear density of the material. The time required to perform one test is reduced to approximately 2 or 3 minutes. A discussion of the theory of air flow through porous plugs is out of the scope of this book, and the reader is referred to the reference 3.

The two commercially available air-flow instruments are the Wira Fiber Fineness Meter and the Sheffield Micronair.® The Sheffield Micronaire is used extensively in the United States for testing cotton. The Micronaire instrument is shown in Figure 2.8a and b. A brief description of the instrument and its method of operation follows.

In this instrument air is supplied at a specified pressure (air supply varies from 40 to 125 psi) through porous plug of fiber weighing 3.24 g, which is enclosed into a compression chamber, shown in Figure 2.8b. The construction of the specimen chamber and the compression plunger allows free flow of air through

*ASTM, D1769-72.

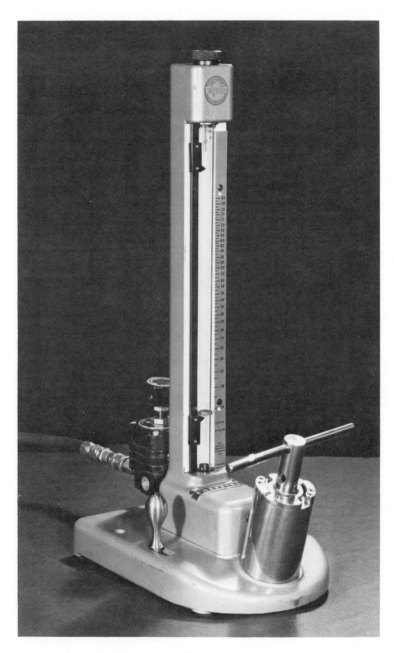

Figure 2.8(a). Sheffield Micronaire.® (Courtesy, The Bendix Corp.)

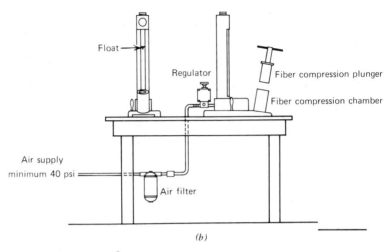

Float

Regulator

Fiber compression plunger

Fiber compression chamber

Air supply
minimum 40 psi

Air filter

(b)

Figure 2.8 (b). Micronaire® Gauge.

the fiber plug. The rate of air flow is indicated by a light metal float in the glass column. A special master claibration plug is used to calibrate the instrument by manipulating the regulator handle until the float rises to the 6.0 mark on the scale. The instrument manufacturers recommend the use of International Calibration Cotton samples, which may be obtained from the U.S. Department of Agriculture, Washington, D.C. After the initial procedure of calibration is completed, the master plug is removed, and the instrument is ready for use. The flow meter scale is graduated in units called the "Micronaire reading." The Micronaire reading of "fineness" is reported to the nearest 0.1-scale unit. The property on which the air flow depends is the specific surface (ratio of the perimeter to the cross-sectional area). Thus this instrument measures a value that is dependent both on the intrinsic fineness and the maturity of cotton (the development of wall thickness). Consequently, a sample containing finer fibers and mature fibers exhibits a higher macronaire reading—a higher resistence to air flow (other things being equal). Micronaire reading is very useful in controlling the mixing (blending) of cotton, in predicting the spinning performance (e.g., Micronaire reading below 3.3 for American Upland would produce a neppy yarn and excessive ends-down in spinning), and the strength and evenness of yarns.

Another instrument based on the air-flow principle is the Arealometer.* This instrument works on the principle of Wheatstone bridge. The specimen of raw cotton is used in the form of a cylindrical plug, and its resistance to air flow (at low constant pressure) is measured at two degrees of compression. The instru-

*For further description see reference 3 and A.S.T.M., D 1449-72.

ment is calibrated to give direct readings of specific surface (square millimeter per cubic millimeter). The two values of specific surface (at two degrees of compression) are then used to calculate the mean specific surface, mean fiber weight, and the maturity of the sample. In this instrument the sample chamber is 0.8 cm in diameter, and the volume of cotton plug used in 0.1 cm^3. This volume is attained by taking a quantity of fiber mass (in grams) equal to one-tenth of the density of fiber in g/cm^3. For example, the test specimen mass for cotton is 152 mg. However, in the portable version of the Arealometer, known as Port-Ar, the sample size used is 8 g.

Fineness Measurement by Vibroscope

This is an indirect method of determining the linear density of a fiber and is based on the theory of vibrating strings. The natural frequency of transverse vibration of a perfectly flexible string is determined by its mass per unit length, tension, and length as given by the equation:

$$f = \frac{1}{2\ell} \left(\frac{T}{m}\right)^{\frac{1}{2}} (1 + a)$$

$$m = T \left(\frac{1}{2 \, \ell f}\right)^2 (1 + a)^2$$

Where f = natural frequency of transverse vibration
   ℓ = length
   m = mass per unit length
   T = tension
and a is a correction factor that depends on the elastic modulus of the material; for fibers of circular cross section, a is given by:

$$a = \frac{r^2}{\ell} \sqrt{\left(\frac{\pi E}{T}\right)}$$

where r is the fiber radius and E the Young's modulus.

This is a nondestructive method for determining the linear density of individual fibers. It is widely used in research when it is necessary to measure the tenacity and other tensile properties of single fibers. This method can be used for fibers of irregular cross-sectional shape. This techinque is not suitable for cotton fibers because of the within-fiber variability. The vibroscope method for measuring the linear density of crimped and uncripmed fibers is the subject of ASTM D1577-66.

Cotton fiber appears as a long, thin, irregular, convoluted, and flattened tube (Figure 2.9). The cross section of a mature fiber shows a thin cuticle and primary wall on the outside and a hollow collapsed lumen in the center of the fiber. The rest of the fiber is made up of secondary wall. The secondary wall material is deposited in a series of daily growth rings on the inside of the primary wall at the time of growth. The material within each layer is made up of a large number of fibrils that deposit in the form of a spiral around the fiber axis. These spirals frequently reverse in direction which, incidentally, correspond in the direction of the external convolutions. Chemically, both primary and secondary walls consist of cellulose. Cotton fiber is the purest source of cellulose, containing about 90% cellulose by weight. In flax the content is 60-85% by weight.

The cellulose molecule consists of a series of glucose rings with the formula $C_6H_{10}O_5$ joined together:

Cellulose Molecule

In native cellulose, the degree of polymerization (that is, the number of glucose rings joined together in the chain molecule) is of the order of 10,000. These long-chain molecules are packed in the fiber and are approximately two-thirds crystalline and one-third noncrystalline material.

Bast Fibers

As the name implies, these fibers are obtained from the inner bark of the stems of plants called *dicotyledenous*. The most important fibers in this group are linen (flax), jute, ramie, hemp, and sunn. These fibers are also composed of cellulose. They are made up of long, thick-walled cells glued together by non-cellulosic materials (lignins and pectins) resulting in long fiber bundles running the entire length of the stem. The amount of noncellulosic material varies considerably from one type of fiber to the other. For example, jute may contain as much as 20% lignin, compared to 8% in flax.

The bast fibers are removed from the woody stems by the process known as "retting." The function of this process is to ferment the noncellulosic material binding the fibers and remove it by washing in water. However, the individual fibers, which are extremely short, are not completely separated from one another, but are extracted in strands to make them viable for textile processing. The strand length varies from one type of fiber to the other, as indicated in Table 2.2. Most bast fibers have very high strength but low elongation-to-break. Bast fibers are generally stronger when wet. The structure of most bast fibers

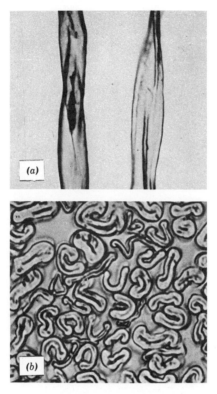

Figure 2.9. Cotton fiber. (*a*) Longitudinal view, × 500, (*b*) cross section, × 500. (1975 AATCC Technical Manual)

resembles that of cotton; however, the spiral angle of lamellar layers is much steeper in the case of the bast fibers. Bast fibers do not have the convolutions characteristic of cotton.

FLAX

This fiber is extracted from the stem of the annual plant *Linum Usitatissimum.* These plants grow in many temperate and subtropical regions of the world. Flax strands are generally processed in lengths varying from 18 in. up to approximately 36 in. This length suffers a loss by the time the strands reach the spinning process. Flax is generally graded on the basis of its color, which is usually yellowish white, but may change depending on the conditions during the process of retting.

Individual flax fibers vary in length from 0.25 to 2.5 in. and are about 25 μ in diameter. The fiber is long, transparent, cylindrical, and has a smooth but sometimes striated appearance (Figure 2.10). It has a narrow lumen running through

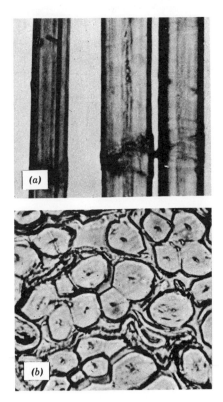

Figure 2.10. Flax fibers. (*a*) Longitudinal view, × 500, (*b*) cross section, × 500. (1975 AATCC Technical Manual)

the center. In cross section, the cell walls appear thick and polygonal in shape. Flax fiber is comparatively stronger than cotton but has very low extensibility. It has an average tenacity of about 60 g/tex and approximately 1.8% extension-to-break. It has a regain of about 12% and is about 20% stronger wet than dry. Linen is mainly used in the manufacture of sail cloth, tent fabric, sewing threads, fishing lines, tablecloths, and sheets.

JUTE

Jute is one of the most important fibers used for industrial applications. It is mostly used in making "sack cloth" and carpet-backing fabric. It is primarily grown in India, Pakistan, and to a certain extent in China. It is a bast fiber and is extracted from the inner bark of plants of the genus *Corchorus*. It is extracted by a retting process similar to the one used for flax. Jute is graded on the basis of its color and string length. Its color varies from yellow to brown to dirty gray, and it is lustrous in appearance. Jute fibers generally have a rough feel; however,

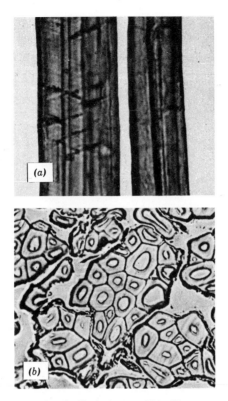

Figure 2.11. Jute fibers. (*a*) longitudinal view, × 500, (*b*) cross section, × 500. (1975
AATCC Technical Manual)

the best quality fibers are smooth and soft. A single jute fiber cell has an average
length of 0.1 in. and a mean diameter of 12 μ. The strand length varies from
approximately 5 to 12 ft. The individual fiber shows nodes and cross markings in
the longitudinal view, and polygonal shapes in the cross section (Figure 2.11).
Jute fibers vary greatly in strength and are not as strong as flax or hemp. They
have an elongation-to-break of approximately 1.7%. Jute is highly hygroscopic in
nature. Besides having many industrial applications, finer quality jute fibers are
utilized in furnishing and curtain fabrics. Bleached jute is sometimes blended
with wool to provide cheap woven apparel fabrics.

HEMP

Hemp comes from the bark of the plan *Cannabis Satina*. It is grown in almost all
the countries of Europe, including the U.S.S.R., and in many parts of Asia. Hemp
fiber is extracted from the woody matter by retting and subsequent breaking
and scutching. It is coarser than flax and has a dark color. The individual fiber

length varies, on an average, between 0.5 and 1 in., and the mean diameter is 17 μ. However, the strand length varies from 4 to 10 ft. The fiber shows joints and cracks on the surface. In cross section, it is seen to be polygonal in shape, with a pronounced lumen in the center. Hemp is primarily used in making ropes and twines and is woven into fabrics used for sack cloth and canvas.

MISCELLANEOUS

In addition to the fibers mentioned above, there are a few more fibers belonging to this class that find varied uses in the textile industry. These include sunn, kenaf, and ramie. These fibers (generally grown in warm climates) are extracted from the bark of plants essentially in the same manner as flax. Each of these fibers has a special place in the region in which it is grown. For example, sunn and kenaf are mostly grown in India and Pakistan and contribute a great deal toward the economy of these countries. Ramie is grown in the United States, U.S.S.R., China, Japan, and in some parts of Europe. These fibers are used mostly in the making of ropes, twines, and sack cloth.

Leaf Fibers

Leaf fibers are obtained from the leaves of *monocotyledenous* plants. These fibers generally find their uses in the making of ropes and cordage. The most important fibers belonging to this category are sisal, henequen, and abaca (Manila), in addition to several other less important ones.

SISAL

Sisal fiber is obtained from the leaves of the plant *Agave sisalana*. It is cultivated in East Africa, Brazil, Mexico, and other parts of Latin America. The mature leaves from the sisal plant are harvested and treated mechanically to separate the fibers from the pulpy material. A single leaf may contain up to 1000 fibers.

A sisal strand consists of a bundle of many individual fibers held together by noncellulosic gummy material, such as lignin, which may amount to approximately 6% based on the dry weight. The individual fiber is approximately 0.1 in. long. The fiber cells tend to be straight and stiff, which affects the flexibility of strands. Sisal fiber has a cylindrical shape in the longitudinal view and has a central canal or lumen that varies in width along the length of the fiber. In cross section, it shows rectangular and polygonal shapes (Figure 2.12). It has good dyeing characteristics and can be dyed with the same direct dyes used for cotton —acid as well as basic. Sisal is used extensively for making binder and baler twines and to a certain extent is converted into mattings and rugs.

HENEQUEN

Henequen fiber is a product of the plant *Agave fourcroydes*. The leaves of this plant are harvested first from the plants when they are between six and seven

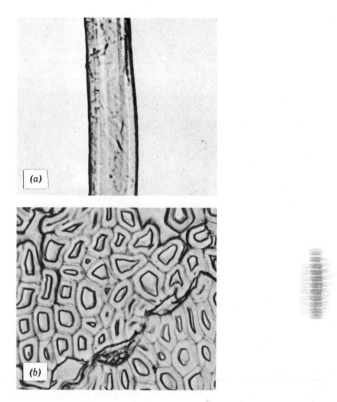

Figure 2.12. Sisal fibers (Java ×). (*a*) Longitudinal view, × 500, (*b*) cross section, × 500. (1975 AATCC Technical Manual)

years old. This procedure is followed at half-year intervals for 15-18 years until the plant flowers and dies. The fiber is extracted by a process similar to the one used for sisal. Its strands are approximately 5 ft. long. Henequen is lustrous and has good color. Its structure resembles that of sisal fiber. It also is used in making twines and coarse canvas fabrics.

ABACA (MANILA)

Abaca or Manila hemp is one of the most important cordage fibers. It is a product of the plant *Musa textilis.* It is indigenous to the Phillipine Islands, which account for nearly 90% of the total world production of Abaca. The abaca plant grows easily in the Phillipines in clusters of sheathlike leaf stalks. The center of the stalk, which is devoid of any fibers, is wrapped with layers of leaf sheaths containing thin layers of fibers. The plant reaches maturity in about two years and yields leaves for a period of up to 15 years. The fibers from the cut leaves are extracted mechanically by a process called decortification, or they may be separated from the ribbons of fibers by scraping the pulp material with a knife. The extracted fibers are then dried in the sun. The strength and the color of the extracted fiber is essentially dependent on the position of the leaf sheaths on the stalk. The outer sheaths produce the strongest and darkest colored and the inner sheaths the weakest and the lightest colored fiber. This difference in the color is due to exposure to sunlight. Fiber quality may also vary according to the maturity of the stalks and the method of extraction employed. The abaca strand is about 15 ft. long and contains many individual fibers held together by gummy cellulosic materials. It contains a very large amount of lignin (approximately 9%). The individual fiber is about 0.25 in long, showing a regular ribbon width with the ends tending to taper to a point. It has thin cell walls and a very large lumen. The fibers are polygonal in cross-sectional shape.

Abaca is mostly utilized in making ropes and cordage. The fiber is strong and moderately flexible. It is resistant to the effects of sea water and, because of its natural buoyancy, is extremely useful for making ship's cables and hawsers. It is also used for making carpets and mats.

Animal Fibers

Although textile fibers of animal origin amount to only 8-10% of the total weight of world fiber consumption, they do, however, play a very important role in the textile industry. These fibers have unique characteristics that make them quite attractive for use as textile materials.

There are two distinct classes within this group. The first type includes the fibers obtained from the fibrous covering or fur of animals, such as wool, mohair, cashmere wool, alpaca, and vicuna, in addition to camel and pig hair and rabbit fur. The second type is silk fiber produced by the silk worm.

Wool and Hair

The most important of these fibers commercially is "wool," the fibrous covering of the sheep, it accounts for nearly 90% of the total world production of animal fibers. It is probably advisable to define the two terms "wool" and "hair" which are most commonly used in the textile industry. Wool is used exclusively to signify the fiber obtained from the covering of sheep. However, both wool and hair are the fibers that form the covering of animals. Wool refers to the fine undercoat; the long and coarse fibers forming the outer coat are termed hair. The chemical structure of all hair fibers is related to that of wool, that is, they are mostly keratin. However, they are all different from wool, as well as from each other, with regard to their structural and physical characteristics.

Wool is obtained by shearing the fiberous covering of sheep and is produced in almost all parts of the world. The shorn wool is known as "fleece" or "clip wool." After shearing, the fleece is "skirted" to remove the soiled wool from the edges of the fleece and then graded by experts who base their judgment on fiber fineness, color, length, and the presence of foreign matter. After grading, the wool is packed in sacks and baled (approximately 300 lb); it is then ready for shipment. The quality of wool depends greatly on the breed of the sheep as well as on the environmental conditions under which the sheep have been reared. The general basis on which the quality of a clip is judged is related to the limiting "size" of the yarn into which it could be spun on the English worsted count system. Thus, for example, an 80^S wool would mean that 1 lb of it would yield 80 hanks of 560 yards each. In recent years, attempts have been made to grade wool on the basis of average fineness or diameter of fibers in microns. However, since there are other factors besides fiber fineness that must be taken into account for quality assessment, a close correspondence between this property and quality grading is not to be expected. Wool may be classified into three broad groups: merino, crossbred, and carpet. There is no international agreement on this classification system, but a widely accepted division in terms of the Bradford system defines the quality as follows: merino wool, 60^S and upward; crossbred wool, 46^S-58^S; and carpet wool, up to 44^S. The first two qualities are used mainly for apparel fabrics; the bulk of the third group is consumed in the production of carpets and felts. There is, in general, a negative correlation between fineness and length of wool fiber as contrasted to the relation observed for cotton. Thus finer wools are usually shorter in staple length. The 80^S Australian wool is about 2.5-3 in. long, 56^S Southdown is 3-3.5 in.; 48^S Romney Marsh 5-6 in; and 36^S Lincoln varies from 10 to 12 in. in length. Finer wools are processed on the worsted system to produce finer and stronger yarns, whereas the coarser wools are spun on the woolen system to yield thick and fuller yarns. The average diameter of a fine quality merino wool fiber is about 17 μ; for medium quality wool, it is about 22-32 μ; and the coarse wools have diameters of approximately 40 μ.

Some idea of the relationship of wool characterisitics to its spinnability (yarn count) is given in the following Table 2.6.

Table 2.6.   Relationship of Wool Characteristics to Its Spinnability

Wool Type	Breed	Average Length (in.)	Averate Diameter (μ)	Count (s)
Fine	Merino	1.5-4	10-30	58-90
Medium	Suffolk Cheviot, etc.	2-4	20-40	46-60
Long	Cotswold Romney Marsh	5-14	25-50	36-50
Crossbred	Corriedale	3-6	20-40	50-60

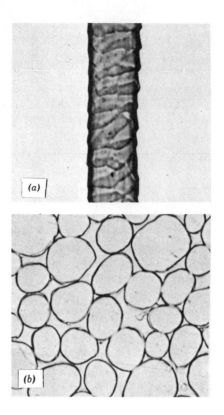

Figure 2.13. Wool Fibers—Merino. (*a*) Longitudinal view, × 250, (*b*) cross section, × 500. (1975 AATCC Technical Manual)

A wool fiber is shown in Figure 2.13. Chemically wool consists of a complex protein called "keratin" with an empirical formula $C_{72}H_{112}N_{18}O_{12}S$. Keratin is composed of a number of α-amino acids with the general formula $NH_2CHR.COOH$, which are linked through their amino and carboxyl groups into a polypeptide chain of the form:

$$\begin{array}{ccccc} & R_1 & & & R_3 \\ & | & & & | \\ & CH & CO & NH & CH \\ CO & NH & CH & CO & \\ & & | & & \\ & & R_2 & & \end{array}$$

Keratin

The polypeptide chains of the keratin molecule are cross linked to adjacent

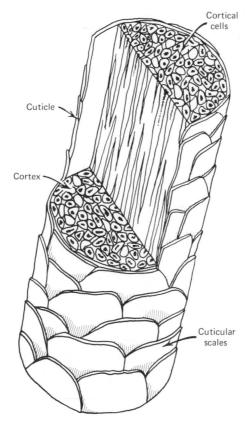

Figure 2.14. Structure of wool fiber.

chains through disulphide bonds (cystine linkages) and salt linkages formed by the combination of free carboxyl and amino groups.

The wool fiber generally appears as a circular cylinder that tapers from the root to the tip. It has a spirally crimped form. When viewed through the microscope, wool fiber shows four distinct regions. They are (1) the outer sheath or epicuticle, (2) the scale-cell layer, (3) the cortex, and (4) the medula, as shown in Figure 2.14.

(1) The outer sheath consists of the nonprotein part of the fiber. It is a thin, water-repellent membrane. It has, however, tiny microscopic pores, through which water vapor may permeate into the internal structure of the fiber. Thus the outer sheath helps wool fabrics to absorb water vapor from the human body, without feeling damp, and release it into the air.

(2) Underneath the epicuticle, at the surface, there are cuticle or scale-like cells. These scales point toward the tip of the fiber. They cause a special direc-

tional frictional effect that has a very important influence on the frictional behavior of wool fibers.

(3) The bulk of the fiber is formed of the cortical cells or cortex, and it is enclosed by the cuticle. Within the cortex there is a fibrillar structure. The cortical cells are 100-200 μ in length and 2-5 μ wide. The cortex of the wool fiber has been shown to have a bilateral structure; one side is called the paracortex and the other orthocortex. The chemical structure of the proteins in the two sections is thought to be different. The paracortex is more stable and is less accessible to dyes than the orthocortex. This bilateral structure gives the fiber a crimped form that is in phase with the mutual twisting of the two sections.

(4) Many coarse wool fibers have a hollow space in the center running along the length of the fiber. This is the medulla, and it may be empty or it may be made up of a different type of cell. The medula is absent in fine wools.

Most finer wools have a white or creamy color, although some breeds of sheep yield brown or black wools. Wool fiber has a natural luster, depending on the type of wool. Merino wools are generally semidull, whereas some other varieties have a silky luster. Wool fiber has a density of 1.32, which makes it slightly lighter than cotton. Wool and other protein fibers with folded molecules are characterized by low tenacity but higher extensibility. This characteristic, nevertheless, does not lower its work of rupture. Wool and other hair fibers are not very good under high stresses, but show large recovery from high strains. For example, these fibers have an elastic recovery of 99% from 2% and 60% from 35% extension, respectively. This makes the wool fibers highly resilient. In other words, wool fibers have a tendency to return completely to their original shape after small deformations, which is a great asset in apparel fabrics.

Wool has an equilibrium moisture regain of 13-19%, depending on the form and condition. When a fiber absorbs moisture, heat is liberated. This liberation of heat has physiologic consequences. For example, the moisture regain of wool changes by about 17% when it is transferred from an indoor atmosphere of 18°C, 45% R.H., to that of 5°C, 95% R.H. outdoors. This change, for a man's suit weighing 1.5 kg. would correspond to the release of 150,000 calories, which is equivalent to the heat produced by body metabolism in as much as 1½ hours. Thus there would be time for the body to adjust to its new environment. The high moisture regain properties of wool also contribute partly toward its non-flammability characteristics. Wool is fairly resistent to mild acids, but it is particularly sensitive to alkaline substances. It is attacked by moth grubs, but is resistant to mildews and bacteria. There are various industrial chemical modification treatments by which wool fabrics can be made shrink and felt proof.

The natural crimp in wool is of great importance, since it results in making a yarn fluffy, thereby trapping air in the interstices between the fibers. This trapping of air helps in forming an insulating layer, thus imparting the characteristic of warmth. It is obvious from the considerations of the physical characteristics of wool fiber that it occupies a unique position in the world textile market. However, the demand exceeds the world supply. This deficiency is remedied to a certain extent by reclaiming the already spun, woven, or knitted and worn wool and reusing it in making new fabrics. The recovered wool is sometimes mixed with fleece wool and used for making low-quality apparel fabrics or blankets. These fabrics are sometimes labeled as "all wool," whereas those made from new material are labeled "virgin wool."

Mohair

In addition to wool, there are a number of other hair fibers obtained from animals of the goat and camel families that are also of commerical importance. Mohair is the product of the Angora goat native to Turkey. Most of the world mohair production now comes from the United States, South Africa, and Turkey, totalling approximately 60 million pounds on a greasy basis.

Mohair fiber has a fine structure similar to wool. However, it appears circular in cross section, with small spots caused by trapped air bubbles. Its staple length varies according to the age of the animal. A six-month-old Angora kid would yield fibers varying from 4 to 6 in., whereas a full-grown goat would produce mohair fibers of 9-12 in. in length each year.

Mohair has physical properties essentially similar to those of wool. In addition, mohair has a very high resistance to wear. It is therefore used in applications such as upholstery where durability is of prime consideration. It is also blended with wool to produce light-weight fabrics for summer wear and other kinds of apparel materials.

Camel Hair

The better variety of camel hair used in the textile industry comes from the Bactrian Camel, found in China, Mongolia, and the U.S.S.R. There are essentially two types of hair obtained from the camel. They are the soft, downy undercoat, which is very fine and light brown in color, varying from 1 to 5 in. in length, and the coarse outer coat, tough fiber ranging in length up to 15 in. The fleece of the camel is not claimed by either shearing or pulling, but is shed naturally in locks. Each animal yields somewhere between 5 and 10 lb. of fiber per year.

The soft fine camel hair fiber is used for making overcoats, dressing gowns, and knitted fabrics. The coarse fibers are used for making belting, ropes, and blankets.

Cashmere Wool

This fiber is obtained from the cashmere goat found in Tibet, Northern China, Mongolia, Northern India, Iran, and Afghanistan. The Cashmere goat is covered with an outer coat of straight, coarse, long hair approximately 2-5 in. long, under which there is a downy undercoat of very fine and soft valuable fibers. The goat sheds the undercoat and some of the outercoat of hair through molting. At this time, the goat is combed and the two types of fiber are separated. The yield of both types of fiber amounts to approximately 8 oz. per animal. Cashmere hair is extremely fine, varying in diameter form 15 to 17 μ. Pure cashmere is often used in the production of such high-quality fabrics as the famous Indian shawls, which have a beautiful drape and soft handle.

Alpaca, Llama, and Vicuna

These animals are the inhabitants of the high mountainous regions of South America. They are found chiefly in Peru, Bolivia, Equador, and Northwestern Argentina. The llama or South American camel is mainly a beast of burden that provides a fleece of thick, coarse fibers about 10-12 in. in length. The fleece is a mixture of fine hair and kemp. It is usually brown in color and is used mainly locally for producing handmade fabrics, rugs, and carpets.

The Peruvian alpaca, a close relative of the llama but a much smaller animal, produces a fine soft fleece that may yield fibers of up to 16 in. in length. It is suitable for spinning on the worsted system. Its color may vary from black or brown to fawn or white. Yet another animal, the huarizo, a cross between llama and alpaca, produces a moderately uniform but less fine fleece than that of the alpaca.

A native of Peru and Bolivia, the vicuna (belonging to the llama family) is a much smaller animal. Its undercoat comprises extremely fine (13 μ) and short (about 5 in.) fibers. The fiber is a tawny brown in color and produces very fine and soft fabrics. It is the finest of all wool or hair-like fibers.

Silk

Silk is the only natural fiber that occurs in the form of a fine continuous filament. The silkworm extrudes a fine filament and wraps it around itself in the form of a protective covering or cocoon before it changes into a chrysalis and then a moth. It extrudes two filaments laid side by side and coated with a natural gum called sericin. This gum hardens on exposure to air, leaving the two threads firmly bound together. There are a number of silk-producing worms; the most important is the "Bombyx Mori" or mulberry worm. This worm feeds exclusively on the leaves of *Morus Alba,* a white mulberry. In addition, there are a few other wild varieties such as Eri, Tussa, and Muga, which are mainly found in India and live on the castor oil plant. There are also some varieties of "wild silk" that are grown in the Far East. The main silk-producing areas are the Far East, India, and some Mediterranean countries.

To reclaim silk filaments, the cocoons are soaked in hot water, which softens the sericin gum. Filaments from several cocoons are picked up, assembled,

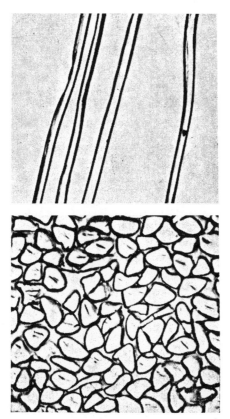

Figure 2.15. Silk fibers. (*a*) Longitudinal view, × 250, (*b*) cross section × 500. (1975 AATCC Technical Manual)

passed through a guide, and made into skeins by the process of reeling. Sometimes a small amount of twist is inserted in the bundle to hold the filaments together. This yarn can then be processed into fabrics before or after degumming. Degumming may be carried out in the fabric form also. The silk that cannot be extracted in the form of continuous-filament yarn by throwsters is salvaged in staple form and is called "waste silk." This is then processed into 100% pure silk spun yarns, or it may be spun in blends with wool, cotton, and other staple fibers.

The silk fiber consists of the condensation product of α-amino acids. The major constituents are glycine, alanine, serine, and tyrosine. Raw silk has a rough and irregular surface appearance, often showing cracks and folds in the outer sericine layer; it has an oval irregular cross section (Figure 2.15).

Silk is essentially used in very expensive luxury goods. It has been able to withstand competition from synthetic fibers in many high-quality textile applications because of its excellent dyeing characteristics, high moisture and absorbancy, and heat-preserving property. It has one major drawback—it does not blend easily with other fibers.

Natural Mineral Fibers

Asbestos

Asbestos is a very important industrial fiber that serves the textile industry in a number of useful ways. It occurs in the form of a natural rock composed of tightly packed fibrous crystals. The fibrous crystals belong to several natural minerals such as anthophyllite, amphibole, and serpentine. These are generally constituted of silicates of either magnesium or its combination with other elements such as calcium, iron, and sodium. Chrysotile (hydrated silicate of magnesium), which occurs in the narrow veins of serpentine rock, forms the major source of the world's supply of asbestos. These fibers may vary in length from about ½ to 14 in. Major asbestos-producing countries of the world are Canada, South Africa, and the U.S.S.R.

Asbestos is highly resistant to heat, acids, alkalies, and other chemicals. It is used primarily in applications such as conveyor belts for transporting hot materials, electrical installations, fireproof clothing, and break linings.

MAN-MADE FIBERS OF NATURAL ORIGIN (REGENERATED)

Cellulosic Fibers

The regenerated cellulosic fibers include viscose, cuprammonium, and acetate rayons. These three forms of rayon together constitute the bulk of world production having a natural polymer (cellulose) base as raw material. Of these, viscose rayon is comparatively easier to produce and accounts for nearly 75% of the world man-made fiber production.

Viscose Rayon

Cotton linters or wood pulp form the basic raw material for the production of rayon. The first step in the production of viscose rayon involves the treatment of bleached and cleaned wood pulp with sodium hydroxide to produce alkali cellulose. After aging, the alkali cellulose is mixed with carbon disulphide to form sodium cellulose xanthate. It is then allowed to ripen for several days at controlled temperatures. During this process, the solution is filtered repeatedly, and when the solution attains a suitable viscosity, it is spun by extruding through the tiny holes of a spinneret. The thread line emerging from the spinneret is passed through a coagulating bath containing a predetermined mixture of sulphuric acid, sodium sulphite, zinc sulphate, and glucose. This process converts the sodium cellulose xanthate into solidified filament cellulose that is insoluble in the coagulating solution. The solidified filament is then drawn and wound on packages. The last stage of the process can be achieved in three different ways. The filament can either be spun by batch process (box and bobbin spinning), or by a continuous spinning process. The filaments may be twisted to form continuous-filament yarns or may be converted into staple after crimping by either mechanical or chemical means.

Viscose fiber may be spun in a variety of cross-sectional shapes. Spun-dyed viscose is produced by mixing pigments in the spinning dope. Titanium dioxide is incorporated in the dope for producing dull fibers. In addition to the regular type, rayon fiber is also produced in various modified forms such as high-tenacity and high-wet modulus or polynosics. High-tenacity rayon such as Tenasco® is produced by applying attitional stretch during drawing (in hot aqueous acid). The purpose of stretch is to increase the degree of alignment of cellulose molecules along the fiber axis and, in addition, causes an increase in the proportion of "skin to core." The skin has a finer texture and is stronger than the core. The high-wet modulus rayon or polynosics have a degree of polymerization around 500, compared to 200 for ordinary rayon. Rayon has a chemical structure similar to that of cotton; both are cellulosic. Regular rayon appears straight and smooth (Figure 2.16). In cross section, it shows striations or channels that run along the length of the filament and a marked difference between skin and core. These striations are caused by the contraction in volume of the filament during coagulation. Ordinary rayon has a dry tenacity of 2-2.5 g/den, compared to up to 4.0 g/den and 5.2 g/den for high-tenacity and high-strength polynosics, respectively. Ordinary rayon has very low wet strength, amounting to approximately 66% of the dry strength. It has high elongation-at-break, stretching up to 25% of its original length when dry. It has a moisture regain of approximately 13% under standard conditions.

$$\left[H \quad -O-CH \begin{array}{c} \overset{\displaystyle OH}{\overset{|}{CH}} - \overset{|}{\underset{OH}{CH}} \\ \diagup \qquad \diagdown \\ \diagdown \qquad \diagup \\ CH \!-\!\!-\!\!- O \\ | \\ CH_2OH \end{array} CH\!- \quad OH \right]_n$$

Viscose Rayon

Rayon is used in a multiciplicity of textile applications. It is cheap compared to

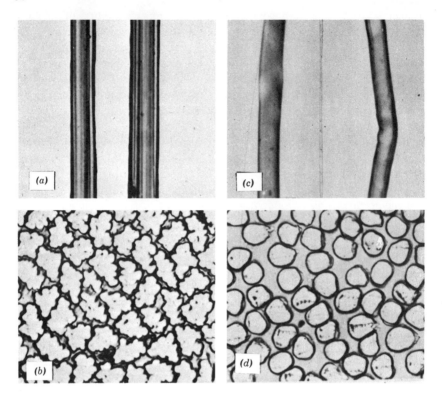

Figure 2.16. Rayon fibers, bright (*a*) Regular, logitudinal view × 250, (*b*) cross section, × 500, (*c*) high-tensity, high-wet modulus longitudinal view, × 500, (*d*) cross section, × 500. (1975) AATCC Technical Manual)

cotton, wool, or any other man-made textile fiber. However, its greatest draw-backs are its loss of strength on wetting and its dimensional instability. Neverthe-less, it has been possible by the use of modern resin finishes to make rayon fabrics more stable when wet. Rayon staple is used extensively in blends with cotton, wool, and other man-made fibers. A large volume of rayon is also used in the disposable nonwoven industry. Rayon fiber is available in a range of deniers in continuous-filament yarn as well as in staple form. The most common deniers are 1.5, 3, 4.5, 8, 12, 15, and 20, with staple lengths ranging from 1¼ to 8 in. The high-tenacity rayons are used in tire cords, conveyor belt fabrics, tarpaulins, and power belt applications. Polynosics are highly desirable for use in blends with polyester fibers because of their high modulus.

Cuprammonium Rayon

Regenerated cellulose fibers produced from the solution of cellulose in a mixture of copper sulfate and ammonia is called cuprammonium rayon or "cupro." The raw material for this type of rayon is similar to that used for the viscose process. Cleaned and bleached cotton linters or wood pulp is mixed with cuprammonium liquor (copper sulfate dissolved in ammonia forms cupritetramino hydroxide and cupritetramino sulfate) at low temperature. To this are added stabilizing agents and an excess of sodium hydroxide, which converts the cupritetramino sulfate to hydroxide. This solution is then filtered through screens, and is deaerated and spun either by a batch or continuous-spinning process. The solution emerging from these spinnerets is passed through a stream of pure water, which dissolves most of the ammonia and some copper. This action coagulates the cellulose, forming it into plastic filaments. The filaments are then drawn to achieve the required denier. They are then passed through a solution of sulfuric acid to remove the remaining copper and ammonia. The filaments are wound into skeins or cakes, washed to remove any trace of ammonium sulfate or copper sulfate, and subsequently dried and lubricated. The cupro filaments produced by continuous-spinning process are generally highly regular and have excellent physical characteristics.

Cupro filaments are produced in very fine deniers, usually 1.3 denier. In cross section, they appear round. They have a silk-like appearance and a tenacity of 1.7-2.3 g/den in the dry state and 1.1-1.3 g/den when wet. Other physical characteristics are similar to those of viscose.

Cupro is a comparatively expensive fiber, and it is used in the production of chiffons, satins, and nets, etc. It is often converted into novelty yarns such as slub, nub, knot, and gimp yarns, which are used in dresswear, drapery, and sportswear fabrics.

Acetate and Triacetate

The raw material for the production of acetate fibers is also the cellulose that is either cotton linters or wood pulp. The purified cotton linters or wood pulp are first steeped in glacial acetic acid and then transferred to a closed reaction vessel containing a mixture of acetic acid and acetic anhydride in a predetermined proportion. A small quantity of sulfuric acid dissolved in glacial acetic acid is added to start the acetylation process. Because of the exothermic nature of the reaction, the temperature of the vessel is maintained at $20°C$ by cooling and is allowed to go up to $25°C$ after about an hour. The acetylation reaction then continues for a further period of eight hours. This reaction yields the product called primary acetate (cellulose triacetate), in which the three hydroxyl groups on each glucose unit have been replaced by the acetate groups. The cellulose triacetate is then partially hydrolyzed by allowing it to stand in water. The hydrolysis is allowed to proceed until a degree of substitution of approximately 2.4 is reached. This acetylated cellulose, called secondary acetate, is precipitated from an aqueous solution and then dried. The flakes are ground and used in spinning filaments by any of the three spinning processes, namely, dry spinning,

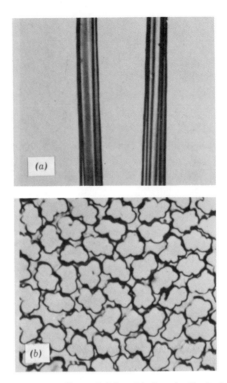

Figure 2.17. Secondary acetate fibers, bright. (*a*) Longitudinal view, × 250, (*b*) cross section, × 500. (1975 AATCC Technical Manual)

wet spinning, or melt spinning (discussed in Chapter 11). In addition to the regular form, acetate fibers are also spun in various modified forms, such as spundyed, delustered, and in various cross-sectional shapes.

Acetate is produced in the form of continuous-filament yarns as well as in staple form in a variety of deniers. The staple fiber generally contains mechanical crimp. Staple lengths of 1.5-3.0 in. are produced for processing on the cotton system, whereas for the woolen and worsted and the spun silk machinery, the most common staple lengths produced are 3-5 in. and 5-7 in., respectively. The most common deniers produced are between 1.4 and 5.0, but sometimes deniers as high as 20 are manufactured.

Acetate fiber has a low tenacity of about 1.1-1.3 g/den, with an elongation-at-break varying from 23 to 30% when dry. Its wet strength varies from 0.60 to 0.75 g/den, with a corresponding elongation of 35-45%. Acetate has generally moderate to good recovery characteristics. Cellulose acetate is a thermoplastic fiber and softens at about 200°C and melts at 232°C. In cross section, its outline appears to have a number of rounded lobes. In the long direction, it has folds and ridges (Figure 2.17).

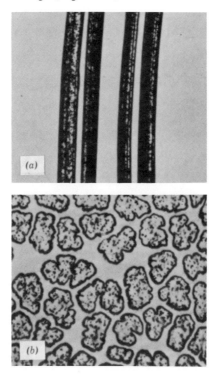

Figure 2.18. Triacetate fibers, dull. (*a*) Longitudinal view, × 250, (*b*) cross section, × 500. (1975 AATCC Technical Manual)

Acetate fiber can be dyed and handled in much the same way as other rayons and natural fibers. It has a soft handle and drapes well. It is used mostly in satins and taffetas and other apparel fabrics and applications such as underwear and lingerie that are worn next to the skin.

The triacetate fiber is produced in much the same way as acetate fiber except that the cellulose triacetate formed after acetylation is dissolved in either acetic acid or methylene chloride and allowed to retain the shape of the original cellulose. This is spun by any of the spinning processes, namely, dry, wet, or melt spinning. It is drawn and may be produced in continuous-filament yarns or as staple, in which case it is first mechanically crimped and then cut in short lengths.

Triacetate fiber shows striations along its length and appears multilobal in cross section (Figure 2.18). It resembles acetate fiber in most of its physical characteristics, with the exception of its handle and thermal characteristics. It lacks the soft handle of acetate; it has instead a crisp firm handle. The heat-setting treatment causes a change in the crystallinity and molecular orientation of the triacetate fiber. This characteristic is a great asset in that it protects triacetate fabrics from shrinkage and imparts better dimensional stability. Triacetate is used in a multiplicity of textile applications, such as knitted and woven undergarments, because of its good shape retention, and in easy-care garment applications, since it has drip-dry properties and needs no ironing. It is also used in blends with cotton, wool, linen, and rayon for such applications as skirts and slacks material, tablecloths, and furnishing fabrics.

Protein Fibers

Protein molecules are formed by linking or polymerization of amino acids by peptide links (CO—NH). The fibrous proteins (such as found in wool and silk) consist of more or less extended long-chain molecules. There are other natural long-chain molecules that may coil up into a compact ball held together by internal chemical cross links. These are called globular proteins. Such globular protein molecules may be unfolded or denatured by either heating or chemical means, and can then be converted into fibers. There is an abundance of certain proteins that form the by-products of industrial processes. These are casein, zein, arachin, and soy bean protein.

Casein is obtianed from skimmed milk. Casein is dissolved in caustic soda, allowed to ripen to a suitable viscosity, and then wet spun. The fibers are hardened by treatment with formaldehyde before use. It is then mechanically crimped and cut into staple of various lengths. Casein is generally blended with cotton, rayon, wool, and linen.

Casein filaments have a smooth surface and nearly round cross section. They have a low tenacity of about 1.1 g/den when dry and suffer a loss in strength of up to 65% when wet; they have a very high elongation of 60-70%. Casein fiber resembles wool in most of its physical characteristics. As a result, the bulk of casein fiber production is used in blends with wool. It has a soft and warm handle, good thermal insulation, and good recovery properties. Casein-wool blends are used extensively in knitting yarns, felt hats, carpets, and as resilient fillings.

Fibers obtained from the protein of ground nuts have almost wool-like characteristics. These are primarily used in blends with wool for making sweaters, blankets, carpets, and felts. However, blends with cotton or wool are sometimes used in the production of fabrics for shirting, pajamas, and dress materials.

Zein fiber "Vicara[R]" was first produced in the United States. It is obtained from zein, the protein of maize. Zein, soybean, and collagen protein fibers are no longer produced commercially and therefore are of little importance to the textile industry.

Alginate and Natural Rubber Fibers

Alginate Fibers

Alginic acid is a polymer of *d*-mannuronic acid with a molecular weight greater than 15,000. It is a constituent of seaweed and accounts for nearly one-third of the weight of seaweed. Alginic acid and its metallic (calcium, beryllium, and aluminum) salts have long-chain molecules that can be aligned to form fibers. However, alginate fibers are very sensitive to water and mild alkalies; consequently, these fibers have not been manufactured on a large scale for normal textile applications. The only fiber produced at present is calcium alginate by Courtaluds, Ltd., U.K. Calcium alginate yarns are used in very special applications.

Alginate fibers are round to oval with a serrated outline. The striations run along the length of the fiber. They have low tenacity ranging from 1.5 to 2.0 g/den and elogations of 2-6%. Alginate fibers do not burn but decompose when exposed to a flame. Calcium alginate is insoluble in water but attacked by mild alkali.

Because of its nonflammable character, alginate fibers have been used for theater curtains. The water solubility of some alginate fibers has proven to be of immense value in producing woolen yarns with high loft and in continuous production of fully fashioned knitted hosiery such as socks. The alginate yarn is introduced for a few courses between socks and it is later dissolved away, leaving the separated socks intact. Alginate yarn is also used in medical applications such as "styptic" elastic dressings, and because of its nontoxicity and easy absorbtion into the bloodstream, it is also used in dental surgery for plugging cavities.

Natural Rubber Fibers

Rubber is obtained by the coagulation of the latex produced from the rubber tree. In its natural state, rubber is a tough and elastic material. It can be softened by heating and milling, which renders it thermoplastic. It can then be molded or extruded to form sheets or filaments. Rubber loses its thremoplasticity and acquires elasticity when vulcanized. Rubber filaments are produced as ribbons by cutting a sheet or in round cross sections by extrusion. It can be used either in these forms or in the form of a core spun yarn, the rubber filament forming the core and the outer covering made up of spirally wound textile yarns. Rubber has low tenacity, high extension, and excellent elastic recovery characteristics. Its moisture regain is negligible, and it has a very high electrical resistivity. It deteriorates on prolonged exposure to sunlight. It is extensively used in sportswear, hosiery, and women's under- and outerwear to provide support and improve the fit of garments.

Silicate Fibers

These fibers are of commercial importance in high-temperature applications. This group includes fibers such as silicates, spun mineral silicates, or mixtures of minerals containing silicates; silica fibers are spun from silicon dioxide; quartz fibers are spun from naturally occurring silica in the form of quartz; silica (G) is obtained from the treatment of glass fibers. After removing all substances other than silica, the fibers are produced by dispersing silica or its derivatives in viscose dope, then spinning the filaments and subsequently burning the combustible products to leave a fiber composed essentially of silica. Silicate and silica fibers are very strong, nonflammable, and highly resistent to chemicals and solvents. None of these fibers absorbs moisture. These fibers are extensively used in boilers, high-temperature electrical and thermal proofing, in socket insulation, missile technology, etc.

SYNTHETIC FIBERS

Polyamide Fibers

Synthetic polyamide fibers, generally known as nylon, were the first successful commercially produced syntehtic fibers. Today they constitute a major part of the fiber production in the textile industry. Polyamide fibers are obtained by the condensation polymerization of long-chain molecules such as a diamine with dicarboxylic acid, or self-condensation of amino acid or its derivative such as a lactam. There are a number of experimentally produced synthetic polyamide fibers, but the two most important commercial ones that form the bulk of world production are nylon 66 and nylon 6.

Nylon 66 is produced by the condensation of hexamethylene diamine and adipic acid and has a general formula of the form:

Diamine Residue   Dicarboxylic Acid Residue

Nylon 66

The first figure refers to the number of carbon atoms in the diamine, and the second figure denotes the number in the dicarboxylic acid of the long-chain molecule. Another commercially important nylon fiber corresponding to this structural form is nylon 6 10 $[-NH(CH_2)_6NHCO(CH_2)_8CO]$.

The second type of nylon structure (nylon 6) is produced by the self-condensation of an amino acid (ϵ-amino-caproic acid), and therefore is denoted by a single figure giving the number of carbon atoms in the amino acid:

$$NH_2(CH_2)_5COOH \rightarrow \left(\begin{array}{c} \text{NH} \quad CH_2 \quad CH_2 \quad CO \\ \diagdown \diagup \diagdown \diagup \diagdown \diagup \diagdown \\ CH_2 \quad CH_2 \quad CH_2 \end{array} \right)$$

Nylon 6

Other examples of plyamide fibers of this structural type include nylon 11 $[NH(CH_2)_{10}CO-]_n$ (known as Rislan in France) and nylon 7, supposed to be produced in the U.S.S.R.

Even though nylon 66 and nylon 6 do not differ in basic chemical structure, there are certain subtle differences in the physical characteristics of the two types of fibers. Some of the most important are:

1. Nylon 6 has a lower melting point (215°C) than nylon 66 (250°C).
2. Nylon 6 has a slightly greater resistance to the influence of ultraviolet light than nylon 66.
3. Using a given dyestuff, nylon 6 will dye to a shade several times deeper than nylon 66, when dyed together.
4. Nylon 66 has a better resistance to degradation because of prolonged heating than nylon 6.
5. Nylon 6 has better elastic recovery than nylon 66.

Nylon (both types) fibers are produced in a multiplicity of finenesses (deniers) and staple lengths. They are produced as multifilament yarns, monofilament, and tow in bright, semidull, and dull luster. In addition to the circular cross section, some manufacturers make nylon fibers in trilobal and multilobal cross-sectional forms (Figure 2.19). Nylon is thermoplastic in nature, and possesses heat-setting characteristics. It is available in crimped and various other textured (bulked) forms.

Nylon fibers generally have smooth surfaces. They are produced in tenacities varying from 4.5 to 5.8 g/den (low tenacity) to as high as 9.0 g/den (high tenacity). High-tenacity nylon has lower elongation (20-28%) than regular filament or staple (30-46%). Nylon fiber has excellent recovery characteristics. It has a moisture regain of 4-4.5% at standard conditions (72 ± 2°F and 65% R.H.). It is not affected by most chemicals, but concentrated hydrochloric and sulfuric

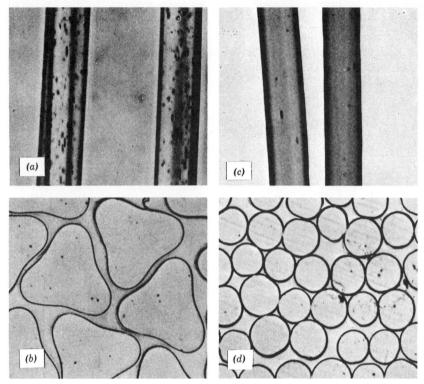

Figure 2.19. Nylon fibers. (*a*) Trilobal, bright longitudinal view, × 250, (*b*) cross section, × 500, (*c*) regular semidull, longitudinal view, × 250, (*d*) corss section, × 500. (1975 AATCC Technical Manual)

acids tend to deteriorate the fiber and cause loss of strength. Also, prolonged exposure to light or ultraviolet light can cause deterioration in fiber tensile properties. Nylon fibers melt when heated to high temperatures. As a result, when nylon fabric ignites, the molten polymer falls away and thus inhibits flame propagation. Nylon has very good dyeing characteristics and may be dyed with a wide range of dyestuffs.

Nylon fibers and filaments are used in a number of textile applications because of their outstanding mechanical properties, excellent recovery, and high resistance to abrasion and wear. In addition to apparel and household textile applications, they are extensively used in carpets and industrial applications such as tire yarns, hose, parachutes, belting and filter fabrics, to name a few.

Polyester Fibers

Polyester fibers form the second largest group of fibers that are used to satisfy the demands of the textile industry. Polyester polymers are produced by a condensation reaction, and the linkages between the monomers are produced by the formation of ester groups. There are a number of different types of poly-ester polymers, but the two most important used in the production of textile fibers are (1) polyethylene terephthalate fibers (PET polyester fibers, e.g., Dacron®) and (2) poly-1,4-cyclohexylene-dimethylene terephthalate fiber (PCDT polyester fiber, e.g., Kodel®). PET-polyester fibers are formed by the condensation polymerization of ethylene glycol with terephthalic acid itself or with demethyl terephthalate:

$$nHO(CH_2)_2OH \quad + \quad nHOOC\langle\!\!\!\!\bigcirc\!\!\!\!\rangle COOH \rightarrow$$

Ethylene Glycol   Terephthalic acid

$$HO\left[-OC-\bigcirc-COO(CH_2)_2O-\right]_n H + (2n-1)H_2O$$

Polyethylene Terephthalate (Dacron)

The second type of polyester (PCDT) is made by condensing terephthalic acid with 1,4-cyclohexanedimethanol

$$[-OC-\langle\!\!\!\!\bigcirc\!\!\!\!\rangle-COOCH_2-\bigcirc-CH_2O]$$

poly-1,4T cyclohexanedimethanol terephthalate

Polyester fibers have a smooth surface and generally have a circular cross section, except for some special types produced in trilobal form (Figure 2.20). Polyester is produced in a wide range of deniers and staple with bright, semidull, or dull luster. It has excellent tensile properties, good recovery from stretch, low moisture content and high abrasion resistance. It has a relatively high coefficient of friction and smooth surface. It is a thermoplastic fiber and melts when heated to high temperatures. It fuses and forms a hard bead when ignited. In all practi-cal senses, polyester fiber is an inert fiber, highly resistant to most of the common organic solvents. Polyester fibers are hydrophobic in character. They are therefore dyed by dyestuffs, such as disperse dyes, and so on, which are insoluble in water.

Polyester fiber or fabric may be heat set to impart dimensional stability which may be affected by any subsequent heat treatments or finishing processes. Polyester filament yarns may be texturized to produce bulk and stretch charac-teristics in woven or knitted fabrics.

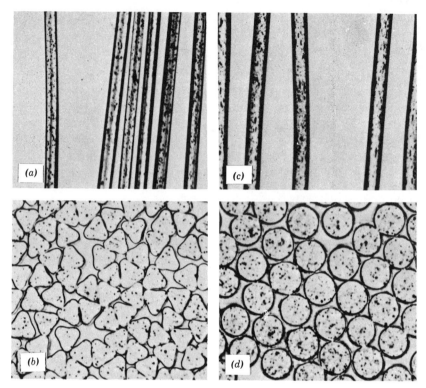

Figure 2.20. Polyester fibers. (*a*) Trilobal, semi-dull, longitudinal view, × 250, (*b*) cross section, × 500, (*c*) regular, semidull, longitudinal view, × 250, (*d*) cross section, × 500. (1975 AATCC Technical Manual)

Polyester fibers are used in many textile applications including carpets, upholstery, and industrial uses. A very large proportion are produced in staple form, which is used in blends with wool, cotton, viscose rayon, and linen. The inclusion of polyester fibers in blends improves the wear and abrasion resistance and the "ease of care" characteristics of fabrics. Other areas of polyester fiber use include fillings in pillows and quilts, sewing thread, tire yarns, conveyor belts, and ropes and twines.

In today's commercial markets, there are several modified forms of polyester fibers being produced, such as bicomponent fibers, fibers with multilobal cross section, and chemically modified forms.

Polyvinyl Derivatives

When vinyl chloride or its derivatives that contain a double bond are polymerized, they form long-chain molecules by the process known as addition polymerization. These polyvinyl compounds have proved to be a source of fibers that are being used as textile raw materials. Some of the most important classes of fibers obtained from polyvinyl derivatives are discussed in the following sections.

Polyacrylonitrile Fibers

These include the fibers spun from polymers or copolymers of acrylonitrile and can be further subdivided into two major groups: (1) acrylic fibers and (2) modacrylic fibers.

ACRYLIC FIBERS

Acrylic fibers (e.g., Orlon®) are spun from polymers consisting at least 85% by weight of acrylonitrile units: $-CH_2-CH(CN)-$. Acrylic fibers may be produced from 100% acrylonitrile or from copolymers, in which case the second component may be incorporated in the form of a normal or a graft copolymer onto the acrylonitrile. Some of the copolymer compounds used commercially are vinyl chloride and methyl acrylate. Acrylic fibers are produced mainly in the form of multifilament yarns, intended for conversion, or as staple fibers. Staple fibers are produced in a variety of deniers and lengths suitable for processing on all spinning systems. Acrylic fibers have extraordinary shrinkage characteristics that enable them to be used along with normal fibers to produce high bulk effects in the resulting product. Acrylic fibers are produced by either dry or wet spinning processes. The tow is drawn when hot, dried, oiled, and crimped, and may be cut into staple after relaxation by heating. The acrylic fibers produced today can be dyed with all types of available dyestuffs.

The acrylic fibers manufactured by different producers have physical properties that vary over a very wide range. They are produced in a variety of cross-sectional shapes depending on the type of spinning system employed. The dry-spun fibers (e.g., Acrian®, Creslan®, and Zefran®) have a round or kidney-bean shaped cross section, whereas the wet-spun fibers (e.g., Orlon®, Darlon®) have a dog-bone or flat cross-sectional shape (Figure 2.21). Acrylic fibers have medium tenacity and relatively high elongation and good recovery from small extensions. They have low moisture regain varying from 1.0 to 3.0%. Acrylic fibers melt on heating and generally exhibit a moderate flame propagation rate. They have good to moderate resistance to alkalies and acids and excellent resistance to sunlight.

Acrylic fibers are used extensively in carpets, furnishing fabrics, apparel (woven and knit fabrics), and nonwoven products.

MODACRYLIC FIBERS

The exact composition of individual modacrylic fibers is rather difficult to obtain. Modacrylic fibers fall under the category of polyacrylonitrile fibers and

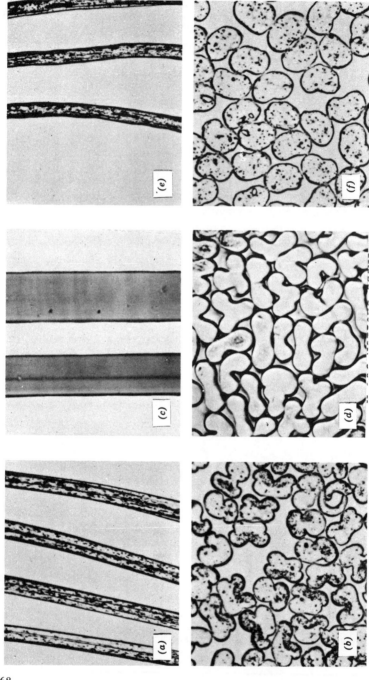

Figure 2.21. Acrylic fibers. (*a*) Modified wet-spun, semidull, longitudinal view, × 250, (*b*) cross section, × 500, (*c*) regular solvent spun, semidull, longitudinal view, × 250, (*d*) cross section × 500, (*e*) bicomponent, semidull longitudinal view, × 250, (*f*) corss section, × 500. (1975 AATCC Technical Manual)

generally contain less than 85% by weight of acrylonitrile. The proportion of the other components may vary widely. VEREL®, fiber produced by Tennessee Eastman Corporation, U.S.A., is one example of this type of fiber and is believed to contain approximately 60% by weight of acrylonitrile. This fiber is produced mostly in staple form and in deniers ranging from 3 to 60.

Modacrylic fibers are easy to dye. They are spun in 100% pure form as well as in blends with wool, cotton, rayon, nylon, and polyester fibers. However, modacrylic fibers generally have slightly higher moisture regain than the ordinary acrylic fibers. They have very good flame resistance characteristics; they self-extinguish and do not ignite easily.

Blends of modacrylic fibers with other fibers are used extensively in knit goods and pile fabrics. They are used in industrial applications such as filter fabrics and protective clothing. Because of their excellent flame resistance characteristics, they are used in drapery and upholstery material. They also find use in carpets because of their good resiliency and rot and stain resistance.

Polyvinyl Chloride Fibers

This class includes fibers spun from 100% vinyl chloride or copolymers of vinyl chloride.

$$CH_2 = CHCl \rightarrow -CH_2-CH-CH_2-CH-$$
$$\qquad\qquad\qquad\quad | \qquad\quad |$$
$$\qquad\qquad\qquad\quad Cl \qquad\quad Cl$$

Vinyl Chloride   Polyvinyl Chloride

PVC (polyvinyl chloride) fibers are produced from the polymerization of 100% vinyl chloride. Rhovyl® and Vinyon®, among others, belong to this class of fibers. These fibers are produced in a variety of staple lengths and in continuous-filament form with a wide range of deniers. Some of the PVC fibers produced commercially have shrinkage characteristics that make them suitable for use in making high bulk fabrics. PVC fibers are hydrophobic in character and may be rendered completely waterproof by suitable coating treatments. They have excellent flame resistance characteristics that make them extremely useful for protective clothing applications.

PVC fibers are smooth and have circular cross sections. They have good tensile properties, and their moisture regain is practically zero. PVC fibers are not affected by acids or alkalies and generally have excellent resistance to a wide range of chemicals. These fibers are used in a variety of textile applications; however, they find their properties exploited most usefully in industrial uses such as filter fabrics, waddings, tarpaulins, glider boats, etc. PVC fibers are also used in blends with wool, cotton, rayon, and nylon for apparel applications.

Textile fibers are also produced from copolymers of vinyl chloride with a smaller proportion of a second compound such as vinyl acetate and acrylonitrile. Vinyon® is an example of a fiber spun from vinyl chloride and vinyl acetate; Dynel® is produced from a copolymer of the former with acrylonitrile. Vinyon HH® is generally used as a binder material in the manufacture of bonded non-woven fabrics such as filter fabrics, felts, and webbings, etc. Dynel® fiber has good tensile and wear properties. It is used in apparel applications in blends with such fibers as wool and rayon.

Polyvinyl Alcohol Fibers

Polyvinyl alcohol fibers, such as Vinal®, are made from polymers or copolymers of vinyl alcohol

$$CH_2 = CHOH \quad \rightarrow \quad -CH_2-CH-CH_2-CH- $$
$$\underset{OH}{|} \qquad \underset{OH}{|}$$

Vinyl Alcohol   Polyvinyl Alcohol

If polyvinyl alcohol fiber is not subjected to heat and aldehyde treatments after spinning, it has a tendency to dissolve in hot water. These water-soluble fibers are used in special applications. Polyvinyl alcohol fibers are produced in filament as well as in staple form. They have excellent tensile properties. These fibers have a smooth surface, and generally have a U-shaped cross section with a flattened tube appearance. The latter characteristic makes them highly flexible, which results in a better feel and handle of fabrics. These fibers are used in all sorts of apparel applications such as suit interlinings, gloves, socks, and intimate garments.

Other fibers that fall under the polyvinyl derivative type are polytetrafluoro-ethylene (e.g., Teflon®), polyvinylidene dinitrile (e.g., Darvan®), and poly-styrene. For detailed information on the above fibers, the reader is referred to the sources of information listed at the end of this chapter.

Polyolefin Fibers

Polyolefin fibers are made from polymers or copolymers of olefin hydrocarbons. The two most important fibers in this category are polyethylene and polypro-pylene.

Polyethylene Fibers

Polyethylene fibers are spun from polymers or copolymers of ethylene:

$$CH_2 = CH_2 \quad \rightarrow \quad -CH_2-CH_2-CH_2-$$

Ethylene   Polyethylene

Polyethylene fibers are very difficult to dye. Consequently, colored fibers are produced in spun-dyed form by incorporating pigments in the molten polymer before extrusion. Polyethylene fibers are produced either in low or high density form. The mechanical and physical properties of polyethylene fibers are highly dependent on the conditions of polymerization because the process of polymerization affects the degree of branching of polymer chains.

Polyethylene fibers generally have a round cross section and a smooth surface. Low-density polyethylene has a low melting point and poor tensile properties. It is used mostly in industrial applications such as ropes, filtration fabrics, and protective clothing.

Because of higher crystallinity, high-density polyethylene fibers have a much higher tensile strength, stiffness, and higher softening point compared to the low-density polyethylene fibers. Nevertheless, they still cannot be used effectively for general textile applications because of their low melting point, and poor resilience and dyeing characteristics. However, because of their strength, rot resistence, and inertness to chemicals and light, polyethylene fibers are used extensively in marine applications.

Polypropylene

Polypropylene fibers are made from the polymers or copolymers of polypropylene.

$$CH_2 = CHCH_3 \quad \rightarrow \quad -CH_2-\underset{\underset{CH_3}{|}}{CH}-CH_2-\underset{\underset{CH_3}{|}}{CH}-$$

Propylene   Polypropylene

The polypropylene molecule consists of a long chain of carbon atoms with methyl groups protruding out from the side of the chain as pendants. This arrangement imparts to the molecule a three-dimensional structure. This structure can occur in a variety of forms with all sorts of spatial arrangements. The regularity of the steric arrangement of the polypropylene molecule has an important influence in determining the physical properties of the fiber. The steric arrangement of the side groups can be classified as (1) tactic, (2) syndiotactic, or (3) atactic. In the case of polypropylene fibers, the isotactic structure would be the one in which all the methyl groups are on one side; in the syndiotactic structure the methyl groups alternate on each side of the plane; in the atactic form the methyl groups are distributed randomly on both sides of the plane along the chain. The tacticity of the polypropylene polymer affects the degree of crystallinity which in turn determines the tensile strength of the fiber. Consequently, polypropylene fibers are produced in a number of forms with a very wide range of tenacities. Polypropylene fibers are smooth and generally round in cross section. Colored fibers are produced in spun-dyed form.

Polypropylene fibers are lighter than water, with a specific gravity of 0.90-0.91. They do not absorb any moisture, and water does not have any effect on their tensile properties. They have excellent resistance to acids and alkalies and very good thermal characteristics.

Polypropylene fibers have excellent processing behavior because of their high fiber-to-fiber coefficient of friction and good crimp retention properties. Polypropylene fibers have a greater covering power, because of their low specific gravity, than any other textile fiber. They have excellent abrasion resistance and flex resistance. Above all, polypropylene fiber is the cheapest of all the synthetic fibers. Polypropylene fibers are extensively used in blankets, upholstery, and carpets. They are also used in ropes and in the manufacture of fishnets, trawls, and lines. They are used in apparel application either in 100% pure form or in blends with rayon.

Polyurethanes

Polyurethanes form the basis of the elastomeric fibers generally known as spandex. These elastomeric fibers are characterized by the high extension and instantaneous recovery normally associated with rubber-like elasticity. Spandex fibers are long-chain synthetic polymers composed at least 85% of the segmented polyurethane. The elastomeric fibers derive their high extension from the unfolding of molecular chains in that segment which is amorphous in character. In order to obtain this soft rubbery region in the polymer, two classes of compounds such as polyesters and polyethers are used. Polyurethane fibers may be made by linking preformed segments through the urethane group, resulting in linear molecules, or they may be branched and/or cross linked to yield three-dimensional structures. The linear polyurethane polymer is formed by reaction of glycol with a diisocyanate:

$$HO(CH_2)_4OH \quad + \quad OCN-C_6H_3(CH_3)-NCO$$

Butane Diol   Isocyanate

$$---O(CH_2)_4O.CONH-C_6H_3(CH_3)-NHCO.O(CH_2)_4O---$$

Polyurethane

The segmented polyurethanes are spun into either monofilaments or multifilament yarns. Sometimes monofilaments are produced by cutting thin ribbons from an extruded sheet. The finest spandex filament produced may be on the order of 40 denier, compared to 150 denier for rubber yarns.

Because of the regular fashion in which the two segments in the polyurethane molecule are arranged and because of the occurrence of tie-points that break and reform during stretching, spandex fibers tend to have superior physical characteristics relative to ordinary rubber fibers. Spandex fibers (e.g., Lycra® and Vyrene®) commonly have a round or dog-bone cross-sectional shape, depending on the method of production. Spandex fibers are generally stronger than ordinary rubber filaments. The tenacity of these fibers varies in the range of 0.55-1.0 g/den as compared to 0.25 g/den for natural rubber. Their breaking elongation may vary up to 700%, and they demonstrate excellent recovery behavior. They have very low moisture regain, which may vary from 1.0 to 1.3%. Spandex fibers have a good resistance to acids, alkalies, and most common chemicals.

Spandex fibers are used in essentially three different forms: (1) uncovered filaments, (2) core spun yarns, and (3) covered yarns. They are extensively used in stretch fabrics, foundation garments, and swimwear.

Aramid

The fiber-forming substance in this class of fibers is a long-chain synthetic polyamide in which at least 85% of the amide ($-C-NH-$) linkages are attached
$$\overset{\|}{O}$$
directly to two aromatic rings. The two aramid-type fibers, Kevlar® and Nomex®, are spun as multifilament yarns and may be cut to produce staple by the process developed by E.I. duPont de Nemours and Co., Incorporated.

This class of fibers has a unique combination of very high strength and roughness never achieved in nature or in any other fibrous material. Some of the important characteristics are:

1. These fibers do not melt; rather they decompose above 380°C.
2. High strength (tenacity varies from 4.8 to 5.8 g/den).
3. Good modulus.
4. Good fabric integrity, particularly at high temperatures.
5. Inertness to moisture.

The above combination of properties makes these fibers particularly suited for end-use applications such as hot air filtration, protective clothing, military applications (helmets and bullet-proof vests), and structural supports for aircrafts and boats. Other uses include ropes and cables, mechanical rubber goods, and marine and sporting goods equipment.

Miscellaneous

There are various other types of man-made fibers such as glass, ceramic, metallic, and carbon, including such generic types as novoloid (Kynol®), anidex, nytril, azlon, etc., that are currently being manufactured. Most of these fibers may be classed as specialty fibers, since their applications are limited to very specific end-uses. Glass fiber is the only exception because of its extensive use in home furnishings, industrial (tire cords), and apparel applications. Glass fibers used for textile purposes are made in two main types, "E" glass and "C" glass. Both types are similar in composition.

Technically, carbon fibers are produced from precursors such as polyacrylo-nitrile or rayon or any other organic base fibers.

Union Carbide was the first company to use viscose-rayon as a precursor to produce carbon fibers commercially. The viscose rayon yarn is heated to about 2000°C and stretched up to 50% while at this brilliant white heat. The Union Carbide fibers are marketed under the trade names of Thornel 25, Thornel 40, and a subsequently improved yarn, Thornel 50.

When polyacrylonitrile (PAN) is used as a precursor, the conversion process is carried out in three successive stages:

1. Oxidation at 200-300°C.
2. Carbonization at about 1000°C.
3. Graphitization at 1500-3000°C, according to the type of PAN.

Carbon fibers are characterized by extremely high modulus and tensile strength (depending on the temperature employed in the graphitization stage), and find extensive uses in industrial applications such as aircraft structures and space applications. For these high-performance structures, carbon fibers are used as reinforcements embedded in a suitable resin (such as epoxide, polyester, phenolic, polyphenylene, and polyimide).

For more detailed information on the production, properties, and applications of these fibers, the reader is referred to the list of reading material at the end of this chapter.

REFERENCES

1. Cook, J. Gordon., *Handbook of Textile Fibers*, 4th ed. Merrow, Watford, Herts, England.
2. Moncrief, R. W., *Man-Made Fibers*, 6th ed. Wiley, New York, 1975.
3. Morton, W. E., and J. W. S. Hearle, *Physical Properties of Textile Fibers*, 2nd ed. The Textile Institute and Heinemann Press, London, 1975.
4. Meredith, R., *J. Text. Inst.*, 1945, 36, T107.

5. Farrow, B., *J. Text. Inst.,* 1956, **47**, T58.
6. Farrow, B , *J. Text. Inst.,* 1956, **47**, T650.
7. Ford, J. E. (Ed.), *Fiber Data Summaries,* Shirley Institute, Manchester, England, 1966.
8. Pierce, F. T., Report of 3rd E.C.G.C. Conference, 1938, 138.
9. Hertel, K. L., and C. J. Craven, *Text. Res. J.,* 1951, **21**, 765.
10. Clegg, G. G., *J. Text. Inst.,* 1932, **23**, T35.
11. Lord, E., *J. Text. Inst.,* 1942, **33**, T205.
12 Lord, E., and C. Underwood, *Emp. Cott. Gr. Rev.,* 1958, **35**, 1.
13. Anderson, S. L., and R. C. Palmer, *J. Text. Inst.,* 1953, **44**, T95.
14. A.S.T.M. Standards on Textile Materials, D1440-72.
15. Webb, R. W., *Proc. ASTM,* 1932, **32**, Parts II, 764.
16. Hertel, K. L., *Text. Res. J.,* 1940, **10**, 510.

3

Yarn Structure

INTRODUCTION

It was mentioned in the introductory chapter that structure plays a major role in the physical properties and performance characteristics of yarns, regardless of fiber content. Consequently, if a particular fiber were processed into a variety of staple and continuous-filament structures, textured and untextured, one could expect substantially different characteristics in each yarn. To fully understand or predict the performance of yarns, it is necessary to explore the structure of yarn in detail.

The study of fiber geometry (spatial configuration of fibers) in yarn was initiated in the early fifties by Morton and Yen (1), who developed a fiber tracing technique to indicate the position of individual fibers relative to the yarn axis. A small proportion (less than 1% by weight) of dyed fiber is thoroughly blended with similar but undyed fiber before conversion into yarn. The yarn is immersed into a liquid media of the same refractive index as that of the fibers, causing the undyed fibers to become practically transparent and allowing the dyed (tracer) fiber to be observed and scanned microscopically.

Successive segments of the fiber or filament being scanned are usually shown to be positioned in various annular zones in the yarn structure from core to surface to core in migrational patterns of various periodicities. The term "fiber migration" is used to describe relative fiber movement during processing and the position of fiber in the final yarn structure. Fiber migration, which is expressed by various parameters and indicies, depends on many fiber properties, characteristics of the fiber assemblages, and processing conditions.

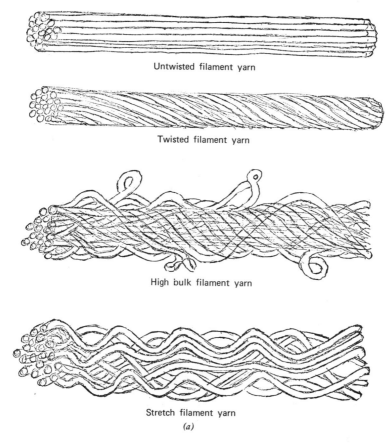

Untwisted filament yarn

Twisted filament yarn

High bulk filament yarn

Stretch filament yarn

(a)

Figure 3.1. Idealized diagrams of (*a*) basic filament and (*b*) spun yarn structures.

The phenomenon of fiber migration in the processing of fibers, filaments, and blends has been studied by many since the earlier work of Morton and Yen. A very thorough treatment of fiber migration in staple and filament yarn and in yarn structure in general is found in *Structural Mechanics of Fiber, Yarns, and Fabrics* by Hearle, Grosberg, and Backer (2). Although a great deal of work on yarn structure has been reported in the literature, much work remains in relating the various yarn structures to performance characteristics other than mechanical properties. The following discussion attempts to simplify some fundamental differences in the structure of various yarns and to relate these differences to performance characteristics. As an aid, idealized models of various structures are shown in Figure 3.1.

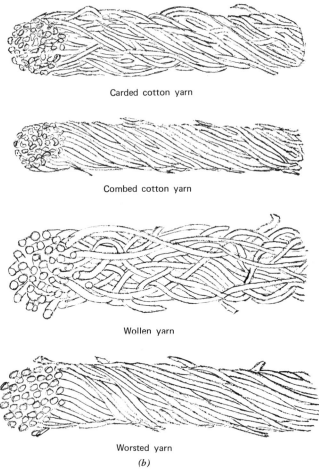

Carded cotton yarn

Combed cotton yarn

Wollen yarn

Worsted yarn

(b)

Figure 3.1. (Continued).

CONTINUOUS-FILAMENT STRUCTURES

In a steel cable or in a rope structure, one strand or component remains in the center, and all other components have a specific and constant radius of spirality (helix), remaining in an annular envelope about the center or cable axis. In examining a segment of the cable, one would find the longest components at the surface and the shortest component in the center. This is possible because each component is individually controlled by tension or rate of feeding during formation of the cable.

During the twisting of multicontinuous filaments, however, there exists no control of the tension or rate of feeding of the individual filaments. Consequently a self-equalization tension phenomenon exists wherein the individual filaments exchange positions radially among the various annular zones in the assemblage. Those filaments or segments of filaments going to the outer layers of the yarn develop more tension and, being tighter, force themselves to the center of the yarn to relieve the tension. The filaments in the core of the yarn that are under less tension, momentarily, are pushed to the yarn surface. This phenomenon is highly desirable in most twisted structures because if one filament breaks it does not continue to unravel indefinitely. This means that every filament is entangled periodically at both the surface and core of the yarn.

Continuous-filament yarns with minimal twist or entanglement have a very long fiber segment length between points of entanglement. With such little entanglement, the filaments are free to spread, and there is no definite or stable yarn cross-sectional shape or diameter. Filaments can be pulled individually away from the main body of the assemblage, creating snagging problems. The yarn tends to flatten out in fabric form and thereby makes possible a tremendous area of contact with any other surface, which is not desirable in most apparel applications. These low-entangled structures do possess the greatest proportion of filament linearity, however, which is very desriable for many industrial applications requiring high strength and low elongation. These structures permit the greatest translation of fiber-to-yarn strength.

As more twist is added to continuous-filament structures, the degree of linearity of filament segment lengths in the yarn decreases slightly, which leads to reduced strength. This reduction in yarn strength occurs because the more linear filaments are strained immediately on loading, whereas the less linear filaments tend to be straightened and then strained, or at least not strained at the same moment and rate. However, greater twist also reduces the average length of filament segments between points of entanglement and the tendency for filaments to spread out and snag. With a sufficient amount of twist, the cross section of a filament yarn can be given enough support so that, under a compressive or bending deformation, an elliptical shape occurs rather than a total collapse or flattening out of the filament assemblage. Increased twist in filament yarn also leads to greater bending stiffness in the yarn structure primarily because of greater entanglement and friction among the filaments.

TEXTURED FILAMENT STRUCTURES

A continuous-filament yarn that is textured into a high-bulk yarn will contain a high degree of nonlinearity in its filaments, with a great amount of twist or entanglement. This is essentially what produces bulkiness with no added extensibility in the yarn. If one were to analyze a segment of high-bulk filament yarn, a tremendous array of filament length would be found. The average length of a filament between points of entanglement, however, would not be great. If these yarns are of the air-jet variety, the tremendous nonlinearity of filaments is in the form of loops. The loopy outer profile of the structure is easily flattened under compressive force (pinching) whereby some of the bulkiness is temporarily removed. The high-bulk yarns resemble the staple yarns in that each structure has a hard-core concentration of fiber and an outer fuzz zone caused by filament loops, in the case of the high-bulk textured yarns (3).

A continuous-filament yarn that is textured into a stretch yarn has a tremendous amount of filament nonlinearity, with minimal, if any, entanglement. This is what permits the extraordinary extensibility and recovery in stretch yarns. Because of the lack of entanglement and the excessive nonlinearity of the filaments, however, the possibility for filament snagging is much greater than for other yarns. Moreover, the stretch yarn has little dimensional stability or structural integrity in either a tensile or compressive deformation. However, the textured features are maintained in normal bending or looping in the absence of a force that would completely linearize the filament assemblage.

STAPLE YARNS

The structure of staple (spun) yarns is an order of magnitude more complex than the structure of continuous filaments. Although a high-quality staple yarn may first appear to be as uniform as a filament yarn, closer examination would indicate a random grouping of fiber and a variation in twist and in linear density, resulting in thick and thin spots along the length of yarn. Also noticeable would be a concentration of fiber forming a core in the center of the yarn and an outer fringe of fiber known as the fuzz zone caused by protruding fiber ends. An untwisting of the spun yarn structure would indicate a more thorough and frequent tying-in of the surface and core layers than found in filament yarns. Examination of low-quality staple yarn would show many localized complex fiber entanglements or hard spots known as neps that are contained in the structure.

A tremendous variation in the quality of staple yarns and uniformity of yarn structure is possible because of the manner in which the fibers are processed. Staple fibers are handled as a mass rather than as individuals and therefore tend to behave and to process in groups and in subgroups. The fibers tend to participate to a much greater extent in their own processing than in the case of filament processing. Each staple fiber has two free ends that may protrude, contributing to fuzz, or may bend, hook, buckle, or roll on themselves or in conjuction with other fibers. Intermediate staple fiber assemblages (sliver and roving) contain a slight internal entanglement caused mainly by the condensation of fiber web into linear assemblages. Also, some twist is normally incorporated into roving structures. In the final stage of staple yarn formation, a rather great amount of twist is superimposed onto the existing looser entanglement. This situation, along with the normal fiber migration pattern, leads to a very complex yarn structure.

In blends of fibers with substantially different processing characteristics, an abnormal fiber migration may occur that is known as preferential radial migration or coring; one component of the blend is found primarily in the core and the other component mostly near the surface. This unusual phenomenon is an abnormal extension or exaggeration of an already highly complex structure.

NON-RING-SPUN STAPLE YARNS

Most yarns made from staple fibers are ring-spun yarn. The twist that provides the final entanglement is inserted into the yarn by the normal technique by which the leading end is attached to a bobbin that is turned by a spindle situated within a ring. The fiber assemblage is delivered rather slowly from a pair of rolls, passes through a clip that travels around the ring, and is wrapped onto a bobbin that rotates at a very high speed. The ratio of spindle speed (revolutions per minute) to yarn delivery speed (inches per minute) determines the yarn twist (turns per inch). Thus the limiting factors in production are the spindle speed and the amount of twist desired.

To overcome this limitation in the production of staple yarns, many ringless systems of spinning have been proposed and developed (4). Commercially speaking, the open-end systems have had the most success. Other systems such as self-twist, thermoplasitically bonded, twistless, and so on, have found limited application at best. To form an open-spun yarn, a strand or linear assemblage of staple fiber is held at both ends, broken in the middle, and reformed in a continuous or accumulative fashion. The rate of production is an order of magnitude greater than in ring spinning.

The open-end spun yarn tends to appear to be no different from ring spun yarns on the surface. However, there exist some rather important differences in the internal structure of the yarns, especially in fiber contiguity (how the fibers come together and entangle or touch one another within the assemblage). The accumulative-entanglement mechanism in open-end spinning leads to more layering of the fibers or fiber segments into specific annular zones in the yarn structure. Moreover, there is less tying-in of fibers or fiber segments from surface-to-core-to-surface than found in ring-spun yarns.

These differences in internal structure are reflected in different performance characteristics. The open-end spun yarns tend to be more uniform in appearance and in linear density than ring-spun yarns. Also, the open-end spun yarns are known to be somewhat more extensible, fuller, and softer. The main disadvantage is that open-end yarns are not as strong as ring-spun yarns. For many applications, however, the advantages of open-end yarns far outweigh the disadvantages. It is pointed out that although significant differences exist between open-end and ring-spun yarns, the differences are not nearly as great as those which exist among various staple and continuous-filament structures.

FUNDAMENTAL STRUCTURAL FEATURES OF YARN

The structural features that are thought to have a major influence on the physical properties and performance characteristics of yarn are (1) volumetric density, (2) fiber segment length between points of entanglement, and (3) mobility of fiber segments between points of entanglement. These three structural features are not easy to measure or even to observe. It is quite necessary, however, to define these structural features and to compare them in a variety of yarns in order to understand the basis for differences in yarn performance.

Volumetric density in yarn is dependent on fiber compactness, the packing density of fiber or filament in the cross section of the yarn in a relaxed state. A great deal of air or empty space is contained in all yarns, but staple and textured filament yarns tend to contain the most. A yarn can be considered dimensionally stable if the fiber packing density is approximately the same in the relaxed state and under low levels of stress. The variation in packing density is also important in relation to aesthetics and yarn quality.

The average fiber segment length between points of entanglement within a yarn structure is sometimes referred to as fiber modular length. In most yarn structures in which twist is the basic mechanism of entanglement, the average fiber or filament segment length between points of entanglement is related to the twist geometry. The orientation of the fiber segment length between points of entanglement with respect to the yarn axis is important also, but to a lesser degree.

The mobility or freedom of movement and direction of movement of fiber segments relative to other fibers and fiber segments in the structure play a major role in yarn performance. Especially important is the lateral movement of fiber segments near the yarn surface on abrasion, snagging, pilling, hand, etc., in fabric structures. Yarn structures that have a minimal degree of fiber segment mobility both longitudinally (in the direction of the yarn axis) and cross-sectionally (normal to the yarn axis) are considered to be dimensionally stable.

If one were to probe various types of yarn with a pin and with the aid of low-power magnification, a qualitative feeling could be gotten for some of the aforementioned yarn structural features such as fiber segment length and mobility. Microscopy would be necessary to observe the orientation of fiber segments and fiber packing density. However, adequate techniques have not been developed as yet to fully characterize the intrinsic fiber contiguity in yarn, especially the internal fiber entanglement.

Yarn structural features depend mainly on the properties of the constituent fibers or filaments and the inherent characteristics of the processing systems. Excluding generic-related parameters (such as fiber friction, modulus, resilience, extensibility, and elasticity), the fiber properties of greatest importance are length, fineness, crimp, and cross-sectional shape. The inherent characteristics of the processing system are fiber orientation and entanglement. Fiber orientation refers to the position of the fiber or filament segments in relation to the yarn axis and, in general, the degree of linearity of the fibers or filaments in a yarn. Fiber entanglement, as used here, relates to both the nature of the entanglement and the frequency or the degree of entanglement.

A comparison of the structural features in various types of yarn and some factors affecting structural features is given in Table 3.1. This tabulation is quali-tative in nature and is presented only as a means of focusing attention on structural differences among the various types of yarn. It is pointed out once again that differences in structure that may appear to be rather subtle to the eye can lead to substantial differences in yarn performance.

STRUCTURALLY RELATED PERFORMANCE OF YARNS

Referring to Table 3.1, it can be seen that, based on the fiber geometric proper-ties and inherent characteristics of the processing systems, staple yarns have sub-stantial fiber density, short fiber segment lengths between points of entangle-ment, and minimal mobility of the fiber segments. This is the basis for the retention of the intrinsic fiber contiguity and thus the excellent dimensional stability of staple yarn structures under low levels of stress. All yarns, except stretch or elastomeric, tend to have excellent dimensional stability when loaded in the direction of the yarn axis. However, staple yarns also have good dimen-sional stability when loaded or deformed cross-sectionally or normal to the yarn axis. This means that spun yarns retain their natural bulkiness, good covering power, and excellent hand under various extensional, compressional, and bend-ing deformations to a much greater extent than would be expected in textured filament yarns.

Again referring to Table 3.1, the untwisted or slightly entangled filament yarns have very long filament segments lengths between points of entanglement and great lateral mobility of the filament segments. This combination of structural features allows the untwisted multifilament yarn to spread and flatten out under normal bending or compressional deformations. The spreading of the filaments changes the yarn cross-sectional shape to a rather flat, ribbon-like structure or

Table 3.1. Comparison of the Structural Features in Various Types of Yarn and Some of the Factors Affecting Yarn Structural Features

	Fiber Geometric Properties			+	Processing Characteristics		=	Yarn Structural Features		
	Length	Fineness	Crimp		Orientation	Entanglement		Density[a]	Segment[b]	Mobility[c]
Staple Yarns										
Carded cotton	short	fine	slight		medium	high		medium	short	minimal
Combed cotton	medium	fine	slight		high	v. high		high	v. short	minimal
Woolen (carded)	medium	coarse	high		low	medium		low	medium	slight
Worsted (combed)	long	coarse	medium		high	v. high		high	v. short	minimal
Filament Yarns										
Untwisted	Continuous	fine	none		v. high	minimal		medium	v. long	great
With twist		medium or coarse	none		high	medium		high	long	slight
High bulk			high		low	high		low	medium	large
Stretch			v. high		v. low	minimal		v. low	v. long	tremendous

[a] Fiber or filament packing density in the yarn cross section.
[b] Average fiber or filament segment length between points of entanglement.
[c] Freedom of movement or mobility of fiber segments held between points of entanglement.

one that is quite elliptical. This collapse of the filament bundle permits a much greater area of contact with other surfaces, resulting in greater friction and discomfort (in apparel applications). Also, the filaments are easily snagged away from the main body of the yarn structure. When the bending, compressional, or snagging stress is removed, the filaments often do not fall back into their original alignment, resulting in a false or pseudoentanglement. With a rather small amount of twist in the filament yarn, however, most of these problems are minimized. With sufficient twist in the filament yarn structure, the problems are completely overcome.

In filament high-bulk yarns, filament segments between points of entanglement are in the form of loops on the surface of the yarn. Whereas filament segments in the yarn core have minimal mobility, the loopy segments on the surface can easily rotate and collapse. This means that the outer profile or loopy surface zone of filament high-bulk yarns can be deformed by a rather slight compressional load, creating a greater area of contact with another surface than a spun yarn of the same size, for example.

In stretch yarns, the packing density of the filaments is very low and the filament segments between points of entanglement are very long. This combination results in tremendous mobility of filament segments along the yarn axis and in any direction away from the yarn axis. The tremendous mobility of the filaments means that the stretch yarn structure is easily deformed and that the yarn has poor dimensional stability, in general. The flattening out of the yarn structure occurs rather easily, causing almost as great an area of contact with other surfaces as found with untwisted, untextured multifilament yarns. In the textured filament yarns, the individual crimped filaments can move laterally, rotate, or can decrimp independent from the other filaments in the structure. Snagging of the individual filaments by a rough surface or edge is greatly facilitated because of the mobility and crimp in the filaments.

In conclusion, it seems that many of the desirable physical properties and performance characteristics of a yarn are related to the dimensional stability of the yarn cross section under various types of deformation. Unfortunately, not enough work has been done on yarn cross-sectional behavior, especially in the development of filament textured yarns. One parameter that can be used to characterize yarn cross-sectional behavior under stress is ellipticity. Ellipticity is the ratio of the maximum yarn diameter to the minimum yarn diameter, where unity indicates a perfectly circular cross-sectional shape. The importance of the stability of yarn cross-sectional shape in relation to tactile and aesthetic properties is discussed in a subsequent chapter.

REFERENCES

1. Morton, W. E. and K. C. Yen, "The Arrangement of Fibres in Fibro Yarns," *J. Textile Inst.*, 1952, **43**, T60.

2. Hearle, J. W. S., P. Grosberg, and S. Backer, *Structural Mechanics of Fibers, Yarns and Fabrics,* Volume I, Wiley-Interscience, New York, 1969.

3. Piller, B., *Bulked Yarns,* SNTL, Prague and The Textile Trade Press, Manchester, England 1973.

4. Lord, P. R., *Spinning in the 70's,* Merrow, Watford, Herts, England, 1970.

SUGGESTED READING

1. Morton, W. E. and K. C. Yen, "The arrangement of fibres in fibro yarns," *J. Text. Inst.,* 1952, **43** T60.

2. Morton, W. E., "The arrangement of fibers in singles yarns," *Text. Res. J.,* 1956, **26** 325.

3. Treloar, L.R.G., "The geometry of multi-ply yarns" *J. Text. Inst.,* 1956, **47** T348.

4. Tattersall, G. H., "An experimental study of yarn geometry," *J. Text. Inst.,* 1958, **49** T295.

5. Riding, G., "An experimental study of the geometrical structure of single yarns," *J. Text. Inst.,* 1959, **50** T425.

6. Gracie, P. S., "Twist geometry and twist limits in yarns and cords," *J. Text. Inst.,* 1960, **51** T271.

7. Hickie, T. S. and M. Chaikin, "The configuration and the mechanical state of single fibers in woollen and worsted yarns," *J. Text. Inst.,* 1960, **51** T1120.

8. Riding, G., "A study of the geometrical structure of multiply yarns," *J. Text. Inst.,* 1961, **52** T366.

9. Hearle, J. W. S. and V. B. Merchant, "Interchange of position among the components of a seven-ply structure—mechanism of migration," *J. Text. Inst.,* 1962, **53** T537.

10. Riding, G., "Filament migration in single yarns," *J. Text. Inst.,* 1964, **55** T9.

11. Wray, G. R. and Q. S. Truong, "A modification of the tracer—fiber technique for yarns," *J. Text. Inst.,* 1965, **56** T157.

12. Iyer, K. B. and R. M. Phatarfod, "Some aspects of yarn structure," *J. Text. Inst.,* 1965, **56** T225.

13. Treloar, L. R. G., "A migrating-filament theory of yarn properties," *J. Text. Inst.,* 1965, **56** T359.

14. Treloar, L. R. G. and G. Riding, "Migrating-filament theory: Apparent variation of twist with radial position," *J. Text. Inst.,* 1965, **56** T381.

15. Hearle, J. W. S., B. S. Gupta, and V. B. Merchant, "Migration of fibers in yarns: Pt. I characterization and idealization of migration behavior, *Text. Res. J.,* 1965, **35** 329.

16. Hearle, J. W. S. and O. N. Bose, "Migration of fibers in yarns: Pt. II A geometrical explanation of migration," *Text. Res. J.,* 1965, **35** 693.

17. Hearle, J. W. S. and B. S. Gupta, "Migration of fibers in yarns: Pt. III A study of migration in staple-fiber rayon yarn," *Text. J. Res.,* 1965, **35** 788.

18. Hearle, J. W. S. and B. S. Gupta, "Migration of fibers in yarns: Pt IV A study of migration in a continuous filament yarn," *Text. Res. J.,* 1965, **35** 885.

19. Hearle, J. W. S., B. S. Gupta, and B. C. Goswami, "Migration of fibers in yarns: Pt. V The combination of mechanisms of migration," *Text. Res. J.,* 1965, **35** 972.

20. Wray, G., "A simple technique for the circumferential viewing of yarns containing tracer fibers," *J. Text. Inst.*, 1966, 57 T42.

21. Hearle, J. W. S. and O. N. Bose, "The form of yarn twisting Pt. II Experimental studies," *J. Text. Inst.*, 1966, 57 T308.

22. Denton, M. J., "The development of false twist in bulking," *J. Text. Inst.*, 1968, 58 344.

23. Hearle, J. W. S. and B. C. Goswami, "Migration of fibers in yarns: Pt. VI The correlogram method of analysis," *Text. Res. J.*, 1968, 38 780.

24. Hearle, J. W. S. and B. C. Goswami, "Migration of fibers in yarns: Pt VII Further experiments on continuous filaments," *Text. Res. J.*, 1968, 38 790.

25. Hearle, J. W. S. and B. C. Goswami, "Migration of fibers in yarns: Pt. VIII Experimental study on a three-layer structure of nineteen filaments," *Text. Res. J.*, 1970, 40 598.

26. Gupta, B. S., "Fiber migration in staple yarns Pt. II The geometric mechanism of fiber migration and influence of roving and drafting variables," *Text. Res. J.*, 1970, 40 15.

27. Balasubramanian, N., "Effect of processing factors and fiber properties on the arrangement of fibers in blended yarns," *Text. Res. J.*, 1970, 40 129.

28. Scardino, F. L. and W. J. Lyons, "Preferential radial migration of fibers in the processing of blends," *Text. Res. J.*, 1970, 40 573.

29. Henshaw, D. E., "A model for self-twist yarns," *J. Text. Inst.*, 1970, 61 97.

30. Henshaw, D. E., "Twist distribution in self-twist yarns," *J. Text. Inst.* 1970, 61 269.

31. Lord, P. R., "The structure of open-end yarn," *Text. Res. J.*, 1971, 41 778.

32. El-Behery, H. M. A. E. and D. H. Batavia, "Effect of fiber initial modulus on its migratory behaviour in yarns," *Text. Res. J.*, 1971, 41 812.

4

Yarn Specifications

INTRODUCTION

For salability or for incorporation into various textile structures, yarns must be spun or produced according to specifications. The specifications may vary considerably, according to the end-use requirements of the yarn. For example, the list of specifications for industrial yarns is much longer and in much greater detail than the list of specifications for yarn used in regular fabrics. Essentially, many of the yarns used in industrial applications are engineered linear fiber assemblies.

Several specifications are common to all types of yarns, mainly for the purpose of identification or designation. Among the more important identifying specifications for yarn are linear density, structural features, fiber content, and an indication of any mechanical or chemical treatments. Spun yarns require several specifications relating to quality parameters. In addition to these descriptive items, performance specifications are often required. Various types of yarns also have peculiar designations and specifications relating to unusual constructional features (e.g., corespun, novelty, combination yarns, etc.).

It is an industrial practice and reality that no standard identifying code exists for the complete designation or description of yarns. This is due to a great extent to the esoteric and fragmentized nature of the textile industry, dating back to the industrial revolution. Different schemes for identification or designation exist among and within each category of yarn structure. Moreover, various countries use different systems for yarn identification. To improve communications in the international marketing of fiber, textile, and apparel products, a universal scheme for the designation of yarn construction has been proposed by the International Organization for Standardization (1). Until such time as the

proposed scheme is adopted, however, it will be necessary for the textile technologist to converse and to relate in all systems of yarn identification.

SPECIFICATION OF YARN LINEAR DENSITY

Many terms are used to designate the variety of yarn sizes. *Yarn count, yarn
number, yarn fineness,* and *yarn size* are all used interchangeably by the textile
industry. The term *yarn linear density* is used by the textile technologist and
textile scientist for the designation of yarn size.

Linear density is defined as the mass or weight per unit length of a material.
Whereas many rigid linear structures are designated in terms of volumetric
density (mass or weight per unit volume), textile yarns are quite flaccid and do
not have a dimensionally stable diameter or cross-sectional shape. Consequently,
rather than specifying diameter or volumetric density, linear density is used as a
more practical and reliable expression of the variety of yarn sizes. The two basic
categories for the expression of linear density of textile yarns are the *direct* and
indirect systems.

The direct systems are based on weight per unit length of linear structures.
Examples of the direct category are the *denier* and *tex* systems. The denier
number is the weight in grams of 9000 meters of yarn. (Example: an 1800-
denier yarn would weigh 1800 grams per 9000 meters or 0.2 grams per meter).
The tex number is the weight in grams of 1000 meters of yarn. (Example:
a 30-tex yarn would weigh 30 grams per 1000 meters or 0.03 grams per meter.)
The tex system has been proposed as the universal system by ASTM (2).
Although the tex system is officially preferred, both these systems can be used
to designate the linear density not only of yarns but of any linear structure (e.g.,
fibers, filaments, slivers, rovings, tapes, ribbons, tow, nonflaccid rods, etc.).
Direct systems are so named because the larger the designated number, the
heavier the yarn. Consequently, with the direct systems, yarn numbers are
directly proportional to yarn linear densities.

Indirect systems for the expression of the linear densities of yarns are based on
a number of standard lengths of yarn per unit weight. Examples of the indirect
categories are the cotton, woolen, and worsted yarn systems for designation of
linear density. The cotton system is based on the number of 840-yard lengths
per pound of yarn. (Example: a 20s cotton yarn, also written 20/1, would have
twenty 840-yard lengths per pound of yarn. One pound of a 20s cotton yarn
would measure 16,800 yards in length). The woolen run system is based on the
number of 1600-yard lengths per pound of yarn. (Example: a 3s run woolen
yarn, also written 1/3, would have three 1600-yard lengths or 4800 yards in one
pound of yarn). The worsted system is based on the number of 560-yard lengths
per pound of yarn. The linen lea system and the woolen cut system are based

on the number of 300-yard lengths per pound of yarn. The metric system is based on the number of 1000-meter lengths per kilogram of yarn. There are several other systems for specialty yarns, all of which can be found in ASTM Standards (3,4). It is common practice in industry to use the same indirect system for specifying the size of rovings as used for designating the size of yarns (2). Regardless of fiber content, yarns are usually numbered according to the system of processing (cotton, woolen, or worsted). Indirect systems are so named because the larger the designated number, the lighter or finer the yarn. Therefore, with the indirect systems, yarn numbers are inversely proportional to yarn linear densities. As an aid in the equivalence or conversion of yarn linear density from one system to another (direct or indirect), it is suggested that one refer to ASTM Standard D2260, Conversion of Yarn Number Measured in Various Numbering Systems (3) and Chapter 5.

SPECIFICATION OF YARN STRUCTURAL FEATURES

The major constructional features that must be specified when describing a yarn are an indication of whether the yarn is basically staple or filament in composition, the amount and direction of yarn twist, and whether the yarn is a singles or plied structure (5).

If the yarn is a staple structure, the system of spinning should be designated (e.g., carded cotton, combed cotton, woolen, worsted), although this is conveniently indicated to an extent by notation in the expression of yarn count. For example, on cotton counts, the singles yarn number is stated first, followed by the ply number. (Example: 30/1 means 30s or a single 30s count cotton systen yarn; 30/2 means two single 30s count cotton systems yarns were plied together resulting in a linear density similar to that of a 15s cotton count yarn, approximately). On woolen and worsted counts, the ply number is given first followed by the singles yarn number. (Example: 1/40 worsted means single 40s count worsted yarn; 2/40 worsted means two single 40s count worsted yarns were plied together resulting in a linear density similar to that of a 20s worsted count yarn, approximately). It is conventional practice in the case of spun yarn designation to use the solidus (/) to separate the singles yarn count from the number of plies. In filament yarns, the ply number is usually placed before the singles component. (Example: 2 X 70 denier or 2/70 denier indicates that two 70-denier filament yarns were combined resulting in a linear density equivalent to a 140-denier filament yarn, approximately).

In the case of plied yarns, it is necessary to know the resultant yarn number (R), which is also referred to as the singles equivalent (SE). The resultant yarn number or singles equivalent is the observed linear density of the plied yarn, cord, or yarn whose original number has been changed significantly by twisting or texturing. In plied staple yarns that are normally numbered on the indirect systems, the resultant is estimated by dividing the singles number by the product of the plies. (Example: the estimated resultant is estimated by dividing the singles number by the product of the plies. (Example: the estimated resultant of a 30/2/3 is a 5s). This is an estimate because contraction of the assemblage during plying has not been taken into account. The actual resultant is usually slightly coarser than the estimated resultant; the more components or the more complex the structure, the greater the difference in estimated and actual linear density of the composite yarn. In filament yarns, which are normally numbered on the direct systems, the resultant is estimated by multiplying the singles number by the product of the plies. (Example: the estimated resultant of a 2 × 3 × 150 denier filament yarn is 900 denier).

Spun yarns always have a substantial amount of twist. Twist is expressed by the number of turns per inch (tpi) or by the number of turns per meter (tpm) and by the direction of twist (S or Z). S refers to the righthand or to the clockwise direction of twist in a yarn; Z refers to the lefthand or counterclockwise direction of twist in a yarn. (Example: 36/1, 24 tpi Z means 36s cotton system yarn with 24 turns per inch in the Z direction). Twist in spun yarn may also be designated in terms of a twist multiplier (TM). The twist multiplier is an expression that relates the amount of twist in a yarn to the square root of the yarn count, thereby indicating the relative degree, level, or angle of twist in the yarn structure. Specifically, twist multiplier equals the turns per inch of twist in a yarn divided by the square root of the yarn count on the cotton system (TM = (tpi) ÷ $\sqrt{\text{cotton yarn count}}$). (Example: 16/1, 5 TM Z means that a 16s cotton system yarn has 20 tpi in the Z direction).

The amount of twist used in filament yarn structure is quite minimal, usually substantially less than that found in staple yarns. An exception to this rule can be found in crepes and other novel filament yarn structures. Quite often, filament yarn is available in 1/4 or 1/2 tpi in a particular direction (S or Z). Zero-twist filament yarn indicates that no special amount or direction of twist has been inserted into the structure. Even in zero-twist yarn, however, some long-term entanglement brought about by filament interchange (radial migration) acts as a localized pseudo twist or pinning phenomenon. This can be demonstrated by running a pin through the filament yarn in the direction of the yarn axis.

SPECIFICATION OF FIBER CONTENT

The Textile Fibers Product Identification Act stipulates that all textile products must be labeled according to generic names and definitions provided by rules of the Federal Trade Commission (6). In the particular case of textile products containing wool fiber, the Wool Labelling Act stipulates that the wool be identified as virgin, reprocessed, or reused (7). When textile products are composed of blends of various fibers, the law also requires that the percentage of each fiber component (greater than 5% by weight) in the blend be stipulated according to its proportional content by weight. Spun yarn products are particularly affected by these laws because of the multitude of blends run on staple systems. The percentages designated must be accurate with ± 2.5%. (Example. 50% polyester-50% cotton, which is also written 50 polyester-50 cotton or 50-50 polyester-cotton, must by proportional weight be between 52.5% polyester-47.5% cotton and 47.5% polyester-52.5% cotton).

In cotton yarns and cotton yarn blends, it is customary to designate some information concerning the properties of the constituent fibers. (Example: for cotton, 1 1/16 in. SLM 4.5 Microgram Reading means that the cotton used as an average staple length of 1 1/16 in., a grade of strict low middling, and a fineness or linear density of 4.5 μg/in.). Experienced textile technologists can assume and predict certain yarn quality and performance standards based on this sort of information. Similarly, in wool yarns and wool-yarn blends, it is customary to state some particulars about the wool properties, mainly fineness and length, which can be used by experienced technologists to predict yarn quality and performance. (Example: for wool, 58s or ½ blood refer to a fiber fineness and spinning quality index based on empirical results developed over the years.)

When man-made fibers are spun on staple yarn systems, the designation of the following fiber properties is usually made: staple length in inches, linear density in denier (den) or millitex (mtex); crimp in terms of regular or super, or in terms of the number of crimps per inch (cpi); the luster factor stated in terms of bright, semidull, or dull, which usually relates to the amount of delusterant incorporated into the fiber. (Example: 1.5 in., 1.5 den. polyester, semidull, 10 cpi). In addition, if the particular fiber variant is identified according to the trademark of the fiber producer, a great deal of information may be assumed about the fiber (e.g., cross-sectional shape and surface characteristics; blending compatability with other specific fibers; breaking tenacity, modulus, and extension under wet and dry conditions; dyeing characteristics; setting and melting temperatures; resistance to attack by various solvents; performance and care characteristics). All this information is provided by the fiber producer for each specific fiber variant marketed.

For filament yarns, the generic name of the constituent fibers must be given. In addition, it is normal practice to state first the linear density of the yarn followed by the number of filaments per yarn, thereby indicating indirectly the linear density of the constituent filaments. The luster factor is usually stated after the notation concerning twist. (Example: the designation 70-34 1/2 Z bright indicates a 70-denier filament yarn with 34 filaments. The linear density of the individual filaments would be 2 denier, approximately). As mentioned in the previous paragraph, if the particular fiber variant is given along with the producer's trademark, a great deal more information may be confidently associated with the designation.

SPECIFICATION OF MECHANICAL AND CHEMICAL TREATMENT

Various mechanical and chemical treatments for improved performance characteristics of textile products have been in existence for many years. Many of these treatments are done in fabric form, but several are performed on yarn structures. Among yarn treatments of major interest are the following: mercerization and slack-mercerization in the case of cotton yarns, shrink resistance for wool yarns, degumming of silk yarns, high bulking of spun yarns, and the texturing of filament yarns for stretch or bulk. Industrial yarns are subjected to a tremendous range of treatments and coatings such as waxing, rubber, thermal plastic resins, electrical insulating finishes, electrical conducting finishes, flame retardants, water repellents, etc., engineered for specific end-use situations. In the case of industrial-product yarns, the treatments and the effects of the treatments on yarn properties are usually well stipulated. With consumer-product yarns, however, the treatments are generally specified, but the effects of treatments on yarn properties are not always stipulated. Particularly, the effect of treatment on yarn linear density is not designated or specified. (Example: a designation such as 70 - 34 - 0 textured polyester indicates that a 70-denier, 34-filament yarn with 0 twist was textured. The designation does not indicate the particular texturing process, the resultant linear density because of texturing, or the change in extensibility possible in the yarn from a relaxed state, if any). Texturing always increases to some degree the linear density of filament yarns in their relaxed configuration. It would be most useful to have pre- and post-treatment designations for the more important yarn properties when significant changes have been made.

In addition to these rather permanent types of treatments, yarns are subjected to a variety of temporary conditionings designed to aid in the conversion of yarn into fabric (e.g., steaming to set twist and overcome internal stresses during spinning and twisting operations, application of sizing to combat chaffing of warp yarns during weaving, application of lubricants to overcome friction and abrasion during processing, etc.). Normally, these chemical processing aids are removed during dyeing and finishing operations, and others are usually added to improve the care and performance of textile products for the consumer. In light of the newly legislated care and performance labeling act for textile products, it is now required by law that many of these mechanical and chemical treatments and their effects be clearly communicated (8).

QUALITY SPECIFICATIONS FOR SPUN YARNS

In subsequent chapters, detailed explanations are given concerning the variation in physical properties that one must expect with spun yarns. Of particular importance is the variation in yarn tensile properties (e.g., breaking strength and breaking extension) and the variation in yarn structure (e.g., linear density, thick and thin segments, fuzz, finish, and other extractible material). If these variations in spun yarns are not minimized, one may expect processing difficulty and off-quality fabric.

Consequently, modern spinning mills have elaborate and comprehensive quality specifications for each spun yarn variant produced. Moreover, most weavers and knitters purchase spun yarns according to well-defined quality specifications. Among the more important quality specifications for spun yarns are the following: grade, according to yarn board appearance standards (9) or classimat ratings; uniformity of linear density (10); nep or imperfection count; single end yarn strength and elongation to break (11); variation in yarn strength and elongation to break; break factor (yarn skein strength X yarn number); level of finish, if any. Several of these quality parameters and their measurements are described in detail in subsequent chapters.

THE PROBLEM OF YARN PACKAGE LABELING

Several aspects of yarn specifications and designations have been reviewed. Only so much information can be placed onto yarn packages to describe the yarn contained. It would be very convenient and beneficial if a universal code were developed that would incorporate and stipulate all of the yarn specifications discussed. This would permit, with a minimal set of numbers, letters, and symbols, the proper and meaningful labeling of every package of yarn produced. Unfortunately, no such code exits.

However, as mentioned earlier in this chapter, a good start is the proposed ISO scheme for deisgnation of yarns (1). This rather comprehensive standard has been adopted to improve communication and to standardize the identification or designation of yarn. Although at first it may appear to be a very elaborate system with academic and technocratic overtones, it is really quite good for reducing to a simple code some rather complex yarn structures. The authors highly recommend that this system be utilized as extensively as possible. Once a universal system for yarn designation is in actual use, perhaps the labeling of yarn packages can be standardized.

REFERENCES

1. Document ISO-1139-1973, Designation of Yarns, International Organization for Standardization, American National Standards Institute, New York, 1973.

2. D861 Use of Tex System to Designate Linear Density of Fibers, Yarn Intermediates, Yarns and Other Textile Materials, Part 24, *Annual Book of ASTM Standards.* American Society for Testing and Materials, Philadelphia, Pa., 1976.

3. D2260 Conversion of Yarn Number Measured in Various Numbering Systems, Part 24, ibid, 1976.

4. D299 Asbestos Yarns and D578 Glass Yarns, Part 25, ibid, 1976.

5. D1224 Designation of Yarn Construction, Part 24, ibid, 1976.

6. Textile Fiber Products Identification Act. U. S. Laws, Statutes, etc., Washington, D. C., 1958, 1965.

7. Wool Products Labeling Act. U. S. Laws, Statutes, etc., Washington, D. C., 1939.

8. Care Labeling of Textile Wearing Apparel. U. S. Federal Trade Commission, Washington, D. C., 1971.

9. D2255 Grading Cotton Yarns for Appearance, Part 24, *Annual Book of ASTM Standards.* American Society for Testing and Materials, Philadelphia, Pa., 1975.

10. D1425 Unevenness of Textile Strands, Part 24, ibid, 1967.

11. D2256 Breaking Load (Strength) and Elongation of Yarn by the Single Strand Method, Part 24, ibid, 1975.

5

Twist in Yarns

INTRODUCTION

The subject of yarn structure is discussed in Chapter 3 of this book. It is pointed out that twist plays an important role in affecting the arrangement of fibers or filaments in the yarn cross section. The study of twist and how it effects some of the important properties of textile yarns is the subject of this chapter.

In staple yarns, twist is essential to hold the fibers together and to impart some degree of cohesiveness to the structure. On the other hand, the filaments in a multifilament yarn would fray away if there were no cohesive forces holding them together. Furthermore, the formation of plied or cabled yarns is also achieved by twisting single or plied yarns together to produce a coherent linear structure. In other words, twist is a means by which a bundle of fibers, filaments, or yarns in a plied yarn is held together so that the ultimate structure is made capable of withstanding the stresses and strains and the chafing action of the many processes involved in the manufacture and use of textile fabrics.

The role of twist is very profound in determining the properties of finished yarns vis-a-vis fabrics. The link between fiber properties and the end-use behavior of a textile fabric can be best explained by the following diagram (Figure 5.1).

When a number of components (fibers or filaments) in a continuous strand are twisted, radial forces develop which in turn affect the relative position of the components in the yarn structure, leading to a close packing of all components in a given cross section. Thus the insertion of twist in fiber assemblies (yarns) affects, in addition to the tensile properties (strength-extension), the diameter and the specific volume (in other words, softness or hardness) of yarns. The change in the fiber packing in turn determines the cover of a fabric and such other properties as warmth, crease recovery, permeability, and various other

96

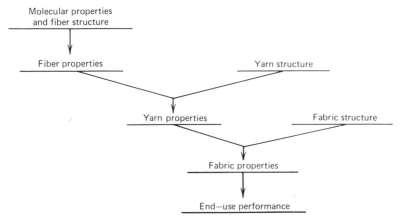

Figure 5.1. Relationships among fiber, yarn, and fabric structure and properties [From Hearle (1)].

related characteristics. Twist also affects the hairiness of yarns, which is a very important property in determining the pilling behavior and the economics of the singeing process.

In his Mather lecture, titled "Twisted Structures," Treloar (2) adequately recognizes the role of twist in yarns and the part it plays in the design of textile structures. He discusses the obvious necessity of twist in the natural and staple fibers and points out:

"Twist is essential to provide a certain minimum coherence between fibers, without a yarn having a significant tensile strength cannot be made. This coherence is dependent on the frictional forces brought into play by the lateral pressures between fibers arising from the application of a tensile stress along the yarn axis. With the introduction of continuous filament yarns, however, the role of twist must be reconsidered. In continuous-filament yarns, twist is not necessary for the attainment of tensile strength (in fact, it reduces it) but it is necessary for the achievement of satisfactory resistance to abrasion, fatigue, or other types of damage associated with stresses other than a simple tensile stress, and typified by the breakage of individual filaments, leading ultimately to total breakdown of the structure. High twist produces a "hard" yarn, which is highly resistant to damage of this kind. The role of twist in continuous-filament yarn is thus to produce a coherent structure that cannot readily be disintegrated by lateral stresses.

From the engineering standpoint, the interesting thing about this structural function of twist is that, in contrast to most structure-building techniques, it produces its effect without significantly increasing the flexural rigidity or resistance to bending of the system. A yarn of 100 filaments has only 100 times the flexural rigidity of a single filament; if the filaments were all cemented together to form a solid rod, it would have 10,000 (100^2) times the flexural rigidity of a single filament.

Extending this line of thought to woven fabrics, we find again that the stiffness of the fabric is of the same order as the total stiffness of all the filaments in a given cross section as shown by Livesey and Owen, and similar considerations apply, no doubt, to knitted or other types of fabric structure. We see, therefore, that the major processes of textile fabrication are concerned with the production of coherent structures having maximum flexibility or minimum resistance to bending stresses, and hence also to compressive or buckling stresses, while retaining, of course, the inherent strength of the original filament material under the action of tensile stresses. This objective is in curious contrast to that normally encountered in engineering structures, where the general problem is to produce maximum resistance to bending and compressive stresses, combined with maximum tensile strength. The engineer achieves his objective of maximizing the rigidity by the introduction of suitably disposed fixed linkages between the various components of the structure. In textile structures, on the other hand, the objective of maximum flexibility is ingeniously achieved by the introduction of geometrical restraints, which, while strongly resisting forces of disruption, do not interfere appreciably with the small relative movements of individual elements associated with bending or other types of lateral deformation. However, a difference of objective does not necessarily imply a difference in method of approach, and there is no reason why the design of a textile structure should not be treated by the same rigorous analytical techniques as the design of any other engineering structure, such as a bridge or an aeroplane. The materials with which the textile engineer works have an inherent strength and other mechanical properties comparable with those of typical structural-engineering materials, and research is continually being concentrated on the improvement of these inherent properties. If these valuable characteristics are to be utilized to the fullest extent, it is equally important to see that the problem of design from the engineering standpoint receives something like the same kind of attention."

The study of twist, therefore, is very important to understanding the structure and behavior of yarns and their ultimate influence on the end-use properties of fabrics.

DEFINITION OF TWIST

When two ends of a straight strand (yarn) are rotated relative to one another, the fibers on the surface of the yarn lie in helices about the yarn axis. In other words, a yarn is twisted when fibers on the surface, which were originally parallel to the axis of the yarn, are now deformed (rotated) so that they make an angle θ with the axis, and the amount of twist is a function of this angle (Figure

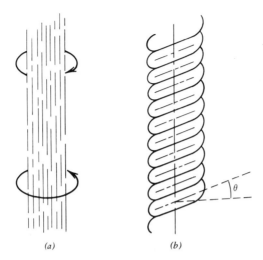

Figure 5.2. (*a*) Parallel bundle of fibers, (*b*) twisted bundle of fibers.

5.2). This definition of twist only applies to the ideal case of an originally straight fiber assembly. However, in actual yarns, variability of yarn diameter, contraction because of twist, migration of fibers from one zone to the other, radial compression of the yarn, and fiber slippage are some of the factors that tend to make the yarn geometry depart from the ideal.

The implications of some of these factors have been taken into account by Woods (3), who studied the kinematics of twist. He introduced the concept of "tortuosity" and stated that when a yarn is distorted its axis lies in a helix rather than in a single plane. On this basis, he has proposed a definition of twist, which is considered valid for all states of a yarn. According to his definition: "The twist is defined as the rotation of the transverse to any chosen generator about the axis, per unit length of the axis; that is to say that as we pass along the axis, the transverses rotate about the axis at a rate which is proportional to the twist in the string at the point considered." The "generator" is defined as any line on the surface that is parallel to the axis when the string is straight and untwisted; the radius of any cross section is the transverse in that section.

DIRECTION OF TWIST

In the designation of yarns, it is essential to specify the direction of twist. Besides its importance in simplifying the trade, it is of great technical significance in designing fabrics. For example, in a twill fabric, the direction of twist in the yarn is of particular importance in determining the predominance of twill

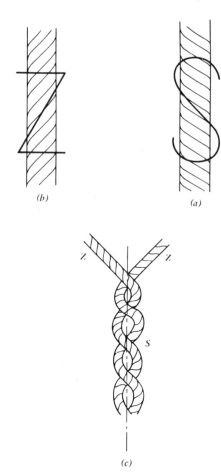

Figure 5.3. Direction of twist. (*a*) Right-handed twist, S, (*b*) left-handed twist, Z, (*c*) twist in folded yarns.

effect. For a right-handed twill, the best contrasting effect will be obtained when a yarn with Z twist is used; on the other hand, a left-handed twist will produce a fabric having a flat appearance.

The direction of twist in a yarn is designated in two ways:

1. Right-handed twist. S twist or clockwise (Figure 5.3*a*).
2. Left-handed twist. Z twist or anticlockwise (Figure 5.3*b*).

Single yarns are generally twisted in anticlockwise directions, whereas the S or right-handed twist is very common in ply yarns.

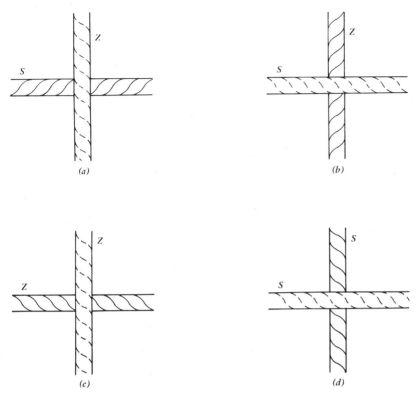

Figure 5.4. (*a*) and (*b*) Conditions that preclude nesting, (*c*) and (*d*) conditions that promote nesting.

Fabric Appearance and Twist Direction

Single warp yarns are generally given Z twist, and the weft or filling yarns, S twist (Figures 5.3*a* and *b*). When the warp and weft yarns are interwoven, they generally lie at an angle approximately $90°$ at the intersection, as shown in Figure 5.4. It is clear from Figure 5.4*a* and *b* that there is no possibility of the warp and weft yarns bedding or nesting into each other. However, if the two sets (warp and weft) of yarn have similar twist, that is, both Z or both S (Figures 5.4*c* and *d*), the fibers at the point of yarn contact will be more or less parallel. It is likely that this situation may induce nesting of yarns into each other. Consequently, the fabric produced will be thinner and firmer in hand than when nesting is prevented.

However, there is one important condition that must be satisfied before the kind of bedding effect described above occurs. This condition, according to Brunnschweiler (4), is that the sum of the angles of twist in warp and filling yarns must approach $90°$. But the angle of twist in normal staple yarns varies between 20 and $30°$. It follows that if the angle of twist in a yarn is $20°$, the fibers will cross at $90-(2 \times 20) = 50°$, which is far from the condition required for bedding. On the other hand, with a $45°$ twist angle (a twist factor far in excess of 77 tex$^{1/2}$ turns/cm used for crepe yarns), the yarn produced would be so compact that it would inherently resist nesting. Nevertheless, the relative direction of twist in the warp and filling yarns with normal twist does affect the fabric structure and the bedding behavior, as has been pointed out by Backer (5, 6). He gives the following reasons: (1) Yarns tend to flatten at the crossover points, and this causes an increase in twist angle. (2) Yarn crimping also increases the angle of twist on the inside of the bend.

Sometimes yarns with opposite twist directions are used to produce special surface texture effects in crepe fabrics. For example, the use of 1S:1Z crepe yarn twist arrangement in both warp and filling produces the finest figure effect in a crepe fabric.

The relative direction of twist has very important implications in determining the distinctness of the diagonal lines in a twill fabric. There are three major considerations that must be taken into account when designing twill constructions. These are:

1. The twist direction in relation to the twill hand. The diagonal twill effect will be predominant if the yarn twist (warp or filling) opposes the twill direction.
2. The relative direction of twist in the warp and filling. The use of warp and filling yarns with the direction of fibers opposing each other will result in a distinct twill effect. In other words, both sets of yarn should have the same twist direction.
3. Nesting or bedding of yarns. This factor is of minor consequence. However, if nesting is prevented by using warp and filling yarns twisted in opposite directions, the yarn floats become prominant, thus emphasizing the twill line.

Similar twist-twill relationships are valid in the designing of satin and sateen weaves for which either the suppression or prominance of the incipient twill lines is a requirement.

Determination of Twist

Accurate determination of twist, particularly in spun yarns, is important, since it is related to strength and other physical characteristics of yarns and fabrics. There are essentially two methods that are used in the industry for determination of twist in single yarns. These are 1. direct counting method, and 2. indirect untwist-twist method.

In the direct counting method, twist can be determined either by directly counting the number of turns in a unit length under a magnifying glass or by the parallel-fiber method in which the specimen is untwisted to zero twist until all the fibers are parallel. Because of the large variation of twist between short segments along the length of the yarn, it is necessary to take a very large number of observations to obtain a representative test average. In the parallel fiber method, the length of the specimen also plays an important role. If the specimen is longer than the staple length of the fiber, there is a possibility of the entanglement of fibers, which makes observation of when the twist is removed more difficult. On the other hand, specimen lengths shorter than or equal to the staple length of the fiber are easier to handle.

The untwist-twist method is much simpler and easier to perform. It is based on the assumption that when the twist in one direction is removed and retwisted by an equal number of turns in the opposite direction, the initial conditions of length and tension will be achieved. Thus the average turns per unit length is obtained. Worner (7) points out some of the limitations of the direct counting and the indirect untwist-twist methods for determination of twist in carded cotton yarns. He reports that there is a good correlation between the nominal (machine) twist and that obtained by the direct counting method. However, the correlation between the machine twist and the values obtained by the untwist-twist method is governed by the degree of tension applied to the test specimens; the best agreement is obtained at the highest tension used in the experiment. He has also shown that the number of turns per unit length obtained by this method is somewhat lower than that observed by the direct method.

GEOMETRY OF TWISTED YARNS

The translation of the physical properties of textile fibers into textile structures such as yarns and fabrics is a function of both the fiber properties and the geometric configuration they assume in the ultimate end product. Earlier studies of the relationships between yarn structure and properties have been along empirical or semi-empirical lines. However, many technologists who are concerned with the evaluation of the physical properties of textile structures have since realized that these relationships are governed by the same physical and engineering concepts as those used so successfully in the characterization of the more classic structural materials such as steel and concrete, etc. Hearle and Backer (1) have given an excellent account of the progressive development of the ideas in this field of study. Nevertheless, it is considered appropriate to introduce the reader to some basic concepts that are relevant to the study of yarn structure and properties. One such aspect is the understanding of the geometry of twisted yarns that is so vital in analyzing the stress-strain behavior of yarns in addition to

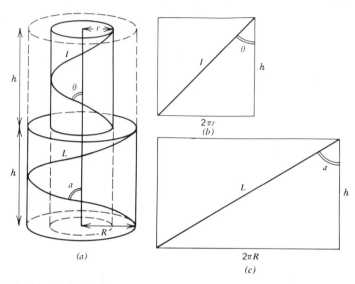

Figure 5.5. Idealized helical yarn geometry. (*a*) Idealized geometry, (*b*) "opened-out" diagram of cylinder at radius *r*, (*c*) "opened-out" yarn surface. [From Hearle (1)].

various other physical properties. A partial list of the pertinent physical characteristics that are influenced by twist will demonstrate its extreme importance:

> bending behavior
> resistance to creasing
> resistance to abrasion
> drapability
> elastic performance
> impact strength
> stress distribution and analysis.

Idealized Helical Geometry

In defining the geometry of a single yarn, the model usually adopted is that of an ideal physical form. This is the coaxial-helix model illustrated in Figure 5.5 (1). The assumptions underlying the model are characterized by the following postulates:

1. The yarn is circular in cross section and is uniform along its length.
2. It is built up of a series of superimposed concentric layers of different radii in each of which the fibers follow a uniform helical path so that its distance from the center remains constant.

3. A filament at the center will follow the straight line of the yarn axis, but going out from the center the helix angle gradually increases, since the twist per unit length in all the layers remains constant.
4. The axis of the circular cylinders coincides with the yarn axis.
5. The number of filaments of fibers crossing the unit area is constant, that is, the density of packing of fibers in the yarn remains constant throughout the model.
6. The structure is assumed to be made up of a large number of filaments; this will avoid any complications arising because of any descrepancies in packing of fibers.

This idealized model can be used to derive some very useful geometrical relations. Let us first define the parameters involved in characterizing the idealized geometry:

R = yarn radius (cm)
r = radius of cylinder containing the helical path of a particular fiber (cm)
T = yarn twist, turns per unit length (cm^{-1})
h = length of one turn of twist (cm)
α = surface angle of twist (the angle between the axis of a fiber on the surface and a line parallel to the yarn axis (degrees)
θ = helical angle at radius r (degrees)
l = length of fiber in one turn of twist, at radius r (cm)
L = length of fiber in one turn of twist, at radius R (cm)

It is clear from the model that

$$h = \frac{1}{T} \tag{5.1}$$

If the concentric cylinders in Figure 5.5a are cut along a line parallel to the yarn axis and opened out flat, they will appear as shown in Figure 5.5b and c. From the geometry of the opened-out diagrams one can obtain the following relations:

$$1^2 = h^2 + 4\pi^2 r^2 \tag{5.2}$$

$$L^2 = h^2 + 4\pi^2 R^2 \tag{5.3}$$

$$\tan \theta = 2\pi \ r/h \tag{5.4}$$

$$\tan \alpha = 2\pi \ R/h \tag{5.5}$$

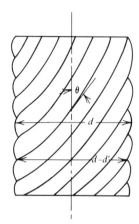

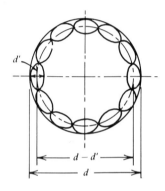

Figure 5.6. Multifilament yarn, showing difference between outer diameter and the diameter $(d\text{-}d')$ effective in measuring twist angle.

In practice the relation in equation 5.5 does not apply exactly in actual yarns with a finite number of fibers. This complexity of twist geometry has been discussed by Schwarz (8) and later by Woods (3) and Treloar (9). Schwarz (8) argued that, in the measurement of yarn diameter, the value obtained corresponds to the diameter of the circle (d), which circumscribes the outer layer of fibers, as shown in Figure 5.6. But in the measurement of twist angle (by a microscopic technique), it is the edges of the fibers in the outer layer that are observed. Therefore, the effective twist angle measurement is made at a diameter represented by the cylinder containing the centers of fibers in the outer layers. The modified form of equation 5.5 then becomes:

$$\tan\ \alpha\ =\ \pi(d\text{-}d')/h \tag{5.6}$$

$$=\ \pi\, dkT \tag{5.7}$$

where d = yarn diameter
d' = fiber diameter

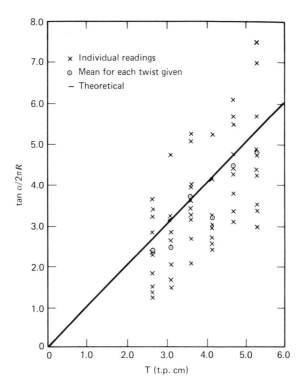

Figure 5.7. External filament angle as function of twist 1650-den. Tenasco. tension = 100 g wt. (Model uptwister) From Riding (11)

and $k = (d\text{-}d')/d$ = Schwarz's constant.

The value of k approaches unity for large numbers of fibers. Schwarz has reported the values of k for single, plied, and cabled yarns and has pointed out its usefulness in twist analysis.

In real yarns, assumptions (2) and (3) made above are incompatible in view of the observations made by Morton (10) on "fiber imigration" in staple yarns. His observations suggest that the path of a fiber in a yarn (continuous filament as well as staple yarn) is in fact not a cylindrical helix, but one whose radius changes along the length of the yarn.

Riding (11) has reported an experimental study in which he examined the validity of equation 5.5 in determining the relationship between twist and yarn structure. He points out that the agreement between experiment and theory is very good for continuous-filament yarns twisted by the continuous method. Figure 5.7 shows the values of tan $\alpha/2\pi R$ plotted against T, turns of twist per unit length for a 1650-den Tenasco yarn. The external filament angle and the yarn radius were measured with the aid of a microscope. The value of α and R

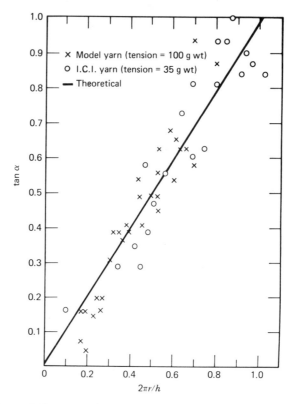

Figure 5.8. Internal filament angle as function of twist factor. (Model uptwister) From Riding (11)

were measured at ten points along the yarn. The theoretically calculated line is also plotted. It is clear from this figure that the experimentally measured values of twist angle are in good agreement with the predictions of the theory (equation 5.5). In the same study, Riding has also examined the dependence of filament angle on pitch and its radial position in the yarn cross section. To measure the filament angle at various radial positions along the length of the yarn, the tracer fiber technique developed by Morton and Yen (12) was used. (This is discussed briefly in Chapter 3.) Figure 5.8 shows the values of tan α plotted against $2\pi \, r/h$. It can be seen that "except for small values of r/h (i.e., small amplitudes), where the experimental errors are relatively large, there is no systematic deviation from the theoretical formula."

YARN SIZE AND TWIST MULTIPLIER

It was pointed out in Chapter 4 that textile yarns are generally specified by their count or size, namely, mass per unit length or linear density. The reason for this is that the diameter (or radius) of a yarn is very difficult to define because of its hairy nature and an indefinite packing density; on the other hand, linear density is easier to measure and control during spinning. The various traditional yarn count systems (direct and indirect) are discussed thoroughly in Chapter 4.

Twist multiplier or twist factor is a measure of the "twist-hardness" of a yarn; it is given by the product of yarn twist and square root of yarn size in the direct system, or the division of the turns per unit length by the square root of the count in an indirect system. Expressed in mathematical terms:

1. Direct system (tex system)–turns per centimeter multiplied by

$$\sqrt{\text{count of yarn, tex}} = \text{turns/cm} \times \text{tex}^{1/2}$$

2. Indirect system (cotton or worsted, etc., count system)–turns per inch divided by

$$\sqrt{\text{count, (cotton or worsted) of yarn}} = \text{tpi/count}^{1/2}$$

Some examples of the major yarn count systems and the twist multipliers and their relationships (conversion of one system to the other) are given in Table 5.1.

Table 5.1. Some Major Yarn Count Systems

Name of System	Unit of Count	Conversion to tex	Unit of Twist Multiplier	Conversion to tex$^{1/2}$ turns/cm
Direct Systems:		Multiply by:		Multiply by:
Tex	g/km	1	tex$^{1/2}$ turns/cm	1
Denier	g/9000 m	1/9 = 0.111	–	–
Indirect Systems:		Divide into:		Multiply by:
Cotton count	840 yd hanks/lb	590.5	tpi/count$^{1/2}$	9.57
Worsted count	560 yd hanks/lb	885.8	tpi/count$^{1/2}$	11.72
Metric number	km/kg	1000	–	–

The idealized yarn geometry can be used to express the relation between yarn count and twist multiplier as derived in the following manner:

Volume of unit length of idealized yarn $= \pi R^2$ and if v_y is the specific volume of the yarn expressed in cm^3/g, then its mass is $\pi R^2/v_y$.

From this it follows that:

$$\text{Yarn count} = C = \text{mass of 1 km (in tex system)}$$

$$= (\pi R^2/v_y) \times 10^5 \ \text{tex} \qquad (5.8)$$

Rearranging equation 5.8:

$$R = (v_y C/10^5 \pi)^{\frac{1}{2}} \qquad (5.9)$$

and combining this with equations 5.1 and 5.5, we get:

$$\tan \alpha = 10^{-3} (40\pi \, v_y)^{\frac{1}{2}} C^{\frac{1}{2}} \tau \qquad (5.10)$$

$$= 0.0112 \, v_y^{\frac{1}{2}} \tau \qquad (5.11)$$

where $\tau = C^{\frac{1}{2}} T \, \text{tex}^{\frac{1}{2}}$ turns/cm (twist multiplier).

The amount of twist needed in a yarn depends on the yarn size, the staple length of the fibers used, and the desired end-use application. A yarn may be soft-twisted (such as knit yarns) or it may be hard-twist (crepe yarns); the factor used to achieve these desired characteristics is the "twist multiplier." It is used in the industry in such a way that it is independent of the yarn size. This is made clear by equation 5.11, which shows the relation between twist multiplier and twist angle. "(These) two quantities are directly related in yarns of different counts but of the same twist factor will be geometrically similar differing only by a dimensionless scale factor" as stated by Hearle (1), Hearle has given values for twist factors and twist angles for some typical yarn specific volumes (Table 5.2).

Table 5.2. Twist Factor and Twist Angle

Specific Volume (cm^3/g)	Twist Factor, $\text{Tex}^{\frac{1}{2}}$ Turns/cm						
	0	20	40	60	80	100	120
	Twist Angles (degrees)						
0.5	0	9	18	25	32	38	44
1.0	0	13	24	34	42	48	53
1.5	0	15	29	39	48	54	59

Some typical values of twist multiplier used in the textile industry are given in Table 5.3. Generally, long staple fibers have a tendency to bind themselves together better than short staple fibers. It is therefore usual to use a slightly

Table 5.3. Twist Factors of Typical Textile Yarns

	Twist Factor ($\text{tex}^{1/2}$ turns/cm)	Twist Factor (traditional units $\text{tpi/count}^{1/2}$)
Cotton yarns		
Doubling weft	29-32	3.0-3.3[a]
Ring weft	32-35	3.3-3.6
Ring twist	38-43	4.0-4.5
Voile	49-53	5.5-5.5
Crepe	57-77	6.0-8.0
Worsted yarns		
Hosiery	17	1.4[b]
Soft worsted	20	1.7
Medium worsted	23	1.9
Hard worsted	26	2.2
Extra hard worsted	29	2.5

[a]Using cotton count.
[b]Using worsted count, from Stanbury and Byerley (27).

lower twist multiplier for the former case than that used for the corresponding yarn made from shorter staple fibers. Because of the helical arrangement of the fibers, a yarn suffers some contraction, which may range from 2 to 3% in a soft-twist yarn and up to 10 to 12% for hard-twist yarns. Consequently, there will be a considerable difference between the nominally calculated twist factor and twist based on the zero-twist yarn length and the actual values in the twisted yarns. The effect of contraction is of importance in adjusting draft during spinning of staple yarns. The subject of contraction because of twist is discussed in section 5.6.

Optimum Twist Factor

There is a certain minimum value of twist factor below which it is impossible to spin a staple yarn. Above this minimum twist factor, which is strongly influenced by staple length, fineness, and fiber surface (frictional) characteristics, the cohesiveness (yarn strength) between fibers increases at a fairly rapid rate initially. At low twist factors, the initial increase in yarn strength is determined by the resistance of fibers to slippage.

At high twist factors, the contribution because of resistance to slippage reaches a steady maximum. However, as the twist factor becomes high, the effect of fiber obliquity comes into play, and this has a tendency to cause a decrease in yarn strength. The twist-strength relationship is therefore the result of the simultaneous operation of the two effects discussed above. The twist factor at which maximum strength is achieved in any given staple yarn is sometimes called the "optimum twist factor." The optimum twist factor and the lowest practicable twist will both depend on such fiber characteristics as fiber length, fineness, bending and flexural rigidity, and frictional properties. The optimum twist factor will also vary with the count of yarn being spun, and it is entirely possible for the maximum value of breaking load and breaking extension to occur at different levels of twist. A more detailed discussion of the twist strength relationship in staple yarn is found in Chapter 6.

CONTRACTION BECAUSE OF TWIST

One effect of twist in a yarn is to cause contraction as a result of the longer path (because of helical geometry) followed by the fibers. The theoretical calculation of contraction has been dealt with by Treloar (9), Morton and Hearle (13), and Gracie (14). The theoretical treatment is based on the consideration of an idealized twist geometry and the occurrence of fiber migration in the yarn. The existence of fiber migration accounts for the effective averaging within the yarn of the different fiber lengths at different positions; the deficiency in length for filaments near the surface is made up by the excess occurring when they are near the center.

The yarn contraction can be defined in two ways:

$$\text{contraction factor} = C_y = \frac{\text{length of zero twist yarn}}{\text{length of twisted yarn}}$$

and

$$\text{retraction} = R_y = \frac{\text{length of zero twist yarn} - \text{length of twisted yarn}}{\text{length of zero twist yarn}}$$

In other words,

$$C_y = 1/(1 - R_y) \tag{5.12}$$

The contraction factor is very useful when dealing with staple yarns. Its value ranges from 1 for no contraction to infinity for contraction to zero length. The contraction affects the size or count of the spun staple yarn; therefore, contraction factor indicates the factor by which the draft must be adjusted to prevent the twist contraction from decreasing the yarn size. For example, if a 1.0 hank roving is spun to 18 single yarn 5% contraction is allowed, the draft adjusted at the drafting rollers must be greater than 18; that is, it should be 18.90 instead of 18. The parameter "retraction" is commonly used for continuous-filament yarns, and its value ranges from 0 to 1. The retraction represents the fractional decrease in length or increase in the linear density (denier) on twisting continuous-filament yarns.

The theoretical relationship that expresses retraction in terms of surface twist angle has the following form (9, 13):

$$R_y = (\bar{\ell}\text{-}h)/\bar{\ell}$$

$$= \frac{\sec \alpha - 1}{\sec \alpha + 1} = \frac{1 - \cos \alpha}{1 + \cos \alpha} = \tan^2 (\alpha/2) \qquad (5.13)$$

Where ℓ = mean length of filaments in the length h of twisted yarn, and
$\quad\alpha$ = surface twist angle.

The contraction factor is given by:

$$\text{contraction factor} = C_y = \frac{\ell}{h} = \frac{1}{2} (1 + \sec \alpha) \qquad (5.14)$$

(For derivation of these expressions, the reader is referred to reference 9 or 13).

The theoretical values of contraction factor and retraction for various surface twist angles are shown in Table 5.4.

Table 5.4. Yarn Contraction[a]

Twist Angle (α)	Contraction Factor (C_y)	Retraction (R_y)
0°	1.0	0
10°	1.008	0.008
20°	1.032	0.031
30°	1.078	0.072
40°	1.153	0.132
50°	1.278	0.217

[a]from Hearle (1).

Tattersal (15) and Riding (11) have carried out experimental studies of the effect of twist on the retraction of continuous-filament yarns. They have compared their experimentally obtained results with the predictions of Treloar's

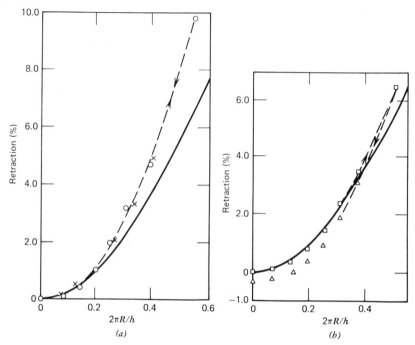

Figure 5.9. Retraction on "static twisting" of 1650 denier Tenasco; Riding (11). (*a*) Twisted under tension of 100 g wt, (x) Twisting, (○) Untwisting, (–) Theroretical; (*b*) Twisted under tension of 500 g wt.

theory (9). The measured retraction values obtained during "static twisting" (a length of yarn mounted between a twist head and a traversing trolley) plotted against $2\pi R/h$, that is, tan α, did not show a good agreement with the theoretical curve. The retraction was found to be dependent on twisting tension, as shown in Figure 5.9. Twisting at low tension gave retraction values considerably higher than predicted but showed perfect reversibility, on the other hand, at higher twisting tension, the retraction was lower than the predicted value and gave negative values on untwisting to zero twist. This latter behavior has been attributed to the irreversible straining of the filaments and kink formation, which was confirmed by visual observation. These kinks are formed during static twisting when the filaments are locked in place and are relatively restrained from migrating freely. The free migration of filaments in a twisted yarn is essential for averaging of path length. However, when retraction values were obtained by untwisting yarns made on either commercial machines or on special laboratory machines (continuously twisted yarns) as described by Riding, the agreement with the theory was found to be excellent. This is shown in Figure 5.10. Twist-

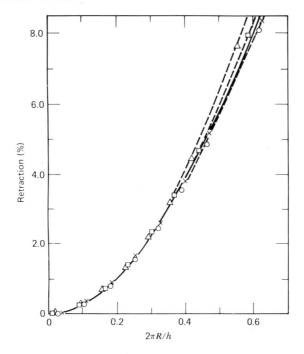

Figure 5.10. Retraction on continuous twisting of 1650 denier Tenasco on a model up-twister, at various twisting tensions. (Riding, 11) (x) 50 g wt tension, (o) 200 g wt tension, (□) 400 g wt tension, (△) 700 g wt tension, (———) theoretical.

ing tension was found to have a negligible effect on retraction values. The retraction values were also found to be independent of the maximum twist inserted and the yarn linear density.

Hearle et al. (16) have suggested the rearrangement of equation 5.14 to the following form.

$$C_y \ (C_y - 1) = 1/4 \ \tan^2 \ \alpha \qquad\qquad (5.15)$$

This equation suggests that a plot of $C_y \ (C_y - 1)$ against $\tan^2 \ \alpha$ should yield a slope of 1/4 and can be conveniently used to check the correspondence of experimental results with the theory. Figure 5.11 shows the results for polyester (Terylene) and viscose yarns. The polyester yarn showed good agreement with the theory. On the other hand, viscose rayon yarn was highly influenced by twisting tension. It showed lower contraction than the predicted value when twisted at high tension. The authors attributed this difference to some permanent straining of the twisted filaments. Figure 5.12 shows the comparison of

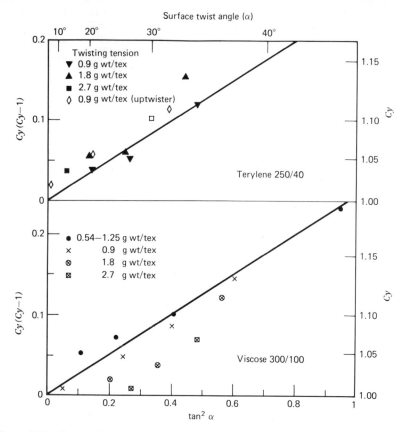

Figure 5.11. Contraction factor results. After Hearle, El-Behery, and Thakur (16). The line is drawn according to equation 5.15.

contraction values for various materials. Viscose rayon and acetate yarns showed good agreement with theory, whereas Tenasco gave anomalously high values.

It is clear from the above discussion that, with minor exceptions, equations 5.14 and 5.15 give good predictions of the contraction and retraction behavior of continuous-filament yarns. The disturbing features that can cause deviations are:

1. irregular twisting of yarn
2. permanent extension of filaments
3. kinking or buckling of the central filaments

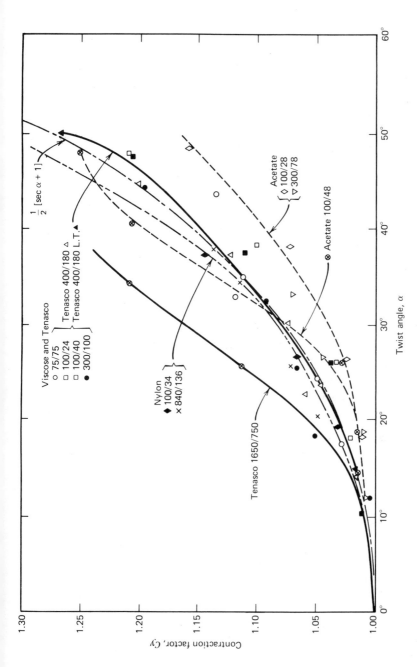

Figure 5.12. Comparison of contraction factors. Hearle, El-Behery, and Thakur (16).

117

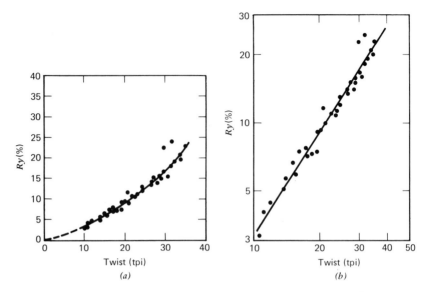

Figure 5.13. Contraction due to twisting of cotton yarn. Landstreet, Ewald, and Simpson (17) (*a*) Plot of retraction against twist, (*b*) log-log plot.

The contraction behavior of spun cotton yarns has been discussed by Landstreet et al. (17). Contraction is plotted against turns per unit length supplied at zero twist for one set of yarns in Figure 5.13. When the data shown in Figure 5.13*a* are presented in the form of a log vs. log plot, as shown in Figure 5.13*b*, there is a good fit with an empirical equation of the form:

$$R_y = k(T)^n \tag{5.16}$$

In practice, the exponent n is found to be of the order of 1.5, which represents the average condition in cotton spun yarns.

TWIST AND FIBER PACKING IN YARNS

Another feature of yarn structure that can be described in a simple idealized form is the packing of fibers in yarns. Schwarz (18, 19) has theorized that the packing of circular fibers can be described by two basic forms: (1) open packing and (2) hexagonal close packing.

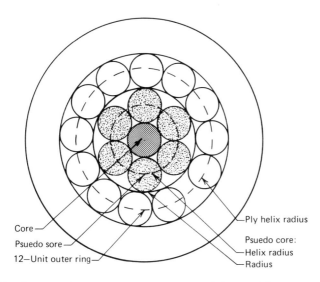

Core
Psuedo sore
12—Unit outer ring

Ply helix radius

Psuedo core:
Helix radius
Radius

Figure 5.14. Concepts of open-packed structure. From Schwartz (19)

Open Packing

In this form, the fibers lie in layers between successive concentric circles, as shown in Figure 5.14. In this assembly, the first layer is a single core fiber around which six fibers are arranged so that all are touching. The third layer, which has twelve fibers, is arranged so that the fibers first touch the circle that circumscribes the second layer. Additional layers are added between the successive circumscribing circles.

The number of fibers in each layer and the total number of fibers in an ideally packed open structure are given in Table 5.4 (for a theoretical explanation reader is referred to reference 1).

Table 5.4. Open Packed Yarn Structure

Layer No.	Maximum No. of Fibers in Layer	Total No. of Fibers
1	1	1
2	6	7
3	12	19
4	18	37
5	25	62
6	31	93

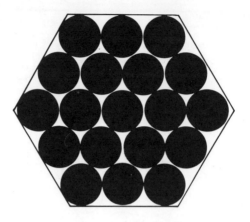

Figure 5.15. Hexagonally close-packed yarn, with three layers.

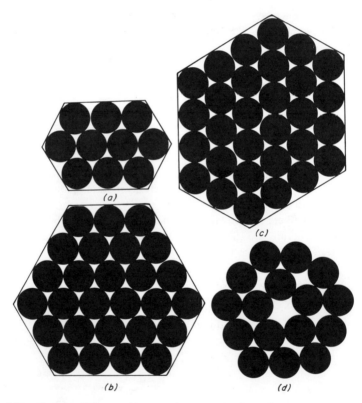

Figure 5.16. Hexagonal close packing on other cores. (*a*) Core of two, (*b*) core of three, (*c*) core of four, (*d*) core of five. From Hearle (1)

Hexagonal Close Packing

The packing of fibers of circular cross section around a single core fiber in a hexagonal configuration leads to what is called "close packing." In this form, all fibers touch each other, as shown in Figure 5.15. Other close hexagonal configurations obtained with cores of two, three, four, and five fibers are shown in Figure 5.16. As the number of fibers in the cross section increases (number of layers), the yarn outline tends to become complicated and deviates from the preferred hexagonal shape. For an ideal close-packed structure, the number of fibers in each layer and the total number of fibers in the cross section are given in Table 5.5.

Table 5.5.   Close Packing of Fibers in Yarn

Layer No.	Number of Fibers in Layer	Total No. of Fibers
1	1	1
2	6	7
3	12	19
4	18	37
5	24	61
6	30	91
7	36	127
8	42	169
9	48	217

Real Yarns

The two ideal forms described above are very rarely achieved in real yarns. There are numerous factors, such as fiber geometry and twist, among others, that influence the yarn configuration obtained in real yarns. Hearle (1) suggests that the factors that operate in determining real yarn structures can be described into two groups, namely, (1) concentrating factors and (2) disturbing factors.

Concentrating factors

(1) One of the concentrating features is the general tendency of fibers to follow the path of minimum energy. In addition, they tend to settle down and cohere like any other materials.

(2) The other factor is twist, which has a stronger effect. Twist causes the development of tangential and radial forces in fibers that tend to bind the fibers together. Any imposed tension in a yarn will generate inward pressure from the outer layers on those inside, leading to a close packing of filaments.

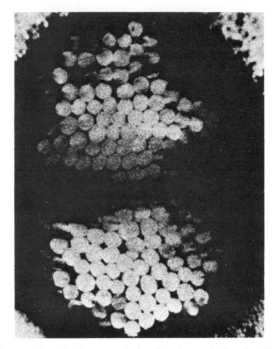

Figure 5.17. Cross section of a hot stretched nylon tire cord, showing filament distortion under pressure. From Wood, Goy, and Daruwalla (26)

High tension and twist can cause deformation in the fiber shape, especially in thermoplastic yarns, as shown in Figure 5.17. This shows a cross section of a hot stretched two-ply nylon tire cord in which the filaments near the center seem to have deformed into polygonal shapes under pressure.

Disturbing factors

(1) Ideal packing is not achieved if the total number of fibers in the yarn cross section differ from the figures given in Tables 5.4 and 5.5. Any deviation from the appropriate number would make the fibers pack unevenly, resulting in an irregularly shaped cross section.

(2) The simple ideal forms discussed earlier are true for a circular fiber shape. Other fiber shapes will modify the ideal packing arrangement.

(3) Twist has a profound effect on the arrangement of fibers in the yarn cross section. Because of twist, the fibers in the outer layers of the yarn follow a helical path; this introduces ellipticity in the cross section of the fibers. The elliptical cross section further affects the packing, thus modifying the ideal arrangement of fibers in the yarn cross section.

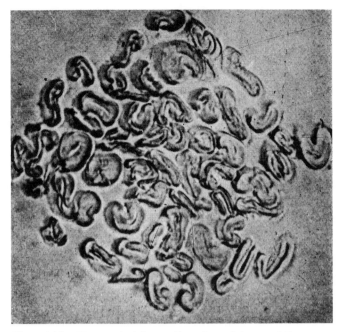

Figure 5.18. Close packed Cotton Yarn. From Schwartz (19)

(4) The other disturbing effect of twist is the variation in the path length between fibers at different radial positions in the yarn cross section. These differences in the path length must be nullified either by buckling of fibers in the center or by an interchange of radial position of fibers (migration), which results in the effective averaging of path lengths.

(5) Sometimes irregularities in the arrangement of fibers prior to yarn formation (such as in roving) may persist in the final yarn structure. Irregularities in the arrangement of fibers can also result from the variations in fiber length supplied to the twisting zone in staple yarn spinning.

(6) Other processing factors, such as the passing of yarns over guides or between feed rolls during winding, may cause changes in yarn shape. In addition, the form of twisting, that is, whether the fibers are presented in a ribbon form or in a cylindrical shape, may introduce asymmetry and irregularity in the shape of the yarn.

Observed Packing of Fibers in Real Yarns

Schwarz (19) has reported a few examples of the kind of fiber packing achieved in actual yarns. Figure 5.18 shows the cross section of a cotton yarn in which

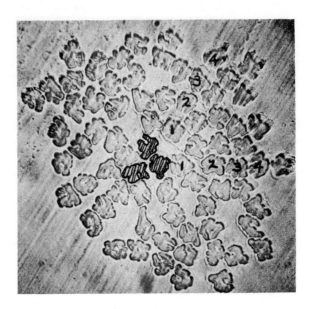

Figure 5.19. 75-filament viscose yarn showing 3 unit core with four outer layers. From Schwatrz (19)

fibers appear to be close packed in a polygonal shape (almost in a square shape). On the other hand, Figure 5.19 shows an example of an open-packed structure for a 75-filament low-twist viscose rayon yarn. This structure has a three-unit core with four outer layers, as identified in the illustration.

Some further examples of the packing of filaments in the cross sections of continuous-filament yarns have been reported by Hearle and Bose (20). These cross sections (Figure 5.20), prepared by embedding yarns in resin and then cutting sections, show that in almost all cases the fibers tend to pack in a close-packed form. There is some evidence of the irregularity and asymmetry in the packing, as pointed out by the authors. The presence of asymmetry has been attributed to the ribbon-twisting effect.

EFFECT OF TWIST ON YARN DIAMETER AND SPECIFIC VOLUME

In woven fabrics, especially in tightly woven structures, the interthread pressures set up during weaving can cause considerable thread flattening. This effect demands considerable attention in understanding the geometry of fabric structures. The changed cross-sectional shape of yarns in fabrics has a significant

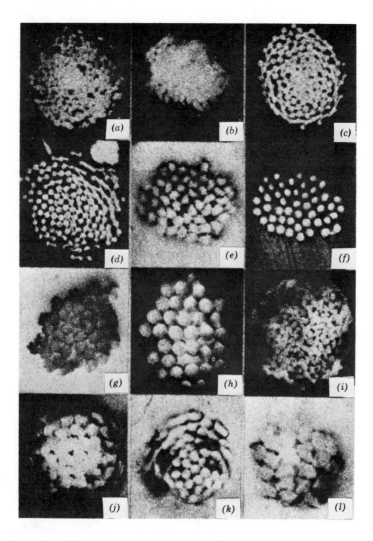

Figure 5.20. Yarn cross sections obtained by Hearle and Bose (20). (*a*) Tenasco 400/180, 10 tpi, uptwisted, (*b*) Tenasco 400/180, 10 tpi, uptwisted, (*c*) nylon 840/136, 10 tpi, uptwisted, (*d*) nylon 840/136, 10 tpi, ring twisted, (*e*) viscose 300/50, 10 tpi, uptwisted, (*f*) terylene 100/48, 10 tpi, uptwisted, (*g*) viscose 100/40, 10 tpi, uptwisted, (*h*) nylon 100/34, 10 tpi, ring twisted, (*i*) Tenasco 400/180, 30 tpi, uptwisted, (*j*) viscose 300/50, 30 tpi, uptwisted, (*k*) Terylene 100/48, 30 tpi, uptwisted, (*l*) viscose 100/24, 30 tpi, uptwisted.

125

effect on the aesthetic as well as some physical characteristics of fabrics. Some of these effects are discussed in a later chapter.

The bulk density or the specific volume of a yarn is a direct manifestation of the way the fibers pack in the yarn cross section. The factors that are likely to affect the yarn density are the fiber type, the density of the fiber material itself, yarn twist, and some external factors such as thread tension and compression encountered during winding, warping, weaving, or twisting. The yarn specific volume is given by the following relationship, which is obtained by rearranging equation 5.9:

$$v_y = \pi R^2/C \times 10^5$$
$$v_y = \tan^2 \alpha/4\pi \; CT^2 \times 10^5 \tag{5.17}$$

With the exception of monofilament yarns, which have a specific volume equal to that of the fiber material, the specific volume of all other textile yarns is determined by the volume occupied by the fibers and the air spaces incorporated into the body of the yarn. Consequently, the yarn specific volume is always more than that of the fibers themselves. This relationship can be analytically expressed in terms of a parameter, packing fraction, ϕ, as follows:

$$\phi = v_f/v_y = \text{fiber volume/yarn volume}$$

or

$$v_y = v_f/\phi \tag{5.18}$$

The specific volumes of textile fibers range from as low as 0.4 for glass fiber to as high as 1.1 for polypropylene.

It must be pointed out that, in addition to the influence of yarn twist, the yarn specific volume is also affected by such factors as the staple length, fiber fineness, the system on which a staple yarn has been spun, and the conditions of tension and compression in different processes.

From equation 5.17 it can be realized that the measurement of the specific volume of a yarn requires the determination of yarn linear density (C, count of yarn) and either yarn diameter of yarn twist and twist angle. Experimentally, the yarn linear density is easily obtained. However, twist and yarn diameter present some difficulty, particularly in the case of staple yarns. Experimental errors in the measurement of twist and diameter of staple yarns result from the diffused nature of their surface outline. Twist measurement is especially a problem with open-end spun staple yarns. Hearle et al. (21) and Hearle and Merchant (22) have used a simple microscopic technique to evaluate the yarn diameter and

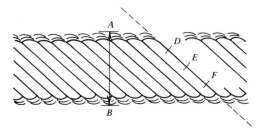

Figure 5.21. Diameter overestimated as AB due to loose fiber on surface. Twist angle underestimated from mean slope of DF instead of true tangent at E. From Hearle (1)

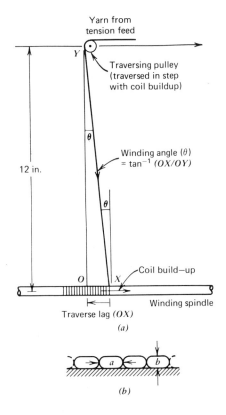

Figure 5.22. Coil winding method of measuring yarn dimensions. From Hamilton, (22) (*a*) Principles of apparatus, (*b*) arrangement of coils on spindle under normal winding conditions.

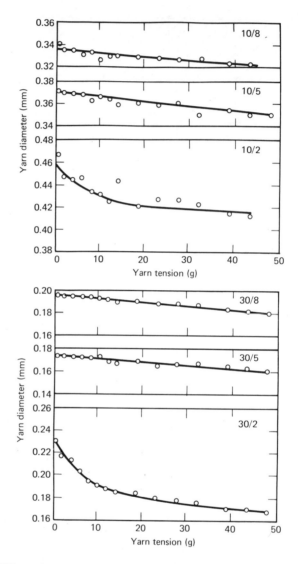

Figure 5.23. Effect of tension on diameter of spun nylon yarns. The figures identifying each diagram refer to cotton count and twist factor, that is, 10/8 is 10s cottons count and twist factor of 8. Hearle and Merchant (22) See overleaf for additional diagrams.

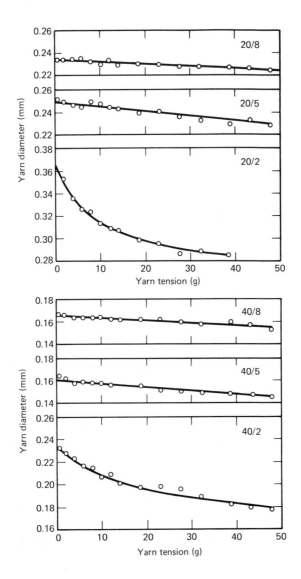

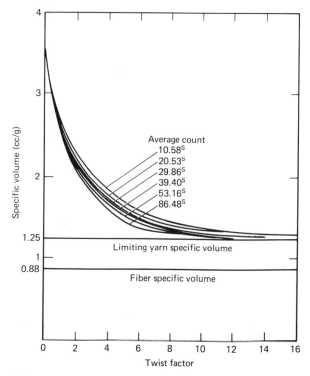

Figure 5.24. Collected curves of specific volume against twist. Hearle and Merchant (22)

twist angle. They observed that the microscopic technique tended to give a rather high value of specific volume, whereas the twist angle appeared to be low. Hearle (1) has suggested reasons for this, as illustrated in Figure 5.21.

Van Issum and Chamberlain (23) have described a photographic technique in which a blurred image of a running yarn is obtained. This tends to average the effect of protruding fibers at the surface. The photograph is then photometrically scanned for diameter values.

However, to understand the structural geometry of fabrics it is essential to consider the thread flattening caused by the interyarn pressure in the fabric. Hamilton (24) has described a method for measuring both the major and the minor diameters of a yarn under conditions of thread tension and compression similar to those which occur during weaving, package winding, warping and knitting. The method consists of winding the yarn in adjacent coils onto a spindle of suitable diameter, as illustrated in Figure 5.22. This method is applicable to both filament and staple yarns over a wide range of yarn sizes. The major diameter a is calculated from the number of turns of coils per unit length,

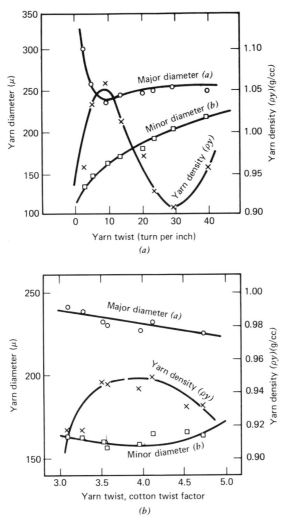

Figure 5.25. Effect of twist on yarn dimensions using coil-winding method. From Hamilton (*a*) 300-Denier continuous-filament rayon yarn, (*b*) 20s staple rayon yarn.

and the minor diameter *b* is measured by pressing a special wheel against the yarn. The feeler wheel is driven by surface contact with the yarn on the winding spindle. There is a provision to change the load on the pressure wheel to achieve desired thread compression. This direct measurement of major and minor yarn diameters, together with the yarn count, allow the determination of the cross-sectional shape and the specific volume. The effect of yarn tension on yarn diameter is shown in Figure 5.23; the dependence of specific volume of spun nylon yarn on twist is illustrated in Figure 5.24 (22). On the other hand, the results of the effect of winding tension and twist measured by the coil winding method are shown in Figure 5.25a and 5.25b.

The values of specific volume reported by various workers are summarized in Table 5.6, 5.7, and 5.8. It is clear that the yarn specific volume, as well as the

Table 5.6. Yarn Specific Volumes (Ignoring Effect of Twist) (van Issum and Chamberlain, 23)

Yarn Type	Specific Vol. (cm³/g)	Packing Fraction (ϕ)
Spun rayon	1.54	0.427
Cotton	1.84	0.358
Worsted	1.70	0.452
Woolen	2.40	0.321

Table 5.7. Dimensions of Staple Fiber Yarns (Cotton-spun; Twist Factor = 33 tex$^{\frac{1}{2}}$ turns/cm) (Hamilton, 24)

Yarn	Diameter (μ) Major	Minor	Linear Density (tex)	Yarn Sp Vol (cm³/g)	Fiber Sp Vol (cm³/g)	Packing Factor (ϕ)
Rayon, 1½ den, 1⁷⁄₁₆ in.	230	162	29.5	1.08	0.66	0.61
Rayon, 3 den, 2½ in.	241	171	31.3	1.11	0.66	0.59
Rayon, 3 den 1⁷⁄₁₆ in.	243	166	30.3	1.14	0.66	0.58
Acetate 3 den, 1⁷⁄₁₆ in.	265	183	31.9	1.30	0.76	0.59
Triacetate, 3 den, 1⁷⁄₁₆ in.	262	179	28.8	1.39	0.77	0.55
Courtelle (acrylic), 3 den, 1⁷⁄₁₆ in.	287	173	27.0	1.61	0.87	0.54
Nylon, 3 den, 1½ in.	276	166	27.4	1.45	0.88	0.60
Terylene (polyester), 3 den, 1½ in.	284	176	33.6	1.30	0.72	0.56
Cotton, Tanguis	273	165	32.0	1.22	0.65	0.53

packing fractions, vary widely for both spun and filament yarns. The packing fractions range from about 0.30 for some spun yarns up to approximately 0.9 for highly twisted filament yarns. The theoretical value for hexagonal close packing is 0.91. In addition to yarn tension and compression, the yarn packing fraction is also affected by fiber variables and blending. These include fiber length, fineness, crimp, and the blending of fibers with different bulking characteristics.

Table 5.8. Specific Volume of Continuous Filament Yarns (Hearle, El-Behery, and Thakur, 21)

Yarn	Twist Factor $(\text{tex}^{1/2} \text{ turns/cm})$	Specific Vol (cm^3/g)	Packing Fraction (ϕ)
100 den/	3.9	1.67	0.40
40 filament/	14.5	1.21	0.54
viscose rayon	29.5	1.11	0.59
	42.7	1.01	0.65
	69.9	0.94	0.70
	102.1	0.90	0.73
100 den/	1.8	2.10	0.36
28 filament/	15.6	1.20	0.63
acetate	27.6	1.12	0.68
	41.3	1.12	0.68
	69.6	1.01	0.76
	101.9	0.96	0.89
100 den/	1.0	3.12	0.25
34 filament/	16.8	1.35	0.65
nylon	26.1	1.20	0.73
	37.0	1.23	0.72
	62.7	1.15	0.76
	97.5	0.99	0.89
100 den/	0.5	2.65	0.27
48 filament/	13.6	1.16	0.63
Terylene polyester	27.6	1.05	0.69
	42.5	1.03	0.70
	57.9	0.98	0.74
	82.9	0.97	0.75
	113.9	0.97	0.75

TWIST IN RELATION TO YARN BENDING

Textile yarns are rarely used in a straight form. They are generally bent into helical forms when twisted to make plied structures and held into torus shapes during the weaving process and into a looped shape when converted into a knitted structure. Yarn bending attains an added dimension during the wear life of a textile structure. Yarns are subjected to bending during draping, creasing (in the bending of a collar or a cuff) or wrinkling of an apparel fabric, in the fluttering of a tent, or the flexing of a rope over a pulley, to name a few.

The mechanical properties of a textile structure during its operational life will be greatly influenced by the kind and the amount of strain deformation that individual components of the structure suffer. The level of deformation in

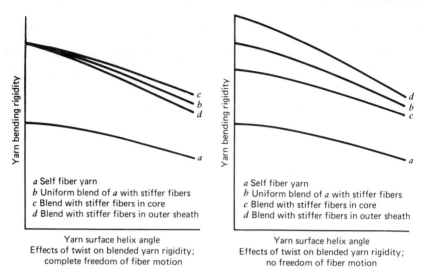

Figure 5.26.   Yarn surface helix angle. From Platt et al. (28)

individual fibers will in turn be affected by the geometric form in which the fiber is made to lie in a yarn and subsequently in a fabric. In a staple yarn, the component fibers do not lie straight but assume a helical configuration imposed by twist. It is therefore essential that a knowledge of the effect of twist on the bent yarn geometry be obtained to help in analyzing the strain and consequently the stress distribution in a textile structure. Backer (25) has reported a geometric analysis of the idealized structure of a bent yarn. The analysis deals particularly with the prediction of local and average fiber strains that occur in highly twisted structures—that are deformed in bending. Specific cases of the relative motion between fibers in both the loosely and tightly twisted structures have been considered. Relationships predicting local fiber curvature because of bending have been derived. This analysis reveals that the level of yarn strain becomes less as the radius of bent yarn (torus) is increased relative to the radius of the yarn. On the other hand, the local strain level decreases as the level of twist in the yarn increases. Furthermore, it is shown that fibers occupying the position at the neutral axis of the yarn experience no strain; the fibers lying at the uppermost and the lowermost positions suffer maximum strain both in the tension and compression modes. Backer has also considered the case of no friction between the fibers in a yarn. This assumption helped in calculating the change in the local helix angle of the fiber at the inside and outside of the bend. This type of prediction can help in the understanding of internal wear in a fabric as well as the crease and wrinkle recovery of textile structures. Platt et al. (28) have reported a theoretical analysis of the effect of fiber properties and yarn structure on singles yarn bending rigidity. The parameters considered are the fiber dimensions, fiber stiffness, torsional to bending rigidity ratio, yarn density, size, twist, fiber clustering and prior relaxation treatments of yarns. The two extreme cases of complete freedom and no freedom of relative fiber movement are analyzed. The theoretically predicted effect of twist on the bending rigidity of the yarn for the two extreme cases is shown in Figure 5.26. This figure also shows how the blend distribution and the stiffness of the fibers being blended affect the bending rigidity of the blended structure. It is obvious that increase in yarn twist has the effect of lowering the bending rigidity. The position (blend distribution) of the stiffer fibers in the yarn cross section influences the yarn bending rigidity differently for the two cases of complete freedom of fiber motion and no freedom of fiber movement. Recently Hunter et al. (29) have shown experimentally that the flexural rigidity of worsted yarns decreased when yarn twist was increased, thus confirming the hypothesis put forward by Platt.

REFERENCES

1. Hearle, J. W. S., P. Grosberg, and S. Backer, *The Structural Mechanics of Fibers, Yarns, and Fabrics.* Wiley-Interscience, New York, 1969.

2. Treloar, L. R. G., *J. Text. Inst.*, 1964, **55**, p. 13.

3. Woods, H. J., *J. Text. Inst.*, 1933, **24**, T317.

4. Brunnschweiler, D., *J. Text. Inst.*, 1953, **44**, p. 114.

5. Backer, S., *J. Text. Inst.*, 1953, **44**, T477.

6. Backer, S., *Text Res. J.*, 1956, **26**, 87.

7. Worner, R. K., *Text. Res. J.*, 1956, **26**, 455.

8. Schwartz, E. R., *J. Text. Inst.*, 1933, **24**, T105.

9. Treloar, L. R. G., *J. Text. Inst.*, 1956, **47**, T348.

10. Morton, W. E., *Text. Res. J.*, 1956, **26**, 325.

11. Riding, G., *J. Text. Inst.*, 1959, **50**, T425.

12. Morton, W. E. and K. C. Yen, *J. Text. Inst.*, 1952, **43**, T60.

13. Morton, W. E. and J. W. S. Hearle, *J. Text. Inst.*, 1957, **48**, T159.

14. Tattersall, G. H., *J. Text Inst.*, 1958, **49**, T295.

16. Hearle, J. W. S., H. M. A. E. El-Behery, and V. M. Thakur, *J. Text. Inst.*, 1960, **51**, T299.

17. Landstreet, C. B., R. R. Ewald, and J. Simpson, *Text. Res. J.*, 1957, **27**, 486.

18. Schwartz, E. R., *Text. Res. J.*, 1950, **20**, 175.

19. Schwartz, E. R., *Text. Res. J.*, 1951, **21**, 125.

20. Hearle, J. W. S., and O. Bose, *J. Text. Inst.*, 1966, **57**, T308.

21. Hearle, J. W. S., H. M. A. E. El-Behery, and V. M. Thakur, *J. Text. Inst.*, 1959, **50**, T83.

22. Hearle, J. W. S. and V. B. Merchant, *Text. Res. J.*, 1963, **33**, 417.

23. van Issum, B. E., and N. H. Chamberlain, *J. Text. Inst.*, 1959, **50**, T599.

24. Hamilton, J. B., *J. Text. Inst.*, 1959, **50**, T655.

25. Backer, S., *Text. Res. J.*, 1952, **22**, 668.

26. Wood, J. O., R. S. Goy, and F. S. Daruwala, *Text. Res. J.*, 1959,**29**

27. Stanbury, G. R. and W. G. Byerley, in *Wool Research 1918-1948*, WIRA, 1949, Leeds, England.

28. Platt, M. M., W. G. Klein, and W. J. Hamburger, *Text. Res. J.*, 1959, **29**, 611.

29. Hunter, I. M., R. I. Slinger, and P. J. Kruger, *Text. Res. J.*, 1971, **41**, 361.

6

Mechanical Properties of Yarns

INTRODUCTION

In addition to the contribution of fiber structure and fiber properties, the behavior of textile materials is significantly influenced by the role played by the yarn structure and yarn properties. This can easily be visualized when one considers the textile material properties that have important technological significance and that determine end-use performance. The textile material properties may be classified into three main groups, as suggested by Backer (1). These are (1) bulk properties, (2) surface properties, and (3) transfer properties. Backer has given a list of material properties that fall under each classification. They are summarized as follows:

1. *Bulk Properties,* which include uniaxial tension and compression; biaxial tension, compression, and shear; bending behavior (including creasing and crease resistance); torsional characteristics; behavior under stress concentration (such as tear resistance or cutting resistance); dimensional stability (shrinkage resistance, pucker resistance, pilling and shedding resistance); and fatigue resistance in tension, compression, and bending.

2. *Surface Properties,* such as hand, roughness, wear resistance, friction, and resistance to stress concentration (snag and pill resistance).

3. *Transfer Properties,* namely, air and water permeability, filtration efficiency, penetration resistance, and heat transfer.

A close look at the above list of properties would clearly indicate that the yarn structure and the mechanical properties of yarns are of great importance in modifying the ultimate behavior of a textile structure. Consequently, to predict and evaluate the behavior of the ultimate textile structure, an understanding of the structure and the mechanical behavior of yarns is essential. Some features of the geometry of yarn structure are discussed in Chapter 3. The intention here is to report and discuss some of the studies dealing with the theoretical prediction of the mechanical properties of yarns, arrived at by the application of simple mechanics, and the actual behavior of these structures.

YARN STRENGTH

There are essentially two ways in which the yarns are characterized as to their strength behavior. These are:

1. Lea or skein test—including the ballistic test.
2. Single thread tests

Lea Strength Test

The first method is commonly used in the older sections of the cotton, wool, and flax textile industries. In this test, a hank or a skein of 120 yards (made into 80 loops with the starting and finishing ends knotted) is broken on a pendulum lever tester from which the lea strength of the hank is obtained. This test does not give a measure of the absolute value of the yarn strength, but yields a relative value that is quite useful for quality control purposes. The strength of the yarn in a lea test is determined largely by the thinnest places in the yarn and the frictional forces between the threads and the hooks on the tester. The strength of the yarn is expressed in terms of the product of lea strength and count, which is called the "count-strength product" or "break factor."

The count-strength product is used for comparing yarns of slightly different counts. This quantity is also used in determining the "spinning efficiency" of a cotton by comparing the break factor of a yarn with a standard value. The count that has a break factor equal to the standard value is called the "highest standard count" and is taken as the measure of the spinning quality of that cotton.

The count strength product is merely the tensile strength divided by the weight per unit length of yarn. This quantity has the dimensions of length and is termed the "breaking length" of a yarn. This is the length of yarn having a weight equal to the breaking load that is observed to cause the yarn to break under free suspension.

The hank of yarn can also be tested under impact loading on a ballistic tester. This test combines the three characteristics of the yarn, namely, breaking load, breaking extension, and work factor. The value obtained from the ballistic test is the amount of energy or work required to rupture a hank of yarn. This test is sometimes used to detect the changes in quality of cotton, machinery performance, and yarn quality.

Single Thread Tests

Single thread tests provide better information regarding the strength characteristics of yarns. They eliminate the disadvantages of the lea test where no measure of the extension of yarns is available. These tests yield load-elongation diagrams that provide a wealth of information regarding yarn properties. There is apparently no simple relation between the lea strength and the single-end strength. However, Gregory (2) reports that the ratio of lea breaking length to single-end breaking length, obtained for a number of experimental yarns, varies from 0.56 to 0.90.

The load-elongation characteristics of yarns are easy to measure by using standard instruments such as the Scott or Uster automatic constant-rate-of-loading inclined plane testers, the Instron constant-rate-of-elongation tester, and the Goodbrand and/or the Cambridge Textile Extensiometer. The experimental conditions such as the rate of extension, rate of loading, the specimen length, temperature, humidity, and the initial tension used to mount the specimen should be controlled and specified when quoting results.

The behavior of a yarn under a gradually increasing applied load is fully expressed by the load-elongation curve including the breakage point (Figure 6.1). The load may be measured in gram weight or dynes and the elongation in centimeters. However, for the purpose of comparing the properties of different types of yarns, quantities that are independent of the dimensions of the specimen must be used. In most physical and engineering applications, load is replaced by stress, which is defined as

$$\text{Stress} = \frac{\text{load}}{\text{area of cross section}}$$

The units of stress are dyne/cm^2 or kg wt/mm^2. In the case of textile materials, it is difficult to measure the cross-sectional area. Moreover, it is the weight and not the bulk of the material that is important. Consequently, the stress is expressed in terms of the mass of the specimen and is called specific stress, defined as:

$$\text{Specific stress} = \frac{\text{load}}{\text{mass/unit length}}$$

The units used for specific stress are g wt/denier and g wt/tex.

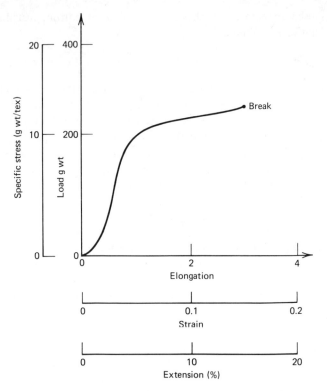

Figure 6.1. Load-elongation diagram for 20cm specimen of 20 tex yarn.

The elongation of the specimen is expressed as tensile strain or percent extension, which takes into account the original length of the specimen:

$$\text{Tensile strain} = \frac{\text{elongation}}{\text{initial length}}$$

The load-elongation curve can thus be converted to a stress-strain curve by changing the units only; the shape of the curve remains the same, as shown in Figure 6.2.

In addition to the abovementioned quantities, there are some other useful parameters that are obtained from the shape of load-elongation curves and the position of the end point where breakage occurs. These are defined as follows:

1. *Strength.* Strength is a measure of the steady force required to break a yarn; for an individual yarn, it is given by the breaking load in g wt or oz wt. To compare different yarns, specific stress at break is used and is called tenacity or

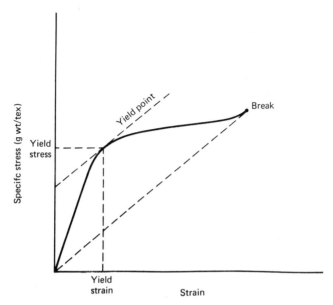

Figure 6.2. Yield point—Meredith's construction.

specific stress; the units for tenacity are g wt/tex or g wt/denier. Alternatively, breaking length may also be used to express the tenacity. When strength is compared on the basis of the area of the cross section, the stress at break is called the ultimate tensile stress, the units for which are kg wt/mm^2 or dyn/cm^2.

2. *Elongation-at-Break.* The elongation-at-break may be expressed by the actual fractional or percentage increase in length and is called the breaking extension.

3. *Work of Rupture.* This is defined as the energy required to break the yarn and is sometimes called "toughness." The work of rupture is given by the area under the load-elongation curve, and its units are dyne cm or g wt × cm.

4. *Initial (Young's) Modulus.* The initial modulus is the slope of the tensile stress-strain curve at the origin. The initial part of the curve is fairly straight, and its slope (ratio of stress to strain) usually remains constant. The modulus is measured in units of stress or specific stress (g wt/tex). Modulus gives a measure of the force required to produce a small extension. A highly inextensible yarn will have a large modulus; an easily extensible material will have a low modulus. The reciprocal of the modulus is called compliance.

5. *Yield Point.* After the initial strength part, the curve tends to bend, and in this region large extensions are produced with relatively small increases in the applied stress, as indicated by the shape of the curve in Figure 6.2 (this is the

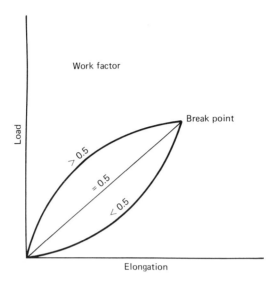

Figure 6.3. Work factor.

usual trend followed by most stress-strain curves). The yield point is located in the bent part of the curve and is determined geometrically. According to Meredith (3), the yield point is defined by the point at which the tangent to the curve is parallel to the line joining the origin and the breaking point. Yield point is also sometimes termed the "limit of proportionality," that is, where the extension ceases to be proportional to stress and "elastic limit." The yield point is characterized by particular values of stress and strain—the yield stress and yield strain.

6. *Work Factor.* For a material obeying Hooke's law, the load-elongation curve would be a straight line from start to the breaking load, and the work of rupture will be given by:

$$\text{Work of rupture} = 1/2 \, (\text{breaking load} \times \text{breaking elongation})$$

Any deviation from the ideal state is defined by a quantity termed the "work factor":

$$\text{Work factor} = \frac{\text{Work of rupture}}{\text{breaking load} \times \text{breaking elongation}}$$

The work factor for an ideal state will be 0.5. If the curve stays above the straight line, the work factor will be more than 0.5, and if below it, less than 0.5 (Figure 6.3).

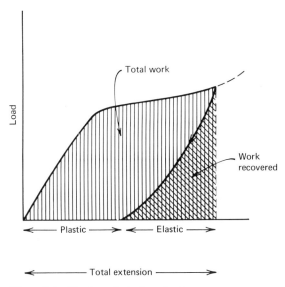

Figure 6.4. Elastic and plastic extension.

7. *Elastic Recovery.* Elasticity, as defined by the American Society for Testing Materials (4), is "that property of a body by virtue of which it tends to recover its original size and shape after deformation." Its opposite is plasticity. The recoverable (elastic) and the nonrecoverable (plastic) part of the deformation observed after stress removal is shown in Figure 6.4. Quantitatively, the elastic recovery may be defined as:

$$\text{elastic recovery} = \frac{\text{elastic extension}}{\text{total extension}}$$

Similarly, work recovery is defined as:

$$\text{Work recovery} = \frac{\text{work returned during recovery}}{\text{total work done in extension}}$$

MECHANICS OF YARN STRUCTURES

In recent years the subject of the mechanics of yarn structure has attracted the attention of many scientists. Backer (1) has given an excellent account of the development of various approaches adopted in analyzing yarn mechanics. Hearle (5) has dealt with the theoretical analysis of the extension behavior of twisted

Table 6.1. Papers on Theory of Tensile Properties of Twisted Continuous Filament Yarns[a]

Name and Reference	Nature of Theoretical Treatment
Gegauff (6)	Theory of spun yarns, but includes basic equations of simplest treatment of filament yarns
Platt (7, 8)	Tensile forces only; includes effects of lateral contraction, large extensions, and deviations from Hooke's law
Hearle (9)	Tensile and transverse forces: Small strains, Hooke's law, no lateral contraction
Hearle, El-Behery, and Thakur (12)	(i) Tensile and trnasverse forces, small strains Hooke's law, with lateral contraction; (ii) Tensile forces only: large strains, lateral contraction, deviations from Hooke's law
Treloar and Hearle (13)	Corrects an error in previous two papers
Treloar (14)	Continuum rubber filament model
Wilson and Treloar (15)	Two-filament rubber model
Wilson (16)	7 and 19 filament rubber models
Treloar and Riding (17)	Energy method—includes effects of transverse forces, constant volume deformation, large strains, deviations from Hooke's law
Symes (18)	Cord properties, with approximations
Kilby (19)	Develops theory to consider effect of equalization or nonequilization of tension in migrating filaments; and effect of bending strains
Treloar (20)	Applies energy method to yarn with migrating filaments
Treloar (21)	Applies energy method to multiply cords
Wilson (22)	Model yarns with five filaments in regular pattern around a core filament
Hearle (84)	Reexamines the energy method and simplifies the treatment of the theory. Suggests four general equations to predict the mechanical properties.
Konopasek and Hearle (85)	Analysis of the mechanics of bending curves
Cheng, White, and Duckett (86)	Apply continuum mechanics and tension matrix analysis

[a]Partially reproduced from Hearle (5).

continuous-filament and staple yarn structures. He has discussed this subject thoroughly, and any attempt to deal with it here would be repetitious. However, to understand the mechanical behavior of actual yarns, a basic knowledge of theories of the mechanics of such structures is considered desirable. A brief discussion of the theories of the load-extension characteristics of continuous-filament and later of staple yarns is presented, this is followed by a discussion of experimental studies reported in the literature.

Most theoretical work on the mechanics of yarn structure found in the literature pertains to continuous-filament yarns. This is because such yarns are easy to manipulate to form twisted yarn structures for theoretical studies and experimental investigation. A summary of published papers on the subject of the theory of the extension of continuous-filament yarns is given in Table 6.1. To work out the analysis of the mechanics, the procedure adopted involves first the calculation of the strains due to an imposed deformation; the next step is to calculate either the stress distribution and equilibrium of forces or the energy due to deformation. Table 6.1 includes the method adopted and the assumptions made by each author in working out the analysis.

Analysis of Tensile Behavior of Continuous-Filament Single Yarns

Hearle (5) has carried out a simple analysis of the mechanics of yarn structure for calculating the modulus of a continuous-filament yarn. He assumed that the yarn has an ideal geometry, as shown in Figure 6.5; the component filaments in

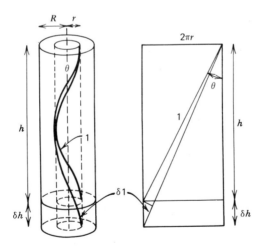

Figure 6.5. Geometry of a yarn subject to extension $E_y = \dfrac{\delta h}{h}$. From Hearle (5)

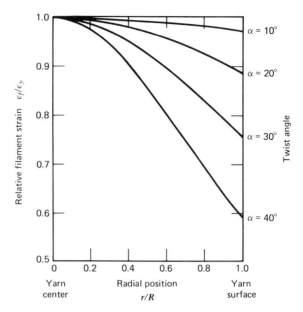

Figure 6.6. Variation of filament strain through yarn. From Hearle (5)

the yarn cross section have uniform properties, and each filament maintains a fixed helical path within the yarn, with no change in radial position. Further, there is no change in yarn diameter during extension, and the packing density is constant.

From the consideration of the yarn geometry and the relationship between the applied yarn strain and axial fiber strain, Hearle reports a relationship worked out by Platt (7):

$$\epsilon_f = \epsilon_y \cos^2 \theta \tag{6.1}$$

where

ϵ_f = filament extension

ϵ_y = yarn extension

θ = helix or twist angle.

This relationship indicates that the extension of the straight filament at the center ($\theta = 0$) of the yarn is equal to the value of yarn extension, and that it decreases to a value $\epsilon_y \cos^2 \theta$ at the yarn surface. This relationship is shown in Figure 6.6, in which the ratio of filament strain to yarn strain (ϵ_f/ϵ_y) is plotted against the radial position of the filament in the yarn (r/R) where r is the radius of a cylinder containing the helical path of a particular fiber, and R is the yarn radius.

Now, considering the forces acting along the filament axis, the expression derived by Hearle for the yarn tension turns out to be of the form:

$$\text{Total yarn tension} = (\pi R^2 \ E_f \epsilon_y / \nu_y) \cos^2 \alpha \qquad (6.2)$$

where

R = yarn radius
E_f = filament modulus
ν_y = yarn specific volume

and

α = surface angle of twist.

This expression can now be used to calculate yarn specific stress and modulus. These quantities are expressed in units of g wt/(g/cm) which is equal to 10^{-5} g wt/tex.

$$\text{Yarn specific stress} = \text{yarn tension} \times (\nu_y / \pi R^2)$$

$$= E_f \epsilon_y \ \cos^2 \alpha$$

and by definition yarn tensile modulus =

$$E_y = \frac{\text{yarn specific stress}}{\text{yarn extension}}$$

$$= E_f \cos^2 \alpha \qquad (6.3)$$

This expression predicts that yarn modulus will decrease with an increase in the value of α (twist angle). This simple relationship has been shown to agree reasonably well with experimental results.

Platt (7) points out that, since the fiber strain at the yarn axis equals the axial strain of the yarn, the rupture will first take place in the central filament, followed by successive rupturing of the remaining filaments terminating in the ultimate yarn break. This kind of phenomenon, in which catastrophic failure occurs because of instantaneous rupture propagation, has been observed; however, this does not apply universally in all cases of yarn breaks. The propagation of failure will depend on factors such as twist, fiber surface properties, blend ratio, etc.

The above treatment of yarn mechanics ignores one very important aspect: the role played by lateral forces (forces acting at right angles to the filament axes) in affecting the load-extension characteristics of yarns, especially spun yarns. The lateral forces are very important in affecting the load transfer from fiber to fiber during the extension of the yarn. The other important parameter is the effect of the reduction in yarn diameter because of contraction. The influence of lateral forces on fiber and yarn in the treatment of the mechanics of twisted yarn structures has been recognized by many research workers. Hearle (9) has worked out the most extensive analysis of the prediction of yarn strength, taking the lateral forces into account.

His procedure was to establish first a relationship between fiber strain and yarn strain, taking into account the yarn contraction ratio σy (Poisson's ratio):

$$\text{Fiber strain} = \epsilon_f = \epsilon_y \, (\cos^2 \theta - \sigma_y \sin^2 \theta) \qquad (6.4)$$

and yarn contraction ratio

$$\sigma_y = \frac{\text{yarn radial contraction}}{\text{yarn axial tension}}$$

He then calculated the fiber strain by assuming that the fiber extension follows Hooke's law (linearly elastic fiber), and reported the following expressions on the basis of elasticity theory:

$$\text{Fiber strain} = \epsilon_f = \frac{X}{E_f} - \frac{2 \, \sigma_2}{E_{f'}} \, (-G)$$

where
$\quad X$ = fiber tensile stress
$\quad E_f$ = fiber tensile modulus
$\quad E_{f'}$ = fiber transverse modulus
$\quad G$ = compressive transverse stress
$\quad \sigma_2$ = transverse Poisson's ratio, that is, (axial strain/transverse strain) for a transverse stress.

He further assumed that the fiber has an axis of symmetry that coincides with its geometric axis; this gives

$$\sigma_2/E_{f'} = \sigma_1/E_f$$

therefore,

$$\epsilon_f = \frac{X}{E_f} - \frac{2\sigma_1}{E_f} \, (-G) \qquad (6.5)$$

where
$\quad \sigma_1$ = axial Poisson's ratio, that is, (transverse strain/axial strain) for a tensile stress
and by combining equations 6.4 and 6.5 we can get the expression for tensile stress X.

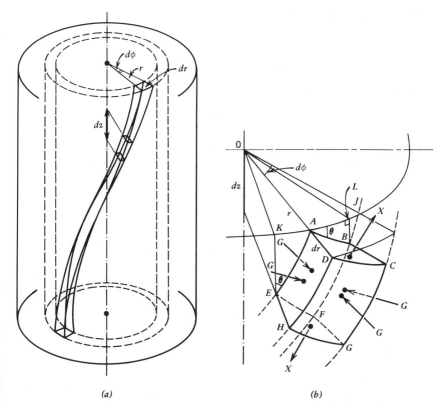

Figure 6.7. (a) Helically twisted yarn model showing element defined by dr, $d\phi$, dz, (b) Enlarged view of element. From Hearle (5)

$$X = E_f \, \epsilon_y \, [\cos^2 \, \theta \, - \, \sigma_y \, \sin^2 \, \theta)] \, - \, 2\sigma_1 G. \qquad (6.6)$$

Hearle normalized the fiber tensile stress X and fiber compressive stress G by dividing each by a factor X_f ($X_f = E_f \epsilon_y$ –which would be present in a filament subjected to yarn axial strain ϵ_y):

$$X = X/X_f$$
$$g = G/X_f$$

Next he established the force acting on the yarn element, shown in Figure 6.7, and derived a differential governing the radial equilibrium in the yarn, after which he solved the differential equation by assuming that the fibers and yarn

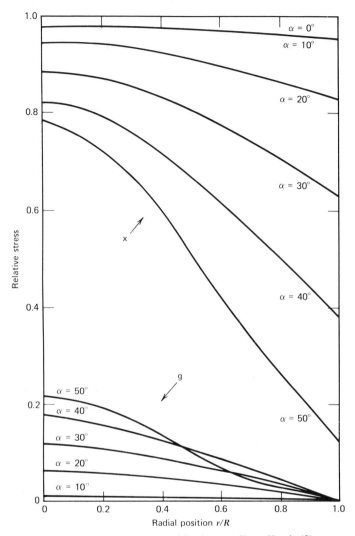

Figure 6.8. Variation of stress with radial position in yarn. From Hearle (5)

deform at constant volume. For the case $\sigma_1 = 0.5$, the expressions for the stresses in a yarn become:

$$x = \tfrac{3}{4} \cos^2 \alpha [1 + (\ell/L)^2] - \tfrac{1}{2} [1 + \ln(\ell/L)] \qquad (6.7)$$

$$g = \tfrac{3}{4} \cos^2 \alpha [1 - (\ell/L)^2] + \tfrac{1}{2} \ln (\ell/L) \qquad (6.8)$$

where

ℓ = length of fiber at radius r

and

L = length of fiber at the surface of the yarn (radius R)

These relations are plotted in Figure 6.8 for various values of twist angle (the relative stresses are plotted against radial position of filament, r/R). Hearle (9) further derived an expression for mean normalized yarn stress for a yarn subjected to small axial yarn strain. The normalized yarn stress is the specific stress in the yarn divided by the specific stress in the individual filaments under small axial yarn strain. This is a useful quantity, since it also leads to the prediction of yarn modulus.

$$\frac{\text{yarn modulus}}{\text{fiber modulus}} = \frac{\text{yarn specific stress at } \epsilon_y}{\text{fiber specific stress at } \epsilon_y}$$

$$= F_y \ (\alpha, \ \sigma_1, \ \sigma_y)$$

Table 6.2 shows the values of yarn parameters reported by Hearle (5) for various values of α, σ_1, σ_y [for $\sigma_y = 0$ (no contraction), and $\sigma_y = 0.5$, constant yarn volume].

Table 6.2. Values of Yarn Parameters

α	$\cos^2 \alpha$	σ_1	G_c/X_f	X_c/X_f	$F_y(\alpha, \sigma_1, \sigma_y)$ with $\sigma_y = 0$	$F_y(\alpha, \sigma_1, \sigma_y)$ with $\sigma_y = 0.5$
0°	1	any value	0	1	1	1
10°	0.970	−0.25	0.0153	1.0076	0.973	0.966
		0	0.0152	1	0.970	0.962
		0.25	0.0151	0.9924	0.966	0.959
		0.5	0.0151	0.9849	0.962	0.955
20°	0.883	−0.25	0.0612	1.0306	0.896	0.867
		0	0.0603	1	0.882	0.854
		0.25	0.0594	0.9703	0.868	0.841
		0.5	0.0585	0.9415	0.855	0.828
30°	0.750	−0.25	0.139	1.0694	0.772	0.713
		0	0.134	1	0.745	0.689
		0.25	0.129	0.9353	0.718	0.665
		0.5	0.125	0.8750	0.693	0.643
40°	0.587	−0.25	0.250	1.1248	0.612	0.522
		0	0.234	1	0.572	0.491
		0.25	0.220	0.8902	0.535	0.462
		0.5	0.207	0.7934	0.502	0.435
50°	0.413	−0.25	0.397	1.1983	0.430	0.321
		0	0.357	1	0.384	0.293
		0.25	0.323	0.8384	0.343	0.268
		0.5	0.293	0.7066	0.309	0.246

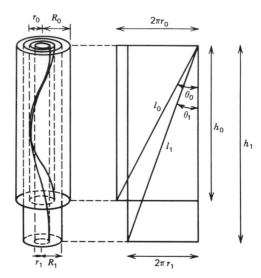

Figure 6.9. Geometry of yarn subject to large extension. From Hearle (9)

It is worth noting that the simple analysis taking the parameter $\cos^2 \alpha$ seems to give reasonably good agreement when compared to the values obtained by the more complex and sophisticated analysis at surface angles below $30°$. The agreement above $30°$ for the $\cos^2 \alpha$ case deviates considerably from the more complex analysis, with differences as high as 25-30% at $40°$ and over 45% at $50°$.

Analyses accounting for lateral pressures have also been reported by Sullivan (10) and Grosberg (11). Grosberg has reported an analysis for predicting resistance to fiber slippage in a low-twist sliver.

Large-Strain Analysis

The analysis discussed in the previous section applies to the case of small strains (such as encountered during draping, wrinkling, etc.) in the filament and yarn. However, if one considers the rupture behavior of yarns in industrial applications, the rupture strains encountered in the filaments are of the order of 15-30%. It would therefore seem desirable that an analysis that takes the yarn strength at break into account should also be carried out. Hearle (5) reports a rigorous analysis and derived the following expression for filament strain from the yarn geometry (Figure 6.9) and large yarn extension:

$$\epsilon_f = \epsilon_y(\cos^2 \theta_1 - \sigma_y\sin^2 \theta_1) - 3/2\ \epsilon_y{}^2(1 + \sigma_y)^2\ \sin^2 \theta_1\ \cos^2 \theta_1$$
$$(6.9)$$

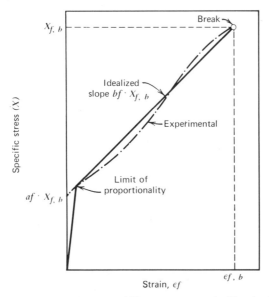

Figure 6.10. Idealized stress-strain curve of filament, compared with a typical viscose rayon curve. From Hearle (5)

Hearle (5) has reported the values of ϵ_f calculated by using various equations. He stated that equation 6.1 gives reasonable agreement for yarn extensions up to 10%, and equation 6.9 (also the one for constant volume deformation not considered here) at 30% yarn extension. He also pointed out that the difference between the use of $\sigma_y = 0.5$ and the correct constant volume equations is small except at high strains (30%) and high twist angles (25°).

For further analysis, Hearle assumed that the stress-strain curve of a filament beyond the yield region up to the breaking point is linear, this idealized stress-strain curve is compared to the actual curve for viscose rayon in Figure 6.10. He then derived the following expression for predicting yarn stress by generalizing the treatment given earlier for large extensions, neglecting the transverse forces but taking the lateral contraction into account and normalizing the non-Hookean region of the stress-strain curve:

$$\frac{\text{Specific stress in filament}}{\text{Filament tenacity}} = \frac{X}{X_{f,b}} \, a_f + b_f \, \epsilon_f \, \text{---} \tag{6.10}$$

where $X_{f,b}$ is the filament tenacity which is related to fiber breakage and a_f and b_f are coefficients. This equation is then combined with the expressions in 6.1 and 6.2 to obtain a detailed expression for the prediction of the stress-strain curve of idealized twisted continuous-filament yarns near the breaking point. Con-

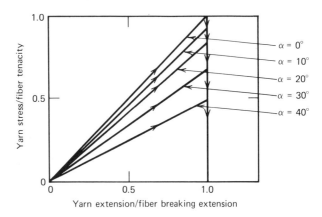

Figure 6.11. Expected behavior of yarns with catastrophic rupture.

sideration of the transverse forces makes the analysis more complex, but Hearle has suggested a generalized form and obtained an expression for the yarn stress by substituting it in the analysis for small strains.

The above relationship for the prediction of yarn stress suggested by Hearle is for large extensions and includes the yarn lateral contraction and non-Hookean behavior. These expressions apply to the yarn stress-strain behavior up to the rupture of the first filament, that which is at the center of the yarn, and the most highly strained with extension equal to that of the yarn extension ($\epsilon_{fb} = \epsilon_y$). This would then be followed by the catastrophic rupture of the whole yarn; if the broken regions in the yarn do not contribute to the tension, the stress-strain curve will drop down linearly, as shown in Figure 6.11.

It should be mentioned that this analysis did not take into account the effect of pressure due to transverse forces. This factor has been introduced by Treloar and Riding (17) in analyzing the mechanics of an extended twisted yarn by the energy method. This method involves the calculation of the energy of deformation, which is obtained by equating the work done by all external forces in extending the yarn to the elastic energy stored in the deformed filaments. This is a simpler method, since the energy is a scalar quantity and can be summed numerically, whereas the stress is a tensor and requires vectorial summation. Hearle (84) has shown that it is possible to work out the essential relations of the energy method in a simpler and briefer form than that reported by Treloar and Riding (17). The following relationships describe the prediction of mean yarn stress for extension of an idealized twisted continuous-filament yarn by the energy method.

$$\text{Specific tensile stress} = f_y = 2 \int_0^1 f_f(\partial \epsilon_f | \partial \epsilon_L) \; x d_x$$

$$\epsilon_f = [(1+\epsilon_L)^2 \; \cos^2 \theta_o + (1 - \nu_y \; \epsilon_L)^2 \; \sin^2 \theta_o]^{1/2} - 1 \qquad (6.11)$$

$$\tan \theta_o = x \tan \alpha_o$$

$$\tan \alpha_o = 2\pi R_o T_o = 2\pi \; V_o^{1/2} \; C_o^{1/2} \; T_o$$

where

C_o = initial yarn linear density
T_o = yarn twist
V_o = yarn specific volume
ν_y = Poisson's ratio

and fiber specific-stress versus strain relation

f_f = function (ϵ_f) subject ot a tensile strain ϵ_L
$x = r_o/R_o$
r_o = radial position
R_o = yarn radius
θ_o = helix angle
α_o = surface twist angle.

The suffix o is used to indicate explicitly that initial values are to be taken. All values are in a consistent set of units.

The validity of equation 6.11 in predicting the stress-strain relationship of twisted continuous-filament yarns is shown in Figure 6.14b.

The case for the inclusion of transverse forces in the analysis for large extension has been dealt with by Treloar and Riding (17) using the energy method. Treloar (20) and Kilby (19) have included the idealized migration pattern of fibers within a yarn and conclude that it has little effect on the tensile properties of yarns.

ACTUAL OBSERVATION OF THE TENSILE BEHAVIOR OF CONTINUOUS-FILAMENT YARNS

Experimental investigations of the tensile behavior of twisted continuous-filament yarns have been reported by many workers and are summarized in Table 6.3. The results reported in the literature by various workers are in general agreement.

Before dicussing the load-extension behavior of twisted continuous-filament yarns, it must be pointed out that their tensile characteristics are profoundly influenced by the method of twisting: this controls the form of the twisted structure and to a certain extent affects the properties of individual filaments.

Table 6.3. Experimental Studies of Tensile Properties of Continuous Filament Yarns[a]

	Modulus
Maginnis (23)	Viscose rayon (sonic and static modulus)
Hamburger (24)	Acetate and nylon (sonic modulus)

	Tenacity and Breaking Extension
Platt (7)	Viscose rayon and acetate
Shrinagbushan (25)	Viscose rayon
Grover and Hamby (26)	Viscose rayon, acetate, nylon, Dacron, Orlon
Taylor et al. (27)	Viscose rayon, nylon, Dacron
Alexander and Sturley (28)	Nylon
Hearle and Thakur (28)	Viscose rayon, acetate, nylon, Terylene
Kilby (30)	Viscose rayon (effect of lubricant)

	Full Stress-Strain Curves
Hearle, El-Behery, and Thakur (31, 32)	Viscose rayon, acetate, nylon, Terylene
Treloar and Riding (33)	Viscose rayon
Riding and Wilson (34)	Rayon, nylon, polyester, triacetate, acrylic, PTFE

[a]From Hearle (5).

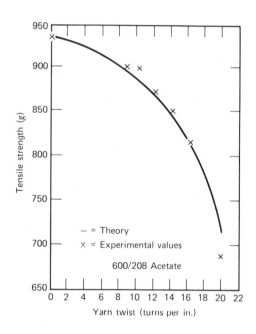

Figure 6.12. Yarn strength vs. twist. From Platt (7)

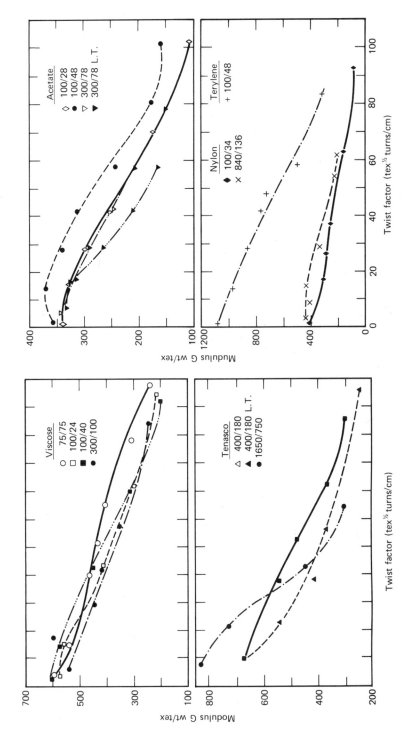

Figure 6.13. Variation of modulus with twist. From Hearle et al (31). Figures below yarn types refer to denier/number of filaments.

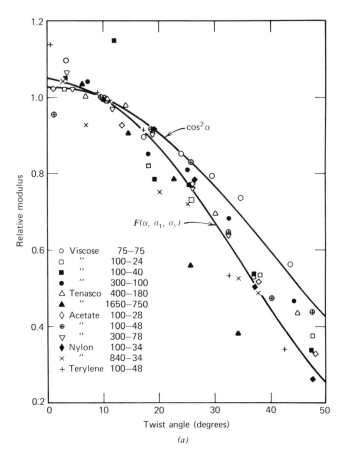

Figure 6.14(a). Comparison of experimental values of modulus with theoretical relations, namely $\cos^2 \alpha$ from equation (6.3) and $F(\alpha, \sigma 1, \sigma_y)$ with $\sigma_1 = 0.5$ and $\sigma_y - 0.5$ from equation for normalized yarn stress. From Hearle and Hearle et al. (5, 31)

The individual load-extension curves also show some variability, as shown in Figure 6.15. For simple comparison, it is a common practice to obtain the average values of the tensile parameters (modulus, tenacity, breaking extension, work of rupture, etc.) and draw an average curve. The load-elongation properties of filament yarns are easily measured using any of the standard instruments mentioned earlier.

The agreement between the theoretically calculated relationship of the tensile strength and twist and the experimentally obtained values of strength has been demonstrated by Platt (7). Figure 6.12 shows a plot of yarn strength (of an acetate continuous-filament yarn) calculated theoretically from Platt's expres-

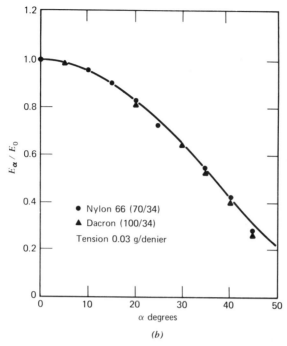

Figure 6.14(b). Experimental and theoretically calculated variation of yarns dynamic modulus with twist. From Zorowski and Murayama (87). E_α and E_0 is the ratio of modulus with surface twist angle α to modulus at zero twist angle. Points are experimental. Line is theoretical.

sion (variation of strength with $\cos^2 \alpha$) and the corresponding experimentally determined values as a function of twist level.

Hearle et al. (31) have reported the results of experimental investigations of the variation of modulus with twist factor for ordinary viscose rayon, high-tenacity viscose rayon, acetate, nylon 66, and polyester yarns. The results, shown in Figure 6.13, exhibit a decrease of modulus with twist as expected on the basis of the theoretically derived expressions. Hearle has also plotted the relative values of modulus (E_y/E_f) against twist angle, as shown in Figure 6.14a.

It is obvious from this plot that the simple expression $(E_y/E_f = \cos^2 \alpha)$ exhibits the general trend, but the more elaborate expressions give the best agreement with the experimental results. A further demonstration of the change (decrease) in modulus with increasing twist for other yarns is shown in Figure 6.15. From observation of these curves, Hearle (5) remarks that the sharpness of yield point is markedly clear for low-twist yarns, whereas the high-twist yarns exhibit a poor transition. This behavior is attributed to the effect of the range of

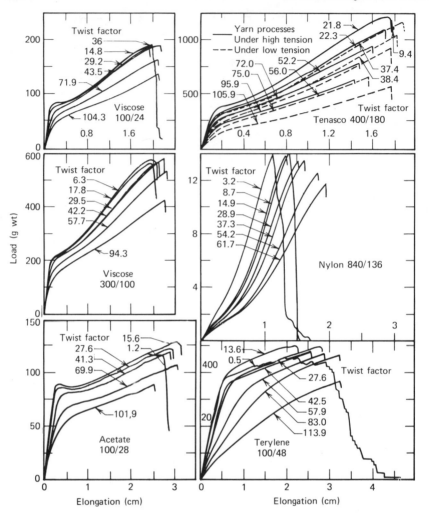

Figure 6.15. Some load-elongation curves of continuous-filament yarns From Hearle et al. (31)

filament extensions occurring in the highly twisted yarns. "When the central filament reaches the yield point, the outer filaments are still at a much lower extension. The sharp bend in the fiber stress-strain curve is thus spread out over a range of yarn extensions." The load-elongation curves for high-twist yarns do not show very good agreement with the predictions made by the theoretical expressions.

However, the observed behavior over the whole range of load-extension curves for yarns shows the best agreement with the predictions made by the energy method developed by Treloar and Riding (17). Their experimental results (33) and those of Riding and Wilson (34) for a range of man-made continuous-filament yarns, for example, high-tenacity viscose, nylon, low-tenacity polyester, high-tenacity polyester, triacetate (tricel), polytetrafluoroethylene (Teflon), and Fortisan, are compared with the theoretically obtained curve in Figure 6.16. Their method of calculation requires the knowledge of geometric factors such as yarn twist and yarn radius for the application of the theory. It can be seen that there is some deviation from the theoretical prediction at low strains and high twist, which has been attributed to the presence of crimp.

RUPTURE BEHAVIOR OF CONTINUOUS-FILAMENT YARNS

The rupture behavior of the whole twisted yarn structure is a complex one and therefore very difficult to treat analytically. The complications arise because of the following:

1. The migration behavior of filaments—affected by the conditions of twisting.
2. Level of local strains during twisting.
3. Local recovery and buckling behavior of filaments.

All these factors will contribute to the variation of the behavior of individual filaments in a given yarn structure. It has already been pointed out that Platt (7) and Hearle (5) have demonstrated a reasonably close agreement between the simple model of yarn tensile behavior and the actual tensile strength of yarn. However, it would be interesting to see whether these theories would explain the actual rupture, that is, the limitations and other related behavioral aspects in a twisted yarn structure.

Platt (7) suggests that, once the first break has taken place in twisted continuous-filament yarn, the rest of the rupture follows suddenly or catastrophically. On the other hand, the hypothesis advanced by Hearle (5) suggests the occurrence of two different mechanisms of rupture. The breakage observed in actual yarns shows that there is a difference between the nature of break in low-twist and high-twist yarns. The rupture in a low-twist yarn occurs in steps, with the individual filaments behaving independently of the others and each filament breaking when it reaches its breaking extension. This independent behavior results from the lack of the development of lateral pressure (a consequence of twist). On the other hand, in highly twisted yarns, the filaments are held together more coherently, resulting in a sharp break, as predicted by the theoretical analysis. Hearle has discussed both these mechanisms and conjectures that, in actual rupture, what happens is that "when a filament breaks, it ceases to be able

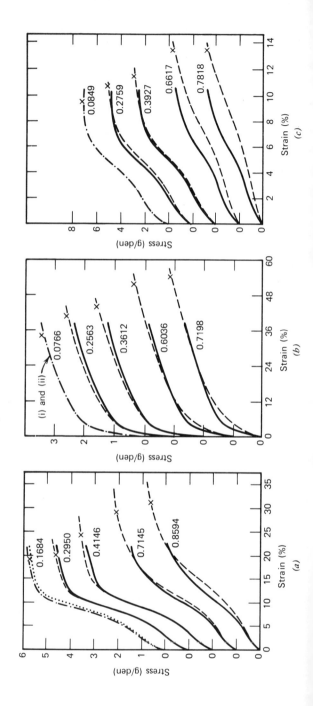

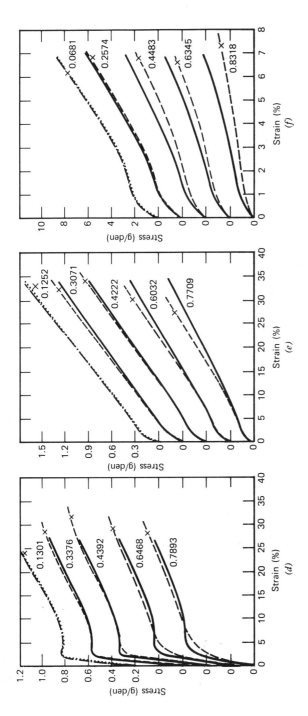

Figure 6.16. Comparison of theoretical and experimental stress-strain curves. From Riding and Wilson (34). i (--) Calculated curve for zero-twist yarn, equivalent to individual filament curve. ii (· · ·) Experimental curve for low (but not zero) twist yarn used as basis for calculation of other curves. (--) Experimental curves. (——) Calculated theoretical curves. (×) Breaking points. Numbers on curves are values of tan α. (a) Nylon, (b) low-tenacity Terylene (polyester), (d) Tricel (triacetate), (e) Teflon (polytetrafluoroethylene), (f) Fortisan.

163

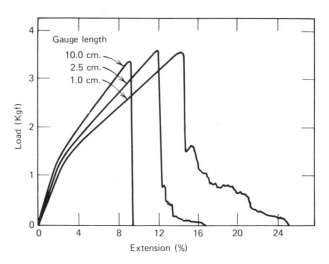

Figure 6.17. Effect of gauge length on load-extension curve of 1650 denier Tenasco yarn at twist factor of 67 tex$^{1/2}$ turns/cm. Rate of Extension of 40% per min. From Hearle and Thakur (29)

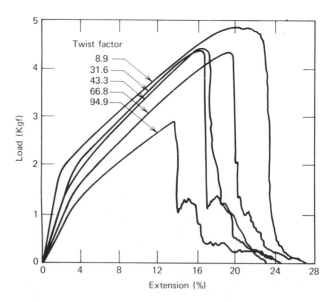

Figure 6.18. Load-extension curves of Tenasco yarns of various twist factors, tested at 1 cm gauge length and rate of extension of 40% per min. From Hearle and Thakur (29)

to support a load at the point of breakage, but, provided there is some friction, it will still remain an effective part of the yarn at positions remote from the point of break." Hearle has reported the effect of yarn twist and gauge length (test specimen), as well as the rate of extension on the observed rupture behavior of continuous-filament yarns. His results on Tenasco 1650/750 yarns are shown in Figures 6.17 and 6.18. It can be observed that, whereas the long specimen (10 cm) shows a sharp break, the 1-cm gauge length specimen depicts a long tail (higher extension), with a part of the yarn breaking sharply. The effect of twist on rupture can be seen in Figure 6.18. Whereas the high-twist-factor yarns (higher than 43 tex$^{1/2}$ turns/cm) show a sharp partial drop, the rupture point in low-twisted yarns has an extended form with complete absence of sharpness at the lowest twist factor (9 tex$^{1/2}$ turns/cm). Tables 6.4 and 6.5 give some additional information on the tensile properties of these yarns.

Hearle also points out that he was able to arrest the process of rupturing midway and stop the progress of the break. This can be seen in Figure 6.19, in which a series of twisted Tenasco yarns with interrupted breaks are illustrated.

On the basis of the observations made from the load-extension curves and the actual rupture behavior of yarns, Hearle states that "whereas for the long specimens and high ratios of extension the breakage is sharp and complete, for short specimens and low rates of extension the break occurs in a series of steps. The change from one mechanism to the other as the gauge length is altered is due to

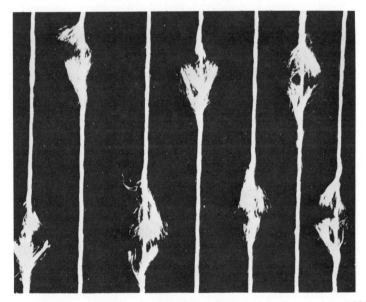

Figure 6.19. Tenasco yarns, tested at 1 cm gauge length and rate of extension of 40% per min, showing interrupted breakage. From Hearle and Thakur (29)

Table 6.4. Breakage of Ring-Twisted Viscose, 300/100[a]

Twist Factor (tex^½ turns/cm)	Twisting Tension (g/tex)	Maximum Load at Break (g), Gauge Length, 1 cm						Load Immediately After Break (g), Gauge Length, 1 cm						Breaking Extension (%)		Breaking Load (g), G.L., 10 cm; Average of 10 Observations
		5 Observations					Average	5 Observations					Average	Average of 5 Observations, G.L., 1 cm	Average of 10 Observations G.L., 10 cm	
7.1	0.9	590	635	600	620	640	617	Continuous drop in load						35.4	22.7	602
18.1	0.9	600	600	610	615	605	606	210	110	120	160	120	144	27.2	24.7	605
17.5	1.8	580	625	610	585	540	599	100	220	130	290	110	170	35.4	25.3	622
17.0	2.7	590	600	580	580	600	590	80	200	160	100	cont.	135	33.5	22.8	609
41.9	0.9	580	620	595	620	580	599	100	180	210	160	120	154	29.5	27.2	598
39.4	1.8	600	615	580	590	600	597	40	110	170	200	200	144	33.6	20.5	569
37.2	2.7	590	585	615	610	605	601	cont.	cont.	cont.	240	200	220	22.5	13.7	592
57.3	0.9	540	540	530	545	540	539	190	280	280	225	80	211	26.7	25.6	532
53.2	1.8	540	555	535	555	585	554	180	260	cont.	290	200	232	27.4	20.7	550
49.9	2.7	560	530	540	520	520	534	120	340	280	240	280	252	26.4	15.8	536
73.7	0.9	450	480	450	480	465	465	190	180	170	200	100	168	27.7	21.6	433
72.4	1.8	470	405	420	490	420	441	280	280	280	220	220	256	22.7	20.7	430
66.5	2.7	500	495	470	500	520	497	200	200	300	240	260	242	26.3	15.7	481

[a]From Hearle (5).

166

Table 6.5. Breakage of Ring-twisted Tenasco, 1650/750, 12 tpi[a]

Rates of Extension	Breaking Extension (%)			Breaking Load (g)			Load Immediately after break (g)		400% per min
	4%	40%	400%	4%	40%	400%	4%	40%	
Gauge length									
	16.4	19.0	18.4	2050	2425	2750	1000	1000	1250
	16.6	20.6	16.0	2275	2700	2525	950	1000	1200
1 cm	22.2	19.4	18.0	2525	2425	2800	1000	1700	1000
Average	18.4	19.7	17.5	2283	2517	2690	983	1233	1150
	14.8	11.6	12.8	2400	2550	2950	1250	850	150
	12.8	13.2	14.4	2600	2900	3100	850	900	150
2.5 cm	12.0	10.4	14.4	2350	2400	3050	750	1100	100
	11.6	13.4		2450	2750		750	450	
	14.4	12.8		2400	3100		1000	750	
Average	13.1	12.3	13.8	2440	2740	3033	920	810	130
	10.2	10.7	12.0	2475	2725	3200	625	200	0
5.0 cm	10.0	10.1	12.4	2625	2600	3260	650	0	0
	10.4	11.6	10.4	2650	2900	1250	75	75	0
	10.1	11.5	10.8	2590	2850	2750	840	150	0
Average	10.2	10.9	11.4	2580	2770	3040	842	142	0
	8.1	8.5	7.6	2700	2900	2750	0	0	0
	7.9	8.5	8.8	2550	3000	2760	0	0	0
7.5 cm	7.7	8.2	8.8	2300	2800	3100	0	0	0
	7.9	7.5		2450	2750		0	0	0
	8.1	8.6		2225	2800		0	0	0
Average	7.9	8.3	8.4	2445	2850	2870	0	0	0
	9.7	10.6	11.6	2600	2600	2950	0	0	0
10.0 cm	9.2	9.9	11.2	2650	2500	2950	0	0	0
	9.1	8.5	11.6	2675	2575	3100	0	0	0
Average	9.3	9.7	11.5	2640	2555	3000	0	0	0

[a]From Hearle and Thakur (29).

the amounts of elastic energy stored in the shorter specimen being less, and insufficient to complete the breakage. Similarly, in slower tests, the stored elastic energy will have decreased due to stress relaxation." He then suggests five different modes of break propagation to explain that the "initial part of the breakage (which) remains steady, and never becomes less than half the whole yarn." These five different possible modes of breakage are shown in Figure 6.20.

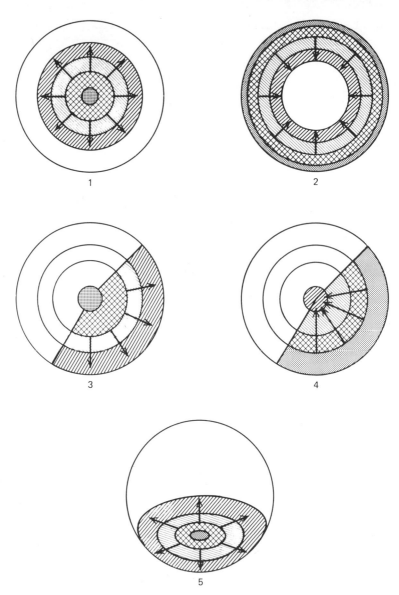

Figure 6.20. Modes of propagation of break, starting at black position and moving out from more heavily to less heavily shaded areas. White portion is left unbroken at end of first stage. From Hearle and Thakur (29)

To support his hypothesis, Hearle further argues in favor of the last mechanism on the basis of the considerations of buckling of the filament, the form of twisting, the change in filament behavior because of migration, and the influence of the twisting process on filament deformation.

The first mode shows the initiation of the break starting from the inside and spreading outward to stop midway through the yarn cross section (according to the first proposed mechanism). The second shows the break starting from the outside and moving inward. Neither of these mechanisms seems plausible, since there is no reason to expect the breaks to stop at the proposed position. The third and the fourth mechanisms, in which the break, assuming the asymmetry of the yarn structure, moves from the center to the outside and from the outside to the center, respectively, cannot be ignored altogether.

However, the fifth mode, in which the break is shown to start in the middle of one-half of the yarn and propagates outward, does show similarity with the observed appearance of a "pseudo-two ply structure" in which one part of the structure (one ply) ruptures and the other stays intact.

Further demonstration of actual rupture behavior has been reported by Backer et al. (35). They made model structures of 91 single twisted continuous-filament yarns under conditions of "no migration" and used mechanical tracer elements to identify the rupture mechanism. The yarns used for this purpose are 70/34 polyester filament yarns and 79s cotton yarns as tracer yarns, the latter dyed with different colors for easy identification. The cotton yarn is used for its low rupture elongation ($\simeq$ 8%) compared to the high rupture elongation of 30-35% for polyester yarns. This system is selected with the idea that the cotton yarn would break in the early stages of straining without influencing the total rupture of the surrounding polyester yarns. Backer et al. also state that this model system permits the study of the movement of rupture from cotton to cotton and from cotton to polyester. The other advantage of using such a structure is that it makes possible the experimental investigation of the distribution of strain and of the lateral pressure within the twisted structure as it is strained axially. The method used to record the location and the frequency of component rupture at each extension level consists of first drawing parallel lines in groups corresponding to the yarns in each layer of the yarn cross section. Each line in the model is assigned a corresponding number according to the position of the yarn in the yarn cross section, as shown in Figure 6.21. The marked number in the cross section refers to the position of cotton yarns; the unmarked refers to that of the polyester yarns. Their results obtained on a 2-cotton/89-polyester component model yarn twisted with a twist multiplier of 2.19 are shown in Figure 6.21. It is clear from the plot that, after straining the yarn to 11%, component #52 (located in the second ring from the core) suffered five breaks in 8 inches, whereas component #8 (in the outer or fifth ring) did not suffer any breaks; at 15% extension #52 and #8 showed 19 and 2 breaks, respectively; at 25% extension

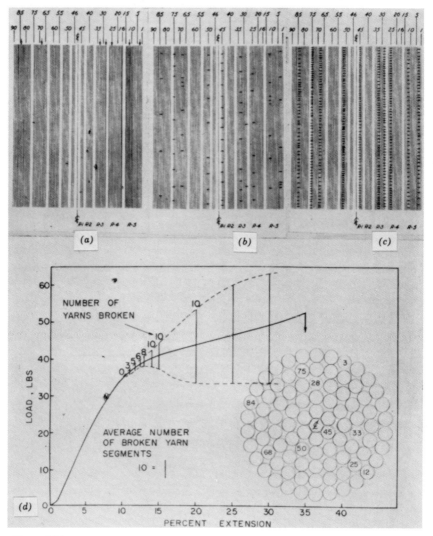

Figure 6.21. Stress-strain-rupture behavior of blended model yarns: 11% cotton/89% polyester, twist multiple = 4.37. (*a*) Rupture count at 11% extension, (*b*) 14%, (*c*) 30%, (*d*) stress-strain-rupture curve. From Backer and Monego (35)

they showed 44 breaks and 13 breaks in #52 and #8, respectively. The investigation also includes studies of the effect of component distribution, twist multiplier, and blend composition in the model yarns on the rupture mechanism. From their experimental investigation of the model yarns, Backer et al. (35) concluded that the first break invariably occurs in the yarns located in the central position. Moreover, the existence of multiple breaks in the cotton yarn invalidates the assumption that the individual filaments behave independently of each other and that the broken filament in a yarn ceases to contribute to the yarn strength.

Effect of Twist on Breaking Extension of Continuous-Filament Yarns

The theory put forward by Hearle (5) suggests that the breaking elongation of continuous-filament yarns remains constant and is not influenced by twist. However, the experimental results reported by Hearle and others do not agree with this hypothesis. Some experimental results for the breaking extension of continuous-filament yarns are shown in Figure 6.22. The results indicate that the breaking elongation of viscose rayon and Tenasco yarns does not show any regular or significant variation with twist. On the other hand, the breaking extension of acetate yarns first shows an increase to a maximum value and then decreases as the twist is increased. The behavior of a 100-denier nylon and Terylene (polyester), however, is just the opposite—it first decreases and then increases.

Hearle (5) has provided various possible explanations to account for the type of behavior mentioned above. These include:

1. The changed behavior of individual filaments caused by the filaments themselves being twisted in the yarn and under large transverse stresses.
2. The alteration of the properties of filaments as a result of the forces imposed on the filaments during the process of twisting; as a consequence, the filaments might suffer permanent extension.
3. Because of the migration of filaments from one helical layer to another, a redistribution of strains within the yarn can occur as a result of slippage when a yarn is stretched; the highly strained regions in the center layer will pull on less strained portions (outer layers) of the same filament. This can result in higher extension of the yarn. This type of behavior is highly likely in the case of low-twist yarns.
4. Twisting tension and the geometry of the twisting triangle plays an influential role in altering the breaking extension of the constituent filaments and eventually the yarn. A large part of the twisting tension is taken up in stretching the outer filaments to accommodate them more easily along the longer paths near the surface of the twisted yarn. After releasing

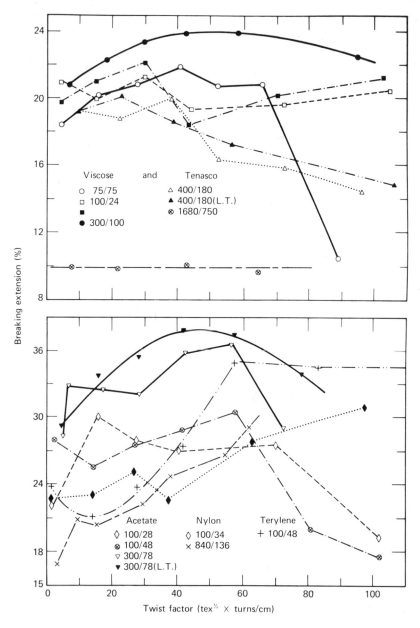

Figure 6.22. Variation of breaking extension with twist. From Hearle et al. (31)

the tension in the yarn, the outer filaments will contract (provided they have good elastic recovery), causing the inner (center) filaments to buckle. This type of mechanism is presumably the one that predominates in increased extension of nylon and polyester yarns made with high twist.

5. Buckling of filaments may occur during the twisting operation provided there is a lack of sufficient migration. However, the extent of the existence of such behavior is not precisely known.

6. There is a discernible variability in the properties of filaments in a yarn, and this will affect yarn breakage behavior, since breakage is determined by the single weakest point in each filament.

Hearle has further discussed this aspect in terms of twist and argues that twist will inhibit the influence (because of increased cohesion) of the weak place in one filament from extending beyond its near neighbor. Second, because of the differences in the path lengths, the outer filaments will be less extended than the center ones. Consequently, the breakage will initiate at or near the center of the yarn. In other words, the extension of the yarn will be determined by the minimum extension of a few filaments at the core, and not by the minimum extension of all the constituent filaments in the yarn cross section. Considering the whole situation statistically, it can be argued that, in general, the breaking extension will be higher than predicted.

On the basis of the above possilbe explanations, Hearle suggests that the increase in the breaking extension with increasing twist, at low twists, is perhaps attributable to the factor of variability. However, at higher twist, the behavior may be explained on the basis of the influence of tension during twisting. Thus fibers having poor recovery characteristics, for example, if they suffer permanent extension, will contribute to decreased breaking extension; on the other hand, fibers with good recovery behavior will cause the central filaments to buckle, thus increasing the breaking extension. Cellulose acetate fibers have poor recovery and therefore fall in the first group; nylon and polyester fall in the second group, and viscose rayon and Tenasco show behavior that falls in both groups.

Hearle has further theoretically analyzed the effects of buckling, deformation, and migration on the breaking extension and has compared the theoretical relationships with the experimental results. His experimental results for viscose, acetate, nylon, and Terylene, along with the theoretically predicted curves, are shown in Figure 6.23. He concludes that nylon and Terylene yarns show maximum buckling, whereas acetate at high twist factors gives evidence of appreciable permanent deformation. However, the viscose yarns behave in a manner that suggests the presence of perfect migration. Nevertheless, he goes on to suggest that this type of behavior could also be expected from a situation that

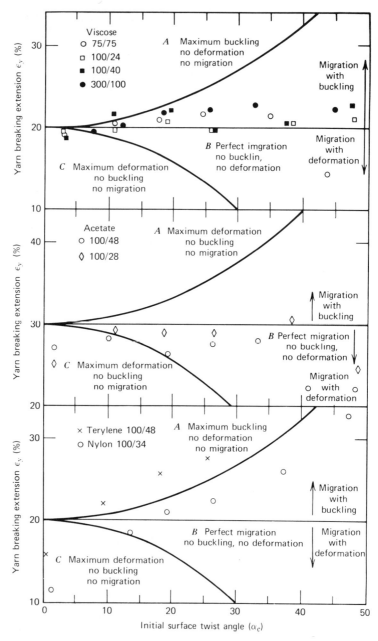

Figure 6.23. Comparison of theoretical and experimental values of breaking extension. From Hearle and Thakur (29)

174

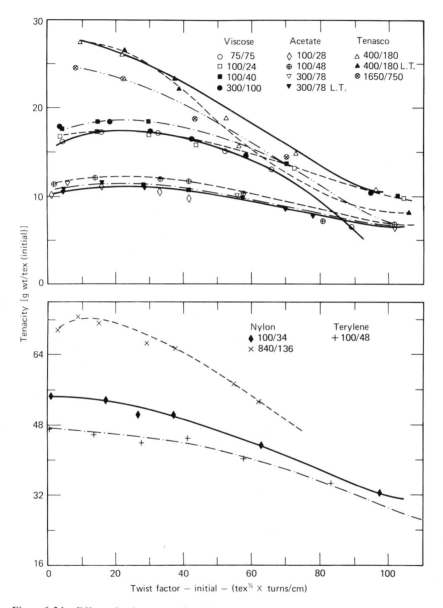

Figure 6.24. Effect of twist on tenacity. From Hearle et al. (31)

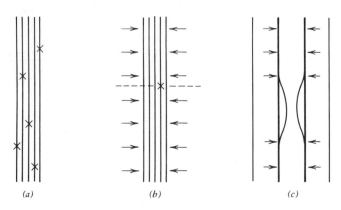

Figure 6.25. (*a*) Choice of weakest points for break of yarn without lateral pressure. Break-ing load < (sum of weakest points of each filament), (*b*) Break located at one position in yarn with transverse forces. Breaking load = (weakest point of whole yarn), (*c*) Weak place in one filament supported by neighboring filaments under action of transverse forces. From Hearle (5)

would exist because of the combination of buckling and deformation even with-out the presence of migration.

Effect of Twist on Tenacity

Figure 6.24 shows the experimental variation of tenacity with twist (29). The tenacity, in general, first increases and then decreases with increase in twist. Hearle attributes the initial rise to the interaction of twist with the variability of the material; the twist will influence the sharing of the load by filaments in the immediate vicinity of the weak places in one filament, and it will also pre-vent the imitation of rupture solely confined to the filaments that happen to be in the highly strained location in the yarn. This phenomenon is shown diagram-matically in Figure 6.25. The decrease in tenacity as the twist is increased might be explained theoretically in terms of the effects of obliquity. Although the agreement between the theoretically predicted values and the experimental results is reasonable in trend, there is no perfect quantitative agreement (Figure 6.24). Some results obtained by Kilby (30) (shown in Table 6.6) indicate this discrepancy; his results indicate that the tenacity of lubricated high-tenacity rayon yarns decreases less rapidly with increase in twist than the trend observed for unlubricated yarns.

EFFECT OF PROCESSING FACTORS ON TENSILE PROPERTIES

In addition to the amount of twist, the tensile properties of yarns are also

Table 6.6. Properties of Lubricated 1650 Denier High-Tenacity Rayon Yarns[a]

Lubricant Content (%)	Tenasco 35			Tenasco Super 70			Tenasco Super 105		
	Twist Angle (degrees)	Relative Tenacity[b]	Breaking Extension (%)	Twist Angle (degrees)	Relative Tenacity[b]	Breaking Extension (%)	Twist Angle (degrees)	Relative Tenacity[b]	Breaking Extension (%)
0.15	0	0.90	9.2	0	0.87	11.0	0	0.85	10.8
	17	0.96	12.2	21	0.93	15.6	23	0.95	14.1
	45	0.35	11.3	46	0.36	14.6	52	0.43	16.0
0.66	0	0.86	8.0	0	0.83	6.8	0	0.86	10.0
	19	0.97	10.6	19	1.00	10.6	22	0.94	12.8
	43	0.55	13.0	46	0.64	16.4	51	0.51	15.6
1.97	0	0.69	6.7	0	0.77	7.0	0	0.89	10.4
	17	0.84	9.2	20	1.00	12.0	20	0.94	12.6
	51	0.41	9.6	41	0.6	15.7	52	0.48	15.5

[a] From Kilby (30).
[b] Relative to value at angle of 12.5°.

influenced by such factors as the method and form of twisting and twisting tension. The form and nature of twisting method affect the arrangement of fibers in the yarn, and they may also modify the properties of the constituent filaments. The methods of twisting include:

1. Static twisting—in this process, the migration of filaments in a yarn is restricted once the yarn has been given a small amount of initial twist.
2. Continuous twisting—here migration will influence the yarn structure. The type of continuous twisting, for example, ring twisting or uptwisting, will also create differences.

Hearle et al. (31, 32) have reported a study dealing with the effect of twisting tension and the operation of twisting on the properties of yarns. Their investigation is based on the following two types of experiments:

1. Yarns twisted on the same machine to the same nominal twist but different tensions.
2. Yarns twisted to the same nominal twist under almost identical tension but on different machines.

The results of the experiments on the effect of twisting tension on mechanical properties of yarns twisted on a ring doubler are shown in Table 6.7. These results indicate that the tensile moduli of viscose rayon yarns increase considerably as the twisting tension is increased (from 0.9 g/tex to 2.7 g/tex). However, nylon and Terylene yarns are very slightly affected, which may be attributed to their higher elastic recovery behavior. On the other hand, Tenasco 1650/750 yarns do not follow the trend mentioned above; the tensile modulus decreases slightly with increasing tension at low twists but shows a slight increase at higher twists. Hearle attributes this observation to the interaction caused by the high linear density of this yarn, since a similar experiment on a Tenasco 400/180 yarn shows behavior identical to that of viscose rayon.

Kilby (30) reports that the tension under which yarn is wound on the bobbin causes a change in the physical properties of the yarn. He finds a considerable difference between the properties of yarn measured immediately after it has been removed from the bobbin and those measured after the yarn has been relaxed for sometime under standard conditions.

EFFECT OF METHOD OF TWISTING ON TENSILE PROPERTIES

The method or the process of twisting can influence the form of twisting (ribbon type or cylindrical form) and the migration behavior of the filaments. These

Table 6.7. Effect of Twisting Tension on Mechanical Properties of the Yarns (Data Taken from Smoothed Curves)[a]

Yarn and Tester	Twisting Tension (g/tex)	Tensile Modulus (g/tex) at Twist Factors ($\text{tex}^{1/2}$turns/cm) of				
		5	10	20	40	60
Viscose	0.9	575	600	605	460	357.5
300/100	1.8	625	638	635	560	460
(Instron)	2.7	650	670	688	650	525
Tenasco	0.9	950	935	868	600	390
1650/750	1.8	965	935	840	590	418
(Instron)						
Nylon	0.9	460	460	450	357.5	230
840/136	1.8	480	490	473	370	250
(Instron)						
Terylene	0.9	1140	1100	1000	740	480
250/48	1.8	1080	1050	975	790	500
(Instron)	2.7	1175	1155	1080	825	—
		Tenacity (g/tex) at Above Twist Factors				
Viscose	0.9	—	17.8	17.8	16.5	14.1
300/100	1.8	—	18.6	18.6	17.0	14.2
(Uster)	2.7	—	17.8	18.8	17.7	14.7
Tenasco	0.9	—	25.5	25.6	23.0	18.4
1650/750	1.8	—	25.5	25.0	22.3	18.0
(Instron)						
Nylon	0.9	73.2	73.8	73.6	70.0	59.0
840/136	1.8	72.9	73.4	73.2	70.5	60.6
(Instron)						
Terylene	0.9	61.2	62.4	62.4	57.2	47.8
250/48	1.8	—	—	63.6	57.6	45.6
(Uster)	2.7	62.8	64.0	63.4	55.2	—
		Breaking Extension (%) at Above Twist Factors				
Viscose	0.9	—	23.7	25.2	22.4	20.8
300/100	1.8	—	20.3	20.4	17.8	16.7
(Uster)	2.7	—	17.8	18.0	14.1	13.6
Tenasco	0.9	—	9.6	10.2	10.8	10.1
1650/750	1.8	—	9.7	10.0	10.5	9.8
(Instron)						
Nylon	0.9	13.8	14.4	15.6	17.0	18.6
840/136	1.8	13.0	13.4	14.3	16.0	17.8
(Instron)						
Terylene	0.9	10.8	11.5	11.9	12.7	14.2
1650/750	1.8	—	—	11.9	12.3	13.2
(Uster)	2.7	10.3	11.3	10.9	11.5	—

[a] From Hearle et al. (32).

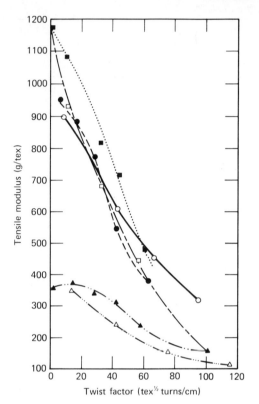

Figure 6.26. Effect of twisting methods on tenxile modulus. Hearle et al. (32) Tenasco 1650/750: (○) Up-twisted; (●) ring-twisted. Acetate 100/48: (△) Up-twisted; (▲) ring-twisted. Terylene 250/48: (□) Up-twisted; (■) ring-twisted.

imposed changes in yarn structure cause differences in yarn properties. Hearle et al. (32) have reported a comparison of the properties of uptwisted and ring-twisted polyester, Tenasco, and acetate yarns. Figures 6.26-6.28 show the effect of the method of twisting on the moduli, breaking extension, and the tenacity of these yarns.

The differences in the moduli and the breaking extensions of the acetate and Terylene (polyester) yarns spun on the uptwister and ring frame are obvious. The ring-twisted acetate and polyester yarns generally show higher values of tensile modulus; the Tenasco yarns, however, do not show any appreciable difference with different methods of twisting.

The uptwisted acetate yarns have generally higher breaking extension than those of ring-twisted yarns (Table 6.5 and 6.8). Tenasco yarns do not show any discernible difference, whereas the Terylene yarns made on the uptwister have a

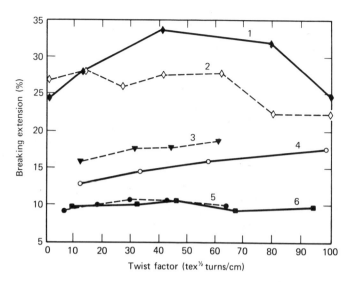

Figure 6.27. Effect of twisting methods on breaking extension. From Hearle et al. (32) Acetate 100/46: (◊) Doubler twisted; (♦) uptwister twisted. Terylene 250/48: (▼) Doubler twisted; (○) uptwister twisted. Tenasco 1650/750: (●) Doubler twisted; (■) uptwister twisted.

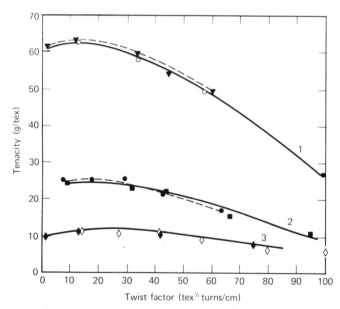

Figure 6.28. Effect of twisting methods on tenacity. From Hearle et al. (32) Terylene 250/48: (▼) Doubler twisted; (○) uptwister twisted. Tenasco 1650/750: (●) Doubler twisted; (■) uptwister twisted Acetate 100/48: (◊) Doubler twisted; (♦) uptwister twisted.

Table 6.8.   Breakage of Uptwisted Tenasco 1650/750[a]

Gauge Length (cm)	Twist Factor (tex$^{1/2}$ turns/cm)	Average Breaking Load (g)	Average Breaking Extension (%)	Load Immediately After Break				
				Sample 1	Sample 2	Sample 3	Sample 4	Sample 5
10	31.6	4650	10.09					
	43.3	4526	10.76			Sharp drop		
	66.8	3413	9.26					
	94.9	2603	9.53					
2.5	31.6	4654	11.89	800	1000	1200	300	900
	43.3	4374	11.61	2000	1500	1000	1100	800
	94.9	2686	10.44	1100	1200	1100	1400	800
1.0	31.6	4445	16.89	1200	cont.	1400	2900	1000
	43.3	4365	18.60	2000	2300	1600	1700	1000
	66.8	3557	13.96	1600	1400	1600	1800	1600
	94.9	2815	13.95	1200	1200	1000	1400	800

[a]From Hearle et al. (32).

lower breaking extension. Hearle (5) points out that "these differences in properties probably arise from differences in yarn structure, such as the form of twisting and the migration behavior of the filaments, as a consequence of the different restraints imposed on the yarn during the two different methods of twisting." In particular, it has been found that some ring-twisted yarns appear to twist as a "ribbon."

STAPLE YARNS

Introduction

The literature is full of reports of experimental investigations on the effects of various physical and mechanical properties of fibers and the yarn processing factors (twist, spindle speed, fiber parallelization, draft ratios, etc.) on the strength of spun yarns. It is not an exaggeration to say that tensile strength has been the most extensively studied property of spun yarns in the textile industry. The strength characteristic has been and is still the primary test used in the textile industry as a guide in comparing the performance efficiency of various processes in quality control, and in determining the end-use behavior of most textile products. This is in spite of the fact that it is difficult to find any correlation between yarn strength and the efficiency of subsequent processes such as winding, warping, beaming, ends down during spinning, slashing, etc. However, the tensile properties of yarns spun from staple fibers, because of their complex structure, are still not fully understood. The complications in understanding the structural mechanics of staple yarns are primarily caused by two factors, namely, discontinuities in fiber length, and the slippage of fibers during extension. Factors such as twist and fiber migration in spun yarns acquire different dimensions, since they are the only reasons why a bundle of short fibers is held together in a linear assembly. The introduction of new types of fibers, the use of various blends compositions, and the new methods of yarn manufacture inject additional complexities and necessitate a thorough understanding of the mechanical behavior of spun yarns if one is to succeed in producing novel structures.

The problem of fiber cohesion was first recognized by Pierce (36), who suggested that the fibers in a yarn are held together because of "random tangle," which was later described by Morton (63) as "fiber migration." The subject of fiber migration in yarns is discussed briefly in Chapter 3.

The earliest attempt made to analyze the mechanics of staple yarns theoretically is attributed to Gégauff (6). He derived an expression for the variation of breaking load in a twisted fiber assembly by assuming that the yarn has a perfect helical geometry and that the twist in the yarn is enough to prevent any slippage. Later, Sullivan (64) carried out his analysis assuming the existence of an arbitrary pressure on the surface to account for the development of transverse presssure. In this analysis, he also introduced the concept of variability of fiber properties. In theory, he established that yarn strength increases first to a maximum value with increase in twist and then falls with subsequent increase in twist. This optimum value of twist is largely determined by the fiber length, fiber fineness, and the fiber coefficient of friction, in addition to the fiber strength. Among the earlier studies, the work of Platt (7), Gregory (65-69), Alexander (70), Shorter (71), and Holdaway (72) should also be mentioned.

Hearle (5) has carried out a more rigorous theoretical analysis of the mechanics of spun yarns. He suggests three levels of approach to studying the problem:

1. a purely qualitative descriptive treatment
2. a rough analysis with many approximations yielding an explicit equation for the variation of staple fiber yarn mechanical properties with twist
3. a full rigorous analysis of the behavior of an idealized model yielding a set of equations that require solving by numerical computation

Of these three approaches, the third is out of the scope of this work, and the reader is referred to the original work by Hearle (73). However, for the benefit of the reader, a simple account of the rough analysis is given here, and the qualitative treatment is discussed at length. The observed behavior of the tensile properties of staple yarns is discussed in terms of the structure (twist, etc.) and properties of the constituent fibers and ultimately linked to the qualitative explanation.

Approximate Theoretical Treatment

In this analysis, Hearle (5) considers yarn as a continuous-filament yarn. By including all other approximations, the derivation of the equation for the variation of modulus with the twist predicts that the modulus varies as $\cos^2 \alpha$ (equation 6.3). Hearle then develops an expression for the reduction in tension in the fibers in the surface layers of the yarn. He obtains an equation for the mean normalized tensile stress in the surface layer resulting from an imposed small axial strain. The final expression has the form:

$$\frac{\text{Yarn modulus}}{\text{Fiber modulus}} = \frac{E_y}{E_f} = \cos^2 \alpha \; [1\text{-}K \cosec \alpha] \qquad (6.12)$$

where

$$K = \frac{\sqrt{2}}{3L_f} \left(\frac{aQ}{\mu}\right)^{1/2}$$

The $\cos^2 \alpha$ term represents the effect of obliquity (twist) and the $(1\text{-}K \cosec \alpha)$ term gives the change in tension caused by fiber slippage; a is the fiber radius, Q is the migration period, L_f is the fiber length, and μ is the coefficient of friction. Hearle points out that, because of the severe approximations, a precise value of k is difficult to obtain; however, this simple treatment does indicate in general how various factors affect the modulus of a staple-fiber yarn.

Qualitative Explanation of the Strength of Staple Yarns

The strength of a staple yarn is determined by various fiber properties and yarn structural and processing parameters. It is known that, as yarn twist increases, yarn strength increases to a manimum value, and any further increase in the former brings about a reduction in the latter. The twist at which yarn strength is maximum is called the optimum twist. One traditional explanation of this phenomenon of change in strength with twist is based on the combination of slippage and breakage of fibers. This is shown diagrammatically in Figure 6.29. It is evident that a staple yarn with no twist (zero twist) has no strength, since the fibers merely slide over one another when stretched. In the initial rising part of the curve, the resistance to slippage increases, but the number of fibers that slip rather than break slowly decreases as the effect of twist in gripping the fibers increases. The downward trend at high twist is obviously caused by fiber obliquity in the yarn; this is identical to the behavior of continuous-filament yarns. A further explanation of this view is given in the following:

1. At low twist, the effect of cohesion between the fibers outweighs the

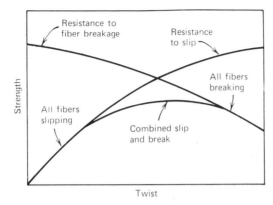

Figure 6.29. Traditional explanations of the effect of twist on strength of staple fiber yarns.

effect of obliquity, giving rise to increase in strength. This increase in strength is slow at first, up to the point at which fibers may just begin to break, and then increases rapidly as larger numbers of fibers break.

2. In the high-twist region, a further increase in cohesion no longer produces an increase in strength, since the majority of fibers break, whereas the increasing inclination of the fibers causes the strength to fall.

However, a close examination of this view indicates that the mechanism governing the twist-strength relationship of staple yarns is not so simple, and this view is incorrect. There will very likely be slippage near the ends of all the fibers, and practically all the fibers are gripped in the central region.

The other explanation, which does not take the effect of migration into account, suggests that the fibers in the outer layers slip, whereas those in the inner layers break, the transition between the breaking and slipping fibers shifting from the inner to the outer zone. This explanation assumes that some arbitrary cohesion in the outer layers causes the development of the transverse pressure. This interpretation does not appear plausible, because the gripping of fibers at low twist will occur if the length of the fibers is reasonably long and there is a high level of migration. Consider an ideal yarn in which all the fibers have similar migration behavior and all will be gripped at some places along their length. Now if the twist is raised to a level at which fibers grip one another, an increase of imposed load will not cause slip because, even though it increases the tension in the fibers that causes slip, it is also effective in the development of transverse forces that increase the frictional hold on the fibers, thus preventing them from slipping. This situation resembles more closely the

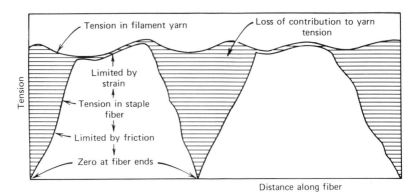

Figure 6.30. Comparison of tension along filament in a continuous-filament yarn and along fibers in the corresponding spun yarn. From Hearle (5)

actual yarn structual behavior, but it still leaves the question of what happens in the regions of fiber ends where slip is liable to occur.

Hearle and El-Sheikh (74) have offered a modified qualitative approach regarding the behavior of spun yarns during extension in terms of the behavior of a continuous-filament yarn. It is assumed that the two structures are identical in all respects (fiber properties, etc.) except for the joining of the discontinuities at the ends of the fibers. Figure 6.30 shows the buildup of tension along the length of the filament when the yarn is subjected to longitudinal extension; the variation in stress is caused by the radial position of the filament imposed by filament migration. Hearle and El-Sheikh suggest that the loss in contribution to the total yarn tension is not so much caused by the slippage of complete fibers as by the failure of portions near the fiber ends to contribute to their fullest extent. Hearle suggests that yarn tension at a given extension or yarn modulus or yarn strength can be expected to vary with twist, fiber length, and fiber friction. Other factors that influence yarn tensile properties are:

1. The rate of fiber migration in yarn: this will have an effect on the rate at which the tension builds up in a fiber.
2. Fiber fineness: the buildup of tension in a fiber is proportional to the area of cross section of the fiber, whereas the frictional resistance to slip is proportional to the circumference and thus to the radius of the fiber. Obviously, the tension causing slippage increases as the square of the fiber radius, whereas the frictional resistance increases as the first power of the radius. Consequently, the greater the fiber radius (coarse fiber), the greater will be the tendency for the fiber tension to overcome the frictional forces resisting slippage. In other words, the finer the fiber, the stronger the yarn.

3. The number of fibers in a yarn cross section: this will affect the pressures generated at the points of interfiber contact in a yarn. Yarn count or size determines the number of fibers in a yarn; thus for a given fiber fineness, yarn size will affect the yarn strength. A very low number of fibers will produce a weak yarn. Goodwin suggests (75) that the minimum number of fibers needed for economical spinning of rayon staple yarn is 64-97, depending on fiber fineness and length. Stanbury and Byerly (38) have suggested a minimum of 20 fibers for worsted spinning.

4. Fiber packing density: this will determine the entanglement of fibers in a staple yarn.

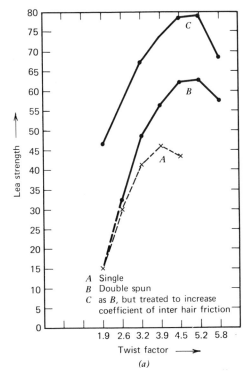

Figure 6.31(a). Reproduction from Studies of Quality in Cotton by Balls (1928). "The inter-relation of twist, adhesion, and strength in singles yarn. Curve A shows the increasing "strength" of a yarn spun with increasing twists, its optimum being at 3.8 twist factor. In Curve B the hair-to-hair adhesion is increased by wet double-spinning for the softest single, thus adding turns up to the same total twists as in A. The optimum rises to 4.9 twist factor. For Curve C the same treatments as in B were preceded by an artificial modification of the hair surface, after the soft single had been spun. The coefficient of friction was thus increased, but the optimum twist is unchanged, though the "strength" is raised throughout."

Table 6.9. Other Experimental Studies of the Strength of Spun Yarns[a]

Reference	Title
Turner and Venkataraman (37)	A study of comparative results for lea, single thread and ballistic tests on yarns from standard Indian cottons
Stanbury and Byerly (38)	The relation between the strength, count, and twist of single worsted yarns
Millard (39)	Nylon staple on the cotton system: some properties of all nylon yarns
Platt et al. (40)	Factors affecting the translation of certain mechanical properties of cordage fibers into cordage yarns
Brown (41)	Correlation of yarn strength with fiber strength measured at different gage lengths
Dickson (42)	Structure and properties of cotton and synthetic single yarn
Ewald and Landstreet (43)	A new method for predicting cotton yarn strength from the observed strength of a single count
Fiori et al. (44)	Effect of single and ply twists on the properties of a 31/2 carded cotton yarn
Fiori et al. (45)	Effect of single and ply twists on the properties of a 15.5/2 carded cotton yarn
Fiori et al. (46)	Effect of cotton fiber strength on single yarn properties and on processing behavior
Fiori et al. (47)	Effect of single and ply twists on the properties of 31/2 or 15.5/2 carded cotton yarns
Platt et al. (48)	Factors affecting the translation of certain mechanical properties of cordage fibers and yarns into cordage strands and ropes
Bogdan (49)	The characterization of spinning quality
Fiori, Brown, and Sands (50)	The effect of cotton fiber strength on the properties of 2-ply carded yarns
Fori, Sands, Little, and Grant (51)	Effect of cotton fiber bundle break elongation and other fiber properties on the properties of a coarse and a medium singles yarn
Virgin and Wakeham (52)	Cotton quality and fiber properties Part IV. The relation between single fiber properties and the behavior of bundles, silvers, and yarns
Coulson and Dakin (53)	The influence of twist on the strength and certain other properties of twofold yarns
Coulson and Dakin (54)	The effects of their physical properties of singles twist; doubling twist, direction of doubling, traveler weight and spindle speed

Table 6.9.   (*Continued*)

Reference	Title
Dakin (55)	The effect of twist and structure on the physical properties of some manifold yarns
Dakin (56)	The effect of twist on some physical properties of an 805/2/3 thread twisted Z/Z/S
Sreenivasan and Shankaranarayana (57)	Twist and tension as factors in yarn characteristics
Bandyopadhyay (58)	The optimum twist factor for jute yarns of various sizes and fiber qualities, and tex-tenacity relations at different twist factors
Nanjudayya (59)	Rupture strength of single yarns measured by ballistic method
Nanjudayya (60)	Strength of a cotton yarn in relation to its structure in the break region

[a]From Hearle (5).

Experimental Studies of Yarn Strength

Several experimental investigations concerning the strength of staple yarns have been reported in the literature. Some of the important studies are listed in Table 6.9. The following discussion is limited to the treatment of results that typify the experimental observations of the strength characteristics of staple yarns.

One of the earliest studies reported in the literature is by Balls (76) in his book "Studies of Quality of Cotton." The three most important chapters are Chapter 10–The Arrangement of Twist, Chapter 11–Hair Adhesion and Yarn Rigidity, and Chapter 12-The Strength of Yarn. These three chapters deal with the role of twist, fiber friction, and the variability in yarn element (yarn structure) on the tensile properties of cotton yarns.

The illustrations in Figure 6.31*a* and *b* show the results of the experiments reported by Balls. Figure 6.31*a* shows the effect of twist and adhesion on the skein strength of yarns, whereas Figure 6.31*b* demonstrates the effect on breaking load of removing twist by untwisting a yarn spun to a high twist. It is clear from the illustrations that the strength of a yarn rises to a maximum value as the twist is increased, followed by a downward trend with further increase. Balls suggests that, for the yarns shown in Figure 6.31*b*, the optimum twist factor ranges between 3.8 and 4.9 (roughly 40 $\text{tex}^{1/2} \times$ t.p.c. and 50 $\text{tex}^{1/2} \times$ t.p.c.), depending on the surface behavior of fibers. Balls also comments that the "initial tension of various yarns was such as to indicate that zero tension would correspond to a twist factor of 1.5." The trend exhibited by these experimental

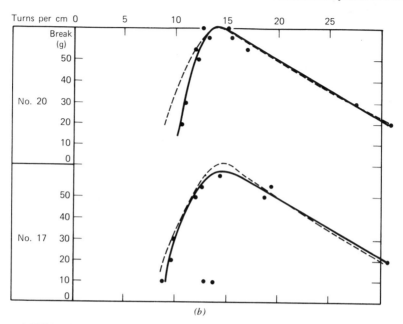

Figure 6.31(b). "Part of figure showing breaking loads of two cotton yarn samples at twists obtained by untwisting of yarn spun to high twist. The measurements were made on the twist-yield tester."

results is in conformity with the general predictions made by the qualitative hypothesis put forward by Hearle.

Balls also interested himself in finding out how fiber properties and yarn structure influence the load-extension and the recovery properties of single yarns. He devised an apparatus that he called an "Elastometer" capable of stretching a yarn at a fixed rate of extension. He studied the elasticity, extension, and permanent set of yarns and made a comparison of the properties of cotton and spun "artificial silk" yarns with the behavior of continuous-filament yarns. The results for extension, elasticity, and permanent set in the region of initial extension for various yarns are shown in Figure 6.32. It is clear that the staple fibers "artificial silk" yarn shows lower tension values and greater permanent set than the corresponding continuous-filament yarns. The effect of increase in twist in cotton yarns on the tensile properties is shown in Figure 6.33. As would be expected, the tension in cotton yarns, at a given extension, increases with increase in twist and then follows a downward trend at the highest twist. With the exception of yarns spun with the two lowest twist levels, the permanent set shows little change. The fibers in the low-twist yarns can easily slide over one another, which accounts for their higher permanent set and low breaking levels.

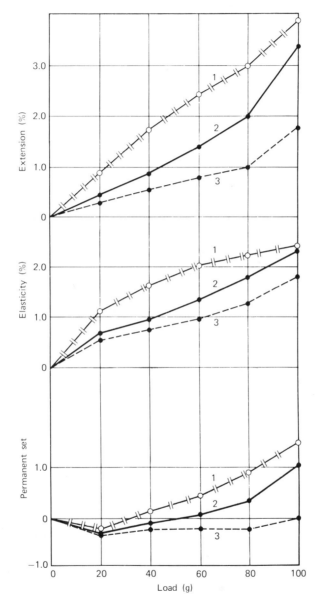

Figure 6.32. Load-extension and elasticity curves for yarns of similar twist and count: 1. cotton yarn; 2. rayon (artificial silk) fibers spun into yarn; 3. rayon (artificial silk) continuous-filament yarn. From Balls (76).

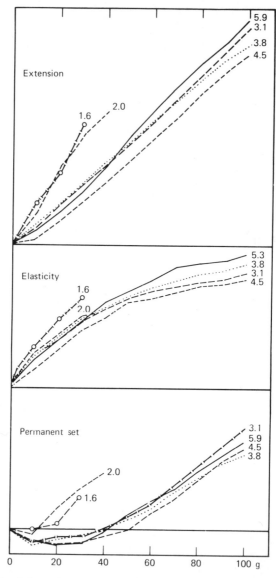

Figure 6.33. Load-elongation and elasticity curves for cotton yarns of different twist factors, expressed in (TPI/count$^{1/2}$). From Balls (76)

Platt (7) has reported a study of the effect of fiber length and twist on the load-extension properties of acetate yarns. Figure 6.34 shows the curves obtained by Platt. An examination of these curves reveals that the staple yarns show a lower tension at all points of extension and a significantly lower strength and breaking extension than the corresponding continuous-filament yarns. For the staple yarns, there is a smoothing out of the yield point. Increase of twist and decrease of staple length have the effect of causing a reduction of tension at a given extension.

Goodwin (75) has reported an experimental study of the effect of twist on the strength of yarns made from three different fiber types. The skein break factor is maximum at the optimum twist factor and decreases gradually to zero at zero twist; however, Hearle (5) has pointed out that "in the absence of an indication of experimental points, it is not certain whether this is a valid deduction from the observations or merely a reflection of the traditional thinking." The elongation-at-break values show a rapid increase with twist (Figure 6.35) in the low twist region, but tend to level off above the optimum twist multiplier.

Another important study that has made a significant contribution to the understanding of the twist-yarn strength relationship is by Landstreet et al. (77). In this study, the authors have dealt with the problem of experimentally defining the overall shape of the twist-strength curve (extending from the low-twist region and beyond the area of maximum strength) for cotton rovings and yarns. Some of their results are shown in Figures 6.36 and 6.37. The results shown in Figure 6.36 are of twist-strength curves for seven rovings ranging from 0.5 to 10.0 hank made from Deltapine, $1\frac{1}{16}$ in. cotton. In addition to the number, the twist-strength relationship is also affected by fiber properties, roving age, position on the bobbin, and method of manufacture. In Figure 6.37 are shown the twist-strength curves for 4.0 hank rovings made from cottons differing widely in fiber properties. It is clear from this illustration that the strength increases slowly at low twist (depicting the behavior of the drafting forces of a low-twist roving), but when the twist is increased, the strength changes very rapidly, exhibiting the behavior of a twisted yarn. The results in Figure 6.37a show twist-strength plots for the rovings and yarns over a wide range of twist; Figure 6.37b shows the twist-strength curve in more detail in the area of maximum strength. These results also show good agreement with the qualitative predictions discussed in the previous section, with the exception of the points of inflection in the maximum strength region.

Holdaway (72) has presented a theoretical model for predicting the strength of single yarns. According to this theory, the mean strength of a yarn can be expressed by the following:

$$F = (\sqrt{M} - 0.0518 \, d_m)^2 \; \psi \, (T_M) \tag{6.13}$$

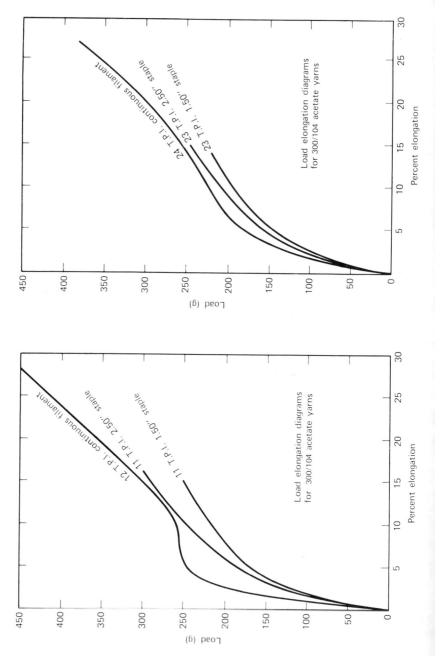

194

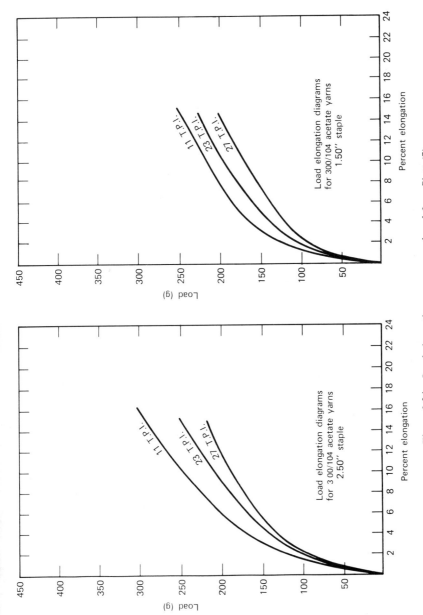

Figure 6.34. Load-elongation curves reproduced from Platt (7).

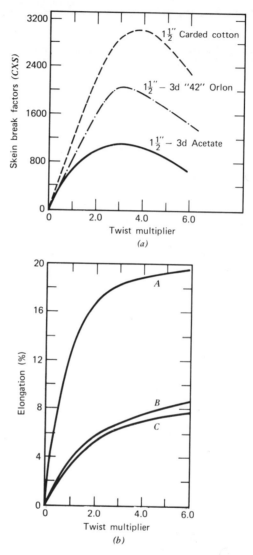

Figure 6.35. (*a*) Effect of twist multiplier on skein break factor of 20s/1 yarn (Courtesy J. A. Goodwin and Lowell Technological Institute), (*b*) effect of twist multiplier on elongation at break. A, 100%, 1½ in. 3 den Orlon 42; B, 100%, 1¼ in. carded cotton; C, 50-50 Orlon-cotton. (Courtesy Lowell Technological Institute.) From Goodwin (75)

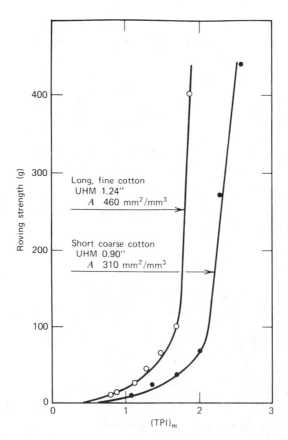

Figure 6.36. Results of Landstreet et al. (77) $(TPI)_m$ is the twist in turns per inch based on yarn length as supplied, $(TPI)_y$ is the twist in turns per inch actual twisted yarn, after twist contraction has taken place. The above diagram shows the twist-strength curves in the region of low twist and depicts the effect of fiber properties on the position of the curves for 2.0 hank roving.

where

F = mean breaking tension in g wt

M = linear density of the pretensioned yarn in tex,

d_m = mean diameter of the single fiber in microns

T_M = the effective twist factor

= $t (\sqrt{M} - 0.0518\, d_m)$

t = turns per inch

and

ψ = a function of T_M.

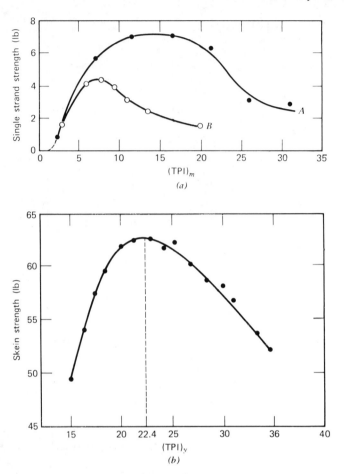

Figure 6.37. From Landstreet et al. (77) (*a*) Twist-strength curves developed over a wide range in twist. Curve A is for a 4s roving given successive increments of twist, becoming coarser due to contraction. Curve B is for yarn spun to an actual count of 4s at all twists, (*b*) twist-strength curve in the region of maximum strength.

In a later publication, Holdaway and Robinson (78) have reported that the theoretical model, in spite of the complexity of spun-yarn structure, is in good agreement with experimentally obtained values of yarn strength. The experimental study was made on worsted single yarns covering a wide range of counts and twists, and the strength and breaking extension measurements were made on a Uster automatic yarn-strength tester. The results of preliminary tests, in which 500 or 600 specimens were tested, demonstrated that the distribution of break-

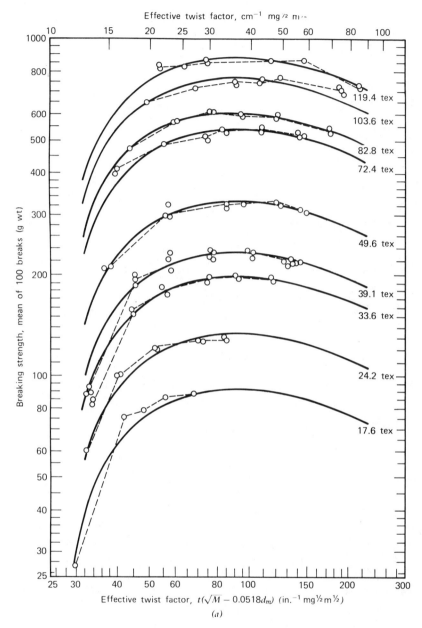

Figure 6.38. Reproduced from Holdaway and Robinson (72) Mean strengths of cap-spun 64s quality wool.

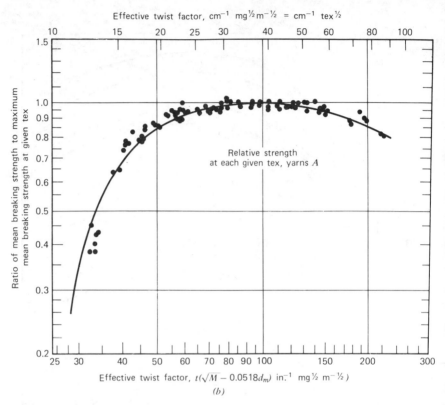

Figure 6.38(b). Reproduced from Holdaway and Robinson (72). Composite curve (data from (a) of relative strength.

ing loads is very close to normal except for two yarns of very low and very high twist that show deviation from the normal distribution.

The expression in equation 6.13 predicts that "for yarns of constant linear density, the mean strength will be a function of the effective twist factor." The actual linear density is replaced by an "effective linear density" given by the term $(\sqrt{M} - 0.0518 \, d_m)^2$ "from which the outer layer of fibers has been removed to a depth of d_m μ"; and the bulk density corresponds to the packing factor of 0.603. This explanation thus differs from the simple view that the breaking load is directly proportional to the linear density. Now if the results according to equation 6.13 are plotted as breaking load F vs. effective twist factor T_M on a logarithmic scale, this should produce curves, for various constant values of $\log_{10} T_m$, that have the same shape but are displaced parallel to the $\log_{10} F$ axis. This is shown in Figure 6.38a. On the other hand, if all the results from Figure 6.38a are plotted as the ratio of the mean strengths to the

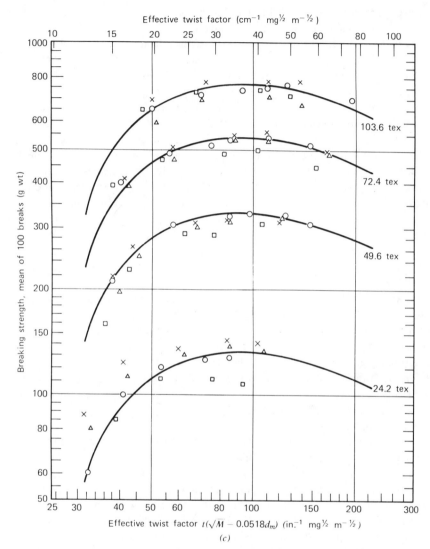

Figure 6.38(c). Reproduced from Holdaway and Robinson (72). Mean strengths of yarns from different wool qualities. (○) 64s (20.68μ); (□) 56s; (×) 64s (22.45μ); (△) 70s.

maximum strength values (relative strength) for each yarn count against the effective twist factor, a single curve is obtained, as shown in Figure 6.38b. This common curve can be used as a basis for predicting the strength of single yarns of a given fiber, provided an allowance is made to take into account the influence of effective linear density. Figure 6.38a demonstrates that the results for yarns made from wool of different qualities are fairly close together when plotted in the same manner as Figure 6.38c.

Holdaway and Robinson report that it has not been possible to give a precise analysis of the factors that influence the extensibility of yarns to rupture. They state that "where fibers tend to slip rather than break, as in yarns of low twist, the mean length of fibers and the distribution of lengths will clearly be important. On the other hand, in yarns of high twist, where fibers mostly tend to break rather than slip, the yield characteristics of the single fibers will be important. With increasing linear density, there will be a statistical effect tending to reduce the frequency of premature breaks due to weaker-than-average fibers, and this will tend to increase the extensibility." Empirically, they demonstrate that yarn breaking extension varies as (0.71 × function of linear density), and this gives a moderately good fit with the experimental results illustrated in Figure 6.39.

There is one obvious factor that has not been taken into account in any of the work reported so far, and that is the "yarn irregularity," which has a profound influence on the strength of a yarn. It has been shown by Fiori et al. (44-46) that the irregularity of a yarn increases as yarn fineness increases, and Morton (79) has reported that an increase in twist increases yarn irregularity. However, it is known that the strength of a yarn is determined by the weakest place; thus an increase in the irregularity would cause the yarn to exhibit lower strength. In other words, for a better understanding of the structural mechanics of yarns, the breaking load should be expressed in terms of the count and twist at the point of rupture instead of the mean values for the whole yarn.

The relationship of yarn strength to its bulk (count) and twist has been very expertly dealt with by Gregory (65-69) in a series of papers on cotton yarn structure. He has tackled the problem of yarn irregularity along the length of a yarn in relation to its strength by examining artificial yarn elements. The strength of the twisted element is governed by its bulk and twist and by the properties of the constituent fibers (fiber length, fineness, and frictional characteristics). The approach adopted to make an artificial yarn element was to grip a length of roving at two points outside the length of the longest fibers separating the grips to produce a thin place at the middle. This method enables the production of yarn elements under controlled conditions and helps in studying the strength properties over a wide range of weights and twists.

Gregory has discussed a procedure to relate the measured breaking load to the weight and twist of the thin place in terms of the weight and twist of the whole yarn specimen between the grips. The method is similar in principle to the engineer's "necked" sample for strength testing. A typical plot of breaking length (count × strength, which is numerically equivalent to tenacity) against twist for yarn elements prepared from combed Giza 12, 6.75 hank rovings is shown in Figure 6.40. This illustration also shows the variation of count with twist. The maximum breaking length is well defined, and the slight reduction in

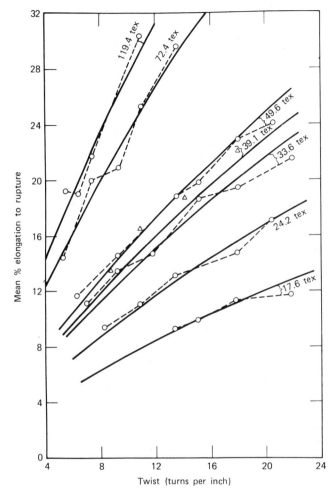

Figure 6.39. Breaking extension vs. twist. From Holdaway and Robinson (72)

count as the twist increases does not obscure the general trend observed for the variation of strength with twist. Gregory also reports that the yarn elements prepared from a particular cotton show a reduction in maximum strength as the count becomes finer. He found that the twist factors for maximum strength vary from 3.9 for the combed St. Vincent cotton to 7.5 for carded Bengals. The twist factors for maximum strength of yarn elements were found to be much greater than the twist factors to which the yarns from the cotton were normally spun.

From consideration of the influence of fiber properties on the twist required to produce maximum strength in twisted yarns, Gregory suggested that the twist

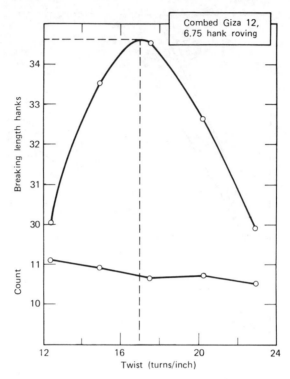

Figure 6.40. Effect of twist on the strength of artificial yarn elements. From Gregory (67)

factor for maximum strength is a function of the product of fiber length, L_1, coefficient of friction, μ, and surface area per unit mass, S. Empirically, the maximum strength is expressed in the form:

$$B = L_1 \ \mu S \qquad (6.14)$$

where L_1 is a biased mean fiber length equal to $L + \sigma^2/L$
where
    L = actual mean fiber length, and
    σ = standard deviation of fiber length.

A plot of B vs. twist factor for maximum strength is shown in Figure 6.41. The dotted line is derived from Sullivan's (64) theoretical curves. The product $B = L_1 \mu S$ defines a critical pressure at which fibers just begin to break.

The concluding part of the study reported in this series of papers deals with a very important aspect that has a great practical value. It covers a very wide area

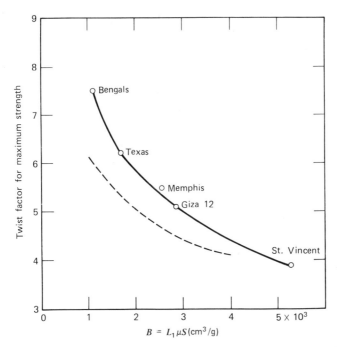

Figure 6.41. Gregory's (68) relationship of twist factor with a function of length, friction, and fineness. The dotted line depicts Sullivan's theoretical relationship.

Table 6.10. Comparative Values of Maximum Breaking Length for Different Cotton Fiber Structures (in kilometers, corrected to 10 sec break)[a]

Structure	Cotton				Range of Ratio St. Vincent Bengals
	St. Vincent	Giza 12	Texas	Bengals	
Single fibers					
1 mm test length	64.0	54.5	41.3	39.6	1.62
1 cm test length	40.7	38.9	27.6	21.1	1.93
Parallel bundle Pressley	47.8	39.7	34.5	32.3	1.48
Twisted yarn element	34.1	26.5	19.1	13.1	2.60
Cabled yarn	32.2	26.2	19.3	12.4	2.59
Cloth weft strip	29.0	20.2	14.0	10.4	2.79
Single yarn					
Single thread test	28.7	19.8	13.2	8.5	3.38
Lea test	24.3	17.8	10.2	5.6	4.34
Range of ratio,					
1 mm fiber/lea	2.64	3.06	4.05	7.07	

[a]From Gregory (69).

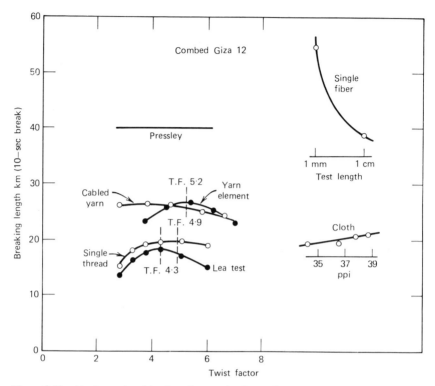

Figure 6.42. Maximum breaking length vs. twist factor for different structures made from cotton–85 count yarn. From Gregory (69)

in which a review is made of the strength of cotton fiber in a variety of textile structures, namely, single fiber, parallel bundles of fibers, twisted bundles of fibers, yarns (both singles and cabled), and cloth. Figure 6.42 shows typical curves for combed Giza 12 yarn spun in 8s ($\simeq$ 74 tex) single count, and the results for the comparison of cotton fiber structures of different types are listed in Table 6.10 and 6.11 and shown in Figure 6.43.

A recent study of the efficiency of fiber-to-yarn strength has been reported by Hurley et al. (61). The yarns were spun to cotton counts of 4.1, 6.2, 9.2, and 14 ($\simeq$ 144, 95, 64, and 42 tex), each with a twist multiplier of 3.5 ($\simeq$ 34 tex$^{1/2}$ $\times$ t.p.c). The results for the breaking force and intrinsic tenacity (yarn tenacity divided by the denier of the thin section in the region of rupture) are shown in Figure 6.44. The denier of yarn in the region of rupture was obtained on an 8-mm length by the Uster Evenness tester. Strength measurements were made on an Instron Tensile Tester and the Uster Single End Strength Tester. The results for the efficiency of fiber-to-yarn strength translation and the strength loss

Table 6.11. Breaking Lengths Expressed as Percentage of Pressley Value[a]

Structure and Text	Cotton			
	St. Vincent	Giza 12	Texas	Bengals
Single fiber: 1 mm	134	137	120	123
Pressley	100	100	100	100
Single fiber: 1 cm	85	98	80	65
Twisted yarn element	71	67	55	41
Cabled yarn	67	66	56	38
Cloth weft strip	61	51	41	32
Single yarn: single thread test	60	50	38	26
8s/3: low-twist single thread	57	47	36	23
Single yarn: lea test	51	45	30	17

[a]From Gregory (69).

factor are listed in Table 6.12. These results are in accord with the accepted view that the strength properties of the yarn are determined by the weakest place.

Other studies dealing with the variation of strength with yarn irregularity have been reported by Hamby et al. (80) and El-Behery and Mansour (62). In these studies, the single-end strength was measured on the Uster Dynamometer, and the yarn irregularity tests were made on the Uster Evenness tester. Some typical

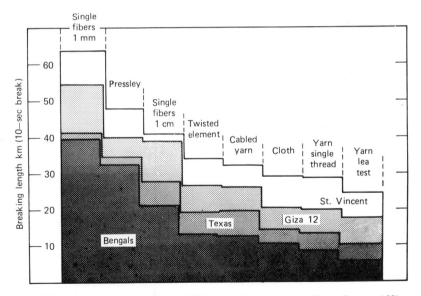

Figure 6.43. Strength relationships of different cotton structures. From Gregory (69)

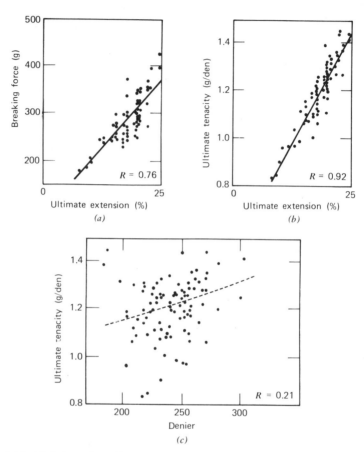

Figure 6.44. Variation of tensile properties of yarns—20-in length specimens broken on Uster Single end strength Tester. From Hurley et al. (61)

results for the variation of yarn strength with mass variation and yarn irregularity index are presented in Figure 6.45 (62).

Hearle and El-Sheikh (74) have reported a study of the tensile properties of twisted worsted and woolen rovings. They have discussed their experimental results in terms of the qualitative explanation of the behavior of spun yarns. A comparison of the experimental results with the equation developed for predicting the tensile properties of spun yarns by Hearle (73) has also been made. The experimental yarns were made by twisting the worsted and woolen roving on a continuous laboratory twister. This method helped to minimize the effects of yarn irregularity.

A typical load-elongation curve and the parameters used to characterize the tensile behavior of spun yarn are illustrated in Figure 6.46. Table 6.13 lists the

Table 6.12. Intrinsic Tensile Properties and Loss Factors for Some Acrylic Yarns[a]

	Specimen Test Length (in.)	
Single-Fiber Properties[b]	0.2	1.0
Denier	5.97 (14)	5.85 (13)
Ultimate Tenacity (g/den)	2.69 (13)	2.58 (15)
Ultimate Extensibility (%)		29 (22)

Yarn properties	Yarn Cotton Count			
	4.1/1	6.2/1	9.2/1	14/1
Skein strength, lb (120-yd skein)	603 (3)	360 (3)	212 (9)	106 (3)
Single end strength (g)	2148 (7)	1302 (9)	813 (10)	447 (18)
Intrinsic tenacity (T_i^c g/den)	2.17 (4)	2.10 (4)	2.07 (8)	2.16 (6)
Skein translation efficiency (TE_{skeins}^c %)	44.6 (4)	42.5 (3)	37.7 (6)	29.7 (3)
Single end translation efficiency (TE_{se}^c %)	60.8 (7)	55.6 (9)	51.6 (10)	43.2 (18)
Intrinsic translation efficiency (TE_i^c, %)	79.3 (4)	76.9 (4)	75.8 (8)	79.1 (6)
Skein loss factor (L_{skeins}, %)	34.7 (7)	34.4 (6)	38.1 (10)	49.4 (6)
Single end loss factor (L_{se}, %)	18.5 (23)	21.3 (25)	24.2 (22)	35.8 (22)

[a]From Hurley et al. (61).
[b]Tests carried out at 70°F, 65% RH and a rate of extension of 100%/minute.
[c]Based on single-fiber tenacity at a gauge length of 0.2 in.
Numbers in ( ) indicate the CV of the measurement.

Table 6.13. Spun Yarn Properties[a]

Twist Angle (degrees)	Linear Density (tex)	Yarn Crimp (%)	Yarn Modulus (g/tex)	At Max. Tension		At Break	
				Stress (g/tex)	Extension (%)	Stress (g/tex)	Extension (%)
10	271	1.5	21	0.4	2.8		
15	308	3.0	98	2.5	3.0		
20	270	2.9	127	5.4	6.8	3.9	15.6
25	285	3.5	124	6.8	30.6	6.6	33.1
30	274	3.3	130	7.5	23.1	7.4	29.4
35	276	4.4	86	7.1	31.5	7.0	32.4
40	330	3.7	55	7.9	47.2	7.9	47.7
50	436	1.3	10	4.6	66.5	4.6	66.6
			Yarn No. 4, Woollen				
15	183	0.9	13	1.1	21.7		
20	185	0.8	41	1.8	15.7	1.8	17.3
25	191	1.8	34	2.5	20.0	2.5	20.7
30	193	1.5	39	3.0	21.7	3.0	22.0
35	202	1.3	41	3.4	25.9	3.4	26.3
40	202	2.3	46	4.1	30.8	4.1	31.3
50	225	3.1	35	3.5	25.3		

[a]From Hearle and El-Sheikh (74).

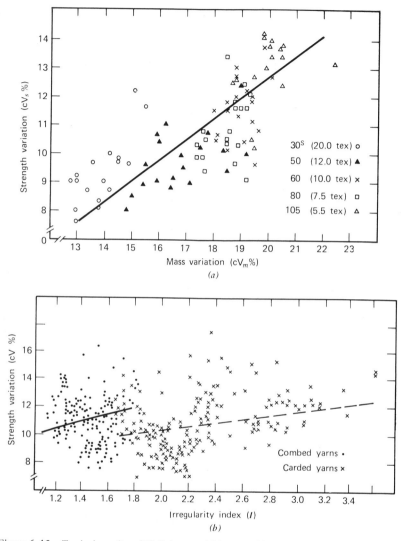

Figure 6.45. Typical results of El-Behery and Mansour (62). (*a*) Strength variation vs. mass variation for combed cotton yarns, (*b*) strength variation vs. yarn irregularity index for cotton yarns.

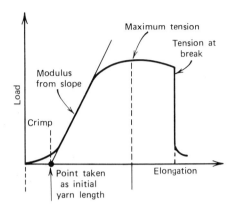

Figure 6.46. Representative load-elongation curve for a yarn. From Hearle and El-sheikh (74)

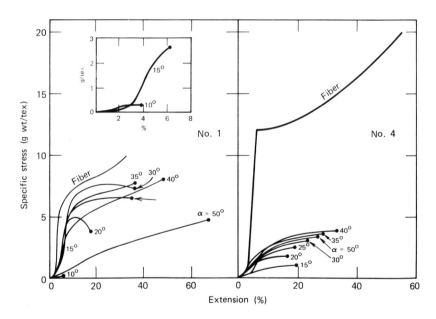

Figure 6.47. Stress-strain curves of yarn and fibers. No. 1. Worsted. No. 4. Woolen. From Hearle and El-Sheikh (74)

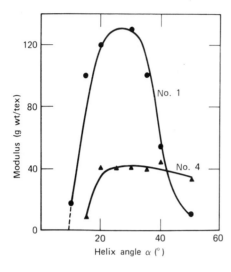

Figure 6.48. Variation of modulus with twist. From Hearle and El-Sheikh (74)

values of these parameters, and the corresponding stress-strain curves are illustrated in Figure 6.47. The plot of modulus vs. twist (twist angle) is shown in figure 6.48. It is evident from these relationships that the general trend of the curves in Figure 6.47 is similar to the form predicted in Figure 6.46.

Hearle and El-Sheikh (74) have presented arguments explaining the experimental results in light of a proposed theory, and for the sake of clarity their discussion is reproduced here.

In the worsted yarn No. 1 there is little resistance to slippage when the twist angle is $10°$, but it subsequently builds up to achieve maximum effect at a twist angle of about $30°$. Thereafter the yarn tension levels fall due to obliquity. The modulus of the woollen yarn No. 4, is much lower and this must be due partly to the more irregular fibre arrangement and partly to the greater proportion of short fibres. The twist angle for maximum modulus in the woollen yarn is nearer $40°$, although the curve is very flat between $20°$ and $50°$. In both yarns the minimum twist angle which is likely to be practically usable is about $20°$.

In both yarns, the initial region of easy extension (called yarn crimp) generally increases in amount as the twist is increased.

Yarn tenacity, whether indicated by the maximum load or the load at which there is a sharp break, rises to a maximum value at around $40°$. Breaking extension is low in the very weak yarns with low twist, and continues to rise steadily as twist is increased. This is similar to the behaviour of polyamide, polyester and acrylic filament yarns where the rising breaking extension has been attributed to the contraction of the yarn following the release of the twisting tension: in parti-

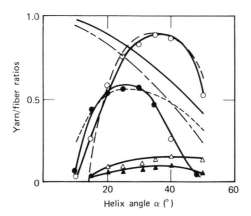

Figure 6.49. Comparison of theory and experiment. (●) No. 1, yarn modulus; (▲) No. 4, fiber modulus; (○) No. 1, yarn tenacity; (△) No. 4, fiber tenacity. (———) $\cos^2 \alpha$; (———)F $(\alpha, \sigma_1, \sigma_y)$; (———) $\cos^2 \alpha$ (1-0.122 cosec α); (- - -) 2.3 $\cos^2 \alpha$ (1-0.0.24 cosec α).

cular, fibres on the surface of the yarn following a longer path are assumed to be under strain during twisting, and then subsequently to contract to their unstrained length with a resultant buckling of the centre fibres. This allows an additional extension of the centre fibres before they come under strain and are extended to break. This behaviour is not shown by filament yarns with poor elastic recovery (where permanent deformation occurs); but in view of wool's good elastic recovery it is not surprising that it is now shown up by wool yarns.

Figure 10 (6.49) shows a comparison between the experimental results and theoretical relations. The variation of modulus of the worsted yarn shows a good agreement with the relation previously discussed, with K equal to 0.133, except for a rapid fall at high twist angle. However, the best relation $F(\alpha, \sigma_1, \sigma_y,)$ for continuous filament yarns, does decrease more rapidly from the $\cos^2\alpha$ relation used in the approximate theory.

The ratio of yarn tenacity to fibre tenacity in the worsted yarn rises to a much higher value than would be expected theoretically. Indeed it approaches a value of 1 at a twist angle of $40°$. This means that the fibres in the twisted yarn are acting as if they were much stronger than appears from single fibre tests. This must be due to the fact that weak places in the fibres are being supported by the grip of neighboring fibres: they will not be extended to the local breaking extension and will be able to contribute to yarn tension. Another way of regarding the mutual support effect is to note that in single fibre tests each fibre can break at its own weakest place, but in a coherent yarn the break must occur at a particular point along the length of the yarn. The theoretical relation can be empirically modified to include this effect by putting:

$$\frac{\text{Yarn tenacity}}{\text{Fibre tenacity}} = A \cos^2 \alpha \ (1 - K \cosec \ \alpha)$$

where A is the ratio of effective fibre strength to measured fibre strength. This equation shows a reasonable fit with A = 2.3 and K = 0.24.

The same arguments do not apply to the modulus which is determined by strain all along the fibre, and not by the effect of an isolated weak place.

The woollen yarns show much lower moduli and tenacities and it is difficult to apply the theory in a meaningful way.

Hearle and El-Sheikh (74) conclude:

... that the tensile behaviour of twist worsted yarns can be explained by the effects of obliquity, which have been worked out for continuous filament yarns, combined with the effects of fibre slippage, which cause a loss in contribution to yarn tension. Yarn properties approach those of filament yarns more closely as yarn twist, fibre length, and fibre friction increase. At very low twists the forces generated are inadequate to cause the yarn to be a self-locking structure. The investigation shows that the idealised cylindrical helical yarn structure forms a

Table 6.14. Twist Factors at Optimum Yarn twist [a] $(\text{tex}^{1/2} \times \text{turns/cm})$

Yarns	Linear Density (tex)	Spinning System	Tensile Test[b]		Fatigue Test[c]	
			Maximum Strength	Maximum Elongation	Maximum Strength	Maximum Elongation
1	35.7	Conventional[d]	50.6	50.6	55.3	42.7
2	35.7	Open-end	54.7	54.7	58.5	47.4
3	25.0	Conventional	49.3	49.3	56.2	40.4
4	25.0	Open-end	57.2	57.2	63.2	47.4
5	20.0	Conventional	49.3	49.3	56.2	40.4
6	20.0	Open-end	56.9	56.9	64.8	53.7
7	16.7	Conventional	48.7	48.7	53.7	42.7
8	16.7	Open-end	55.9	55.9	63.2	52.5
9	31.3	Conventional	47.4	47.4	52.1	39.5
10	28.6	Open-end	55.3	55.3	56.9	53.7
11	20.0	Conventional	47.4	47.4	53.7	39.5
12	20.0	Open-end	54.7	54.7	61.6	50.6

[a] From Barella (83).
[b] Pendulum-type strength tester.
[c] Fatigue tests were carried out at 250 cycles/min.: relative extension 1.66% with a load of 2 gf/tex on a Comptiss apparatus.
[d] Ring spun yarns.

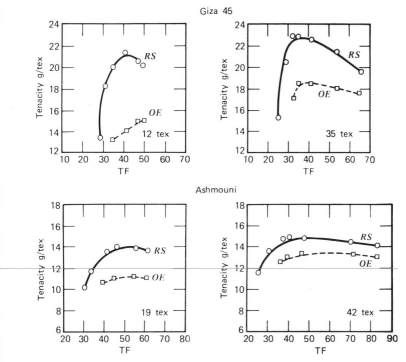

Figure 6.50. Reproduced from Sultan and El-Hawary (82). Yarn tenacity vs. twist (T.F.−tex$^{1/2}$ × turns/cm.)

reasonable model for theoretical study, provided that the important property of fibre migration is included.

This discussion of the structure and the mechanical properties of staple yarns would not be complete without a reference to open-end spun yarns, which form a significant part of the textile scene. The open-end spun yarns have a distinct structure that differs from that observed in ring-spun yarns. The most basic structural differences that are observed are in the fiber extent, fiber migration, and the fiber-packing density, which collectively determine the tensile behavior of yarns. Hearle et al. (81) report that open-end yarns (spun on the air vortex and drum machines) have a relatively high proportion of hooked and entangled fibers, which tends to reduce the fiber extent. This has a detrimental effect on the tenacity of yarns. Because of the mechanism of formation (Chapter 9), fiber migration in an open-end spun yarn tends to be more local, and the yarn is made up of layers. This aspect of the structure affects the degree of fiber interlocking (discussed in the last chapter) within the body of a yarn, which in turn influences the tensile and other physical characteristics of the yarn. Sultan and El-Hawary (82) have reported a study on the comparison of the properties of open-end spun yarns and the ring-spun yarns from two Egyptian cottons. Figure 6.50 shows the twist-strength curves of open-end and ring-spun yarns. It is clear from the results that the twist-strength curves from the open-end spun yarns are flatter in the region of maximum strength. The twist factors for maximum strength for these yarns are generally higher by about 12% than the corresponding ring-spun yarns. These results are in agreement with those reported by Barella (83), as shown in Table 6.14 (in which the results for the optimum twist for maximum strength for open-end and ring-spun yarns are listed). The breaking extension of open-end spun yarns is also found to be 10-40% higher than for the corresponding ring spun yarns made at the same twist factor. These differences in properties result from variations in structural parameters discussed earlier.

REFERENCES

1. Backer, S. in Hearle, J. W. S., P. Grosberg, and S. Backer (Eds.), *Structural Mechanics of Yarns and Fabrics,* Volume 1. Wiley-Interscience, New York, 1969.

2. Gregory, J., *J. Text. Inst.,* 1953, **44**, T515.

3. Meredith, R., *J. Text. Inst.,* 1945, **36**, T107.

4. A.S.T.M., *1971 Book of ASTM Standards,* Part 24, Textile Materials—General Methods and Definitions, American Society for Testing and Materials, Philadelphia.

5. Hearle, J. W. S., in Hearle, J. W. S., P. Grosberg, and S. Backer (Eds.), *Structural Mechanics of Fibers, Yarns, and Fabrics,* Volume 1. Wiley-Interscience, New York, 1969.

6. Gegauff, C., *Bull. Soc. Ind. Mulhouse,* 1907, **77**, 153.

7. Platt, M. M., *Text. Res. J.,* 1950, **20**, 1.

8. Platt, M. M., *Text. Res. J.,* 1950, **20**, 665.

9. Hearle, J. W. S., *J. Text. Inst.,* 1958, **49**, T389.

10. Sullivan, R. R., *J. Appl. Phys.,* 1942, **13**, 157.

11. Grosberg, P., *J. Text. Inst.,* 1963, **54**, T223.

12. Hearle, J. W. S., H. M. A. E. El-Behery, and V. M. Thakur, *J. Text. Inst.,* 1961, **52**, T197.

13. Treloar, L. R. G. and J. W. S. Hearle, *J. Text. Inst.,* 1962, **53**, T446.

14. Treloar, L. R. G., *Brit. J. Appl. Phys.,* 1962, **13**, 314.

15. Wilson, N. and L. R. G. Treloar, *Brit, J. Phys.,* 1961, **12**, 147.

16. Wilson, N., *Brit. J. Appl. Phys.,* 1962, **13**, 323.

17. Treloar, L. R. G., and Riding, G., *J. Text. Inst.,* 1963, **54**, T156.

18. Symes, W. S., *J. Text. Inst.,* 1959, **50**, T241.

19. Kilby, W. F., *J. Text. Inst.,* 1964, **55**, T589.

20. Treloar, L. R. G., *J. Text. Inst.,* 1965, **56**, T359.

21. Treloar, L. R. G., *J. Text. Inst.*, 1965, **56**, T477.

22. Wilson, N., *Brit. Appl. Phys.*, 1965, **16**, 1889.

23. Maginnis, J. B., *Text. Res. J.*, 1950, **20**, 165.

24. Hamburger, W. J., *Text. Res. J.*, 1948, **18**, 705.

25. Shrinagbhushan, *Text. Digest*, 1956, **17**, 4.

26. Grover, E. G., and D. S. Hamby, *Textile Processing of Synthetics–Continuous Filament*, School of Textiles, North Carolina State College, 1956.

27. Taylor, J. R. et al., WADC Technical Report No. 52-55, March 1952.

28. Alexander, F. J., and C. H. Sturley, *J. Text. Inst.*, 1952, **43**, Proc. 1.

29. Hearle, J. W. S., and V. M. Thakur, *J. Text. Inst.*, 1961, **52**, T49.

30. Kilby, W. F., *J. Text. Inst.*, 1964, **55**, T589.

31. Hearle, J. W. S., H. M. A. E. El-Behery, and V. M. Thakur, *J. Text. Inst.*, 1959, **50**, T83.

32. Hearle, J. W. S., H. M. A. E. El-Behery, and V. M. Thakur, *J. Text. Inst.*, 1960, **51**, T229.

33. Treloar, L. R. G. and G. Riding, *J. Text. Inst.*, 1963, **54**, T156.

34. Riding, G. and N. Wilson, *J. Text. Inst.*, 1965, **56**, T205.

35. Backer, S. and C. J. Monego, *Text. Res. J.*, 1968, **38**, 762.

36. Pierce, F. T., *J. Text. Inst.*, 1926, **17**, T355.

37. Turner, A. J. and V. Venkataraman, *J. Text. Inst.*, 1931, **22**, T197.

38. Stanbury, G. R. and W. G. Byerley, *J. Text. Inst.*, 1934, **25**, T295.

39. Millard, F., *J. Text. Inst.*, 1951, **42**, T168.

40. Platt, M. M., W. G. Klein, and W. J. Hamburger, *Text. Rex. J.*, 1952, **22**, 641.

41. Brown, H. M., *Text. Res. J.*, 1954, **24**, 251.

42. Dickson, J. B., *Text. Res. J.*, 1954, **24**, 511.

43. Ewald, P. R., and C. B. Landstreet, *Text. Res. J.*, 1954, **24**, 1064.

44. Fiori, L. A., J. J. Brown, and J. E. Sands, *Text. Res. J.*, 1954, **24**, 267.

45. Fiori, L. A., J. J. Brown, and J. E. Sands, *Text. Res. J.*, 1954, **24**, 428.

46. Fiori, L. A., J. J. Brown, and J. E. Sands, *Text. Res. J.*, 1954, **24**, 503.

47. Fiori, L. A., J. J. Brown, and J. E. Sands, *Text. Res. J.*, 1954, **24**, 526.

48. Platt, M. M., W. G. Klein, and W. J. Hamburger, *Text. Res. J.*, 1954, **24**, 907.

49. Bogdan, J. F., *Text. Res. J.*, 1956, **26**, 720.

50. Fiori, L. A., J. J. Brown, and J. E. Sands, *Text. Res. J.*, 1956, **26**, 296.

51. Fiori, L. A., J. E. Sands, H. W. Little, and J. N. Grant, *Text. Res. J.*, 1956, **26**, 553.

52. Virgin, W. P. and H. Wakeham, *Text. Res. J.*, 1956, **26**, 177.

53. Coulson, A. F. W. and G. Dakin, *J. Text. Inst.*, 1957, **48**, T207.

54. Coulson, A. F. W. and G. Dakin, *J. Text. Inst.*, 1957, **48**, T258.

55. Dakin, G., *J. Text. Inst.*, 1957, **48**, T293.

56. Dakin, G., *J. Text. Inst.*, 1957, **48**, T321.

57. Sreenivasan, K and K. S. Shankaranarayana, *Text. Res. J.*, 1961, **31**, 746.

58. Bandyopadhyay, S. B., *Text. Res. J.*, 1963, **33**, 9.

59. Najundayya, C., *Text. Res. J.*, 1965, **35**, 795.

60. Najundayya, C., *Text. Res. J.*, 1955, **36**, 954.

61. Hurley, R. B., R. G. Duby, and C. R. Pfeifer, *Text. Res. J.*, 1968, **38**, 1174.

62. El-Behery, H. M. A. E. and S. A. Mansur, *Text. Res. J.,* 1970, **40**, 896.

63. Morton, W. E., *Text. Res. J.,* 1956, **26**, 325.

64. Sullivan, R. R., *J. Appl. Phys.,* 1942, **13**, 157.

65. Gregory, J., *J. Text. Inst.,* 1950, **41**, T1.

66. Gregory, J., *J. Text. Inst.,* 1950, **41**, T14.

67. Gregory, J., *J. Text. Inst.,* 1950, **41**, T30.

68. Gregory, J., *J. Text. Inst.,* 1953, **44**, T499.

69. Gregory, J., *J. Text. Inst.,* 1953, **44**, T515.

70. Alexander, E., *Text. Res. J.,* 1952, **22**, 503.

71. Shorter, S. A., *J. Text. Inst.,* 1957, **48**, T99.

72. Holdaway, H. W., *J. Text. Inst.,* 1965, **56**, T121.

73. Hearle, J. W. S., *Text. Res. J.,* 1965, **35**, 1060.

74. Hearle, J. W. S. and El-Sheikh, Papers of Third International Wool Textile Conference, Institute Textile de France, 1965, p. 199.

75. Goodwin, J. A., Press, J. J. (Ed.), *Man-Made Textile Encyclopedia,* Chapter V-3. Textile Book Publishers, New York, 1959.

76. Balls, W. L., *Studies of Quality in Cotton.* Macmillan, London, 1928.

77. Landstreet, C. B., P. R. Ewald, and J. Simpson, *Text. Res. J.,* 1957, **27**, 486.

78. Holdaway, H. W. and M. S. Robinson, *J. Text. Inst.,* 1965, **56**, T168.

79. Morton, W. E., *J. Text. Inst.,* 1930, **21**, T205.

80. Hamby, D. S., W. C. Stuckey, B. Gast, and R. J. Hader, *Text. Res. J.,* 1960, **30**, 435.

81. Hearle, J. W. S., P. R. Lord, and N. Senturk, *J. Text. Inst.,* 1972, **63**, T605.

82. Sultan, M. A. and I. A. El-Hawary, *J. Text. Inst.,* 1974, **65**, T194.

83. Barella, A., *J. Text. Inst.,* 1972, **63**, T226.

84. Hearle, J. W. S., *J. Text. Inst.,* 1969, **60**, T95.

85. Konopasek, M. and J. W. S. Hearle, *J. Text. Inst.,* 1974, **65**, T217.

86. Cheng C. C., J. L. White, and K. E. Duckett, *Text. Res. J.,* 1974, **44**, 698.

87. Zorowski, C. and T. Murayama, *Text. Res. J.,* 1967, **37**, 852.

7

Uniformity Characteristics of Yarns

IRREGULARITY IN YARN

The Incidence of Irregularity and its Effects

All staple-fiber yarns vary in linear denisty, and most problems of yarn quality are related to this basic property.

Irregularity in yarn is recognized in one of a number of ways:

1. variation in linear density
2. variation in thickness as seen by the eye
3. variation in twist
4. variation in strength
5. variation in color

These all arise from the same underlying cause—the uneven distribution of fibers along the length of the yarn. This produces variation in the number of fibers per cross section and in linear density. When twist is inserted in a yarn, it is distributed in such a way that the angle of twist is approximately constant. This means that thin places have more turns per unit length than thick places, and are relatively hard, whereas thick places are soft and comparatively bulky. This accentuates the visual appearance of the irregularities.

The combination of varying numbers of fibers per cross section with varying forces binding these fibers together because of twist variation leads to varying tensile properties. If two yarns are made from the same material to the same count and twist specification, and one is more uneven in linear density than the

219

other, it will be more irregular in strength; it will therefore contain more weak places and places of greater weakness. Such a yarn may suffer more breakages in spinning, winding, warping, weaving, and knitting; this will reduce the efficiency and increase the cost of all these processes.

If the yarn is woven into cloth, it is unlikely that the presence of weak places will be responsible for any serious variations in cloth strength, since the forces binding fibers together are supplemented by the interlacing of threads. In knitted fabrics, thin places in yarn may be the source of holes developing, but in both woven and knitted fabrics, complaints are more likely to arise on the grounds of poor cloth appearance caused by the use of uneven yarns.

Thus for purposes of processing efficiency, and in the interests of cloth appearance, there are levels of unevenness beyond which the yarn is unacceptable; these are not necessarily the same in the two cases. With the advent of suitable testing instruments, the specification of required standards can be made more precise; at the same time, these instruments enable quality control procedures in yarn manufacture to be improved and the requisite standards to be met.

Color variation in single yarn is not usually related to variations in linear density but to inadequate mixing of differently colored fiber components of the blend. When blends of different fibers are used, there is also the possibility of uneven distribution of types of fiber; this can produce color variations on subsequent yarn or piece dyeing if the different types have different dyeing characteristics, and can also be responsible for increased variability of yarn strength.

It will have been noted that, with the exception of color variation, all other forms of irregularity arise from the first, the variation in linear density. In discussing the irregularity of textile products, one often wishes to refer to it as applying in a general way to yarns, rovings, slivers, or tops; the word *strand* is used to imply any or all of these. *Irregularity* is used to imply variation in any property along a strand, and *unevenness* is reserved to mean variation in linear density of a strand or part of it.

Causes of Unevenness in Yarn

Unevenness in yarns produced in roller drafting processes can be assigned to four causes:

1. a random distribution of fibers in slivers by preparatory machinery
2. a wave-like grouping of fibers caused by roller drafting
3. regular periodic variations in linear density caused by mechanical imperfections in drafting mechanisms
4. fortuitous happenings in drafting caused by working conditions and faulty operations

In condenser yarn-making processes, there is no roller drafing; yarn unevenness is caused by:

1. a random distribution of fibers in the card web produced by the card and earlier preparatory machinery
2. irregular division of the card web by the condenser tapes
3. regular periodic variations in linear density caused by mechanical imperfections and faulty operation of the card
4. fortuitous causes due to working conditions

Random Fiber Arrangement

The machinery used to convert a stock of fibers into a yarn does not have the ability to arrange fibers in anything better than a random way even in the most favorable circumstances. Consider what would have to be done to produce a uniform yarn with the same number of fibers at every cross section; the problem is that of arranging the fiber ends in such a way that each time a fiber terminates another is introduced to take its place. If the *fibers* were of equal and uniform linear density, the yarn would also be perfectly uniform.

Textile machinery, however, cannot detect the termination of one fiber to introduce another. Two questions therefore arise:

1. How does machinery normally arrange fibers?
2. What amount of unevenness must be expected from this type of arrangement?

It has been postulated (1) that the best that the early machines can do is to thoroughly mix the fibers so that they come out arranged in a random order in the card sliver. In practice, performance falls short of this best, but otherwise perfect machinery, without powers of discrimination to arrange fibers in a specified order, can do no better than this.

In the case of woolen or cotton waste cards, this would be true of the webs produced, and if the condenser tapes divided up the web without any bias, the slubbings from the condenser would also have a random fiber arrangement.

In the cotton and worsted spinning processes, doubling and drafting occur in the subsequent operations. In doubling there is no way in which thick places can be systematically used to compensate for thin places in adjacent slivers—the operation is random. The order of fiber ends is modified by putting two or more slivers together, but this only produces a thicker sliver with another random order.

Perfect drafting would consist of multiplying the spacing of fiber ends by the draft so that, if strands with a random arrangement of fiber ends were fed into drawing machinery and drafted in this way, the result would eventually be a yarn in which fibers were still arranged at random. The unevenness of such a yarn can be calculated.

Unevenness Caused by Random Fiber Arrangement

The unevenness of a strand with an average of n fibers per cross section in which the fibers are arranged at random is given by

$$\sigma_n = \sqrt{n}$$

where σ_n is the standard deviation of fiber number per cross section.

Therefore, the coefficient of variation of fiber number per cross section is

$$V_n = \frac{100\sqrt{n}}{n} = \frac{100}{\sqrt{n}} \tag{7.1}$$

This equation holds for a strand made of fibers of any length or mixture of lengths. If all the fibers themselves were uniform and of the same weight per unit length, it would also represent the coefficient of variation of linear density of the strand. In the case in which the fibers themselves vary in linear density, the effect of this variation must be added to the variation of fiber number to give the overall variation of strand linear density.

In the case of wool fibers, it is usual to measure fiber diameter and its coefficient of variation V_d. From this the coefficient of variation of linear density of the fiber can be calculated, and it can be shown from equation 7.1 that if V_r is the coefficient of variation of linear density of the strand (r indicating the random arrangements of the fibers),

$$V_r = \frac{100}{\sqrt{n}} \cdot \sqrt{1 + 0.0004\, V_d^2} \tag{7.2a}$$

In the case of cotton and other fibers, the coefficient of variation of linear density of the fibers is usually determined, if this is V_f, by

$$V_r = \frac{100}{\sqrt{n}} \cdot \sqrt{1 + 0.0001\, V_f^2} \tag{7.2b}$$

which for fibers other than wool may be a more convenient form than 7.2a.

In the case of wool fibers V_d is usually of the order of 25% for a 64s quality wool, in which case equation 7.2a gives a value

$$V_r = \frac{112}{\sqrt{n}}$$

According to Foster (2) it is sufficiently accurate for cotton to take

$$V_r = \frac{106}{\sqrt{n}}$$

In the case of a cotton yarn of count C (tex) and fiber linear density H (millitex)

$$n = \frac{1000C}{H}$$

A similar calculation is not so easy in the case of a wool fiber, since the linear density of wool fibers is not usually specified in such terms.

Theoretically, a random distribution of fibers admits the possibility of infinitely thick and infinitely thin places in any of these strands. But textile machinery will not admit of either; the extremely thick will not pass through guides, the extremely thin will break. Therefore, it has been argued (3) that the unevenness of a yarn with a random fiber arrangement will be less than that given by equations 7.2a or 7.2b. This is, however, an academic rather than a practical consideration. What might be of more importance is the effect of autolevellers of modifying thick and thin places in slivers during manufacture.

This theory then states that, because of the inability of fiber processing machinery to do other than arrange fibers in a random order, fiber strands must have at least a minimum variation given by

$$V_r = \frac{100a}{\sqrt{n}} \tag{7.3}$$

where a is a factor characteristic of the unevenness of the fibers concerned.

A yarn with this limiting irregularity is sometimes called an "ideal yarn," because if all other causes of variation could be removed from the yarn-making process it is postulated that this is the most uniform yarn that could be made. This argument has been accepted because the development of the theory is logical and because in practice no yarns produced by conventional methods have ever been produced with a coefficient of variation less than that given by equation 7.3.

The Drafting Wave

The wave-like thickness variations in strands produced by drafting rollers have

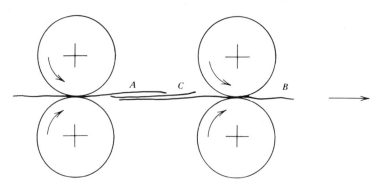

Figure 7.1. Drafting fibers.

been called drafting waves. The way in which they can be considered to arise in the simplest drafting device is shown in Figure 7.1.

In the drafting zone shown in Figure 7.1 there are three types of fibers; A, fibers held by the back rollers that are moving slowly; B, fibers held by the front rollers and being withdrawn relatively quickly; and C, fibers not held by either set of rollers but supported by other fibers. These are called floating fibers.

In this situation, if fiber C is not effectively held by A type fibers, there will be a tendency for B type fibers to drag it from the drafting zone. If this happens, there will be more fibers moving forward quickly, and the effect will become cumulative. Thus a group of floating fibers will be removed to form a thick place in the drafted strand; conversely, it will leave a deficiency of floating fibers in the drafting zone, and a thin place will follow. This thin place will continue until the feed rollers have brought forward a new supply of floating fibers into the drafting zone, at which point the action will recur.

This mechanism has been described by Foster (2) who considered that floating fibers moved at the surface speed of the back rollers until the point at which they were suddenly accelerated to the surface speed of the delivery rollers and were removed. This simple picture of fiber movement has been questioned by Taylor (4), who, by following radioactive tracer fibers, has shown that floating fibers may speed up and slow down more than once before being removed from the drafting zone. More recently Breny (5), Montfort (6), and Vroomen (7) have advanced the view that fibers do not draft as individual units but in bundles or clusters—"grappes." Nevertheless, the conception of a drafting wave given by the simple picture serves to explain in broad terms the irregularities in drafted strands as they are found in practice.

Modern instruments have enabled investigators to study the characteristics of these wave-like variations in linear density. Unlike most waves encountered in scientific and technological studies, as they succeed one another in the records of thickness variation they vary in both amplitude and wavelength. In this they are like many records found in meteorological data; for example, it is generally known that weather tends to vary in cycles of about 10 to 11 years and this is generally related to sunspot activity which varies in a similar way. No one, however, is able to measure accurately either the wavelength or intensity of such activity, or to predict the level or incidence of future activity, because the wavelength and amplitude both vary somewhat. This has been described as characteristic of a system that tends to oscillate with a regular wavelength and amplitude but is continually being upset by random disturbances. This is the picture of the drafting wave as suggested by Foster and Martindale (8). The mechanism of the supply and removal of floating fibers in the drafting zone pictured above provides an oscillating system, and one would think that with fixed parameters— draft, roller setting, and fiber length—it could be expected to oscillate with a steady periodicity and amplitude, but it does not. Rather, it is subject to disturbances perhaps caused by changing interfiber entanglements or the type of irregular fiber movement found by Taylor (4).

Although the wavelength and amplitude of the drafting wave both vary considerably, it has been shown that their mean values are affected by a number of factors.

Since the amplitude of the wave is a measure of the magnitude of the strand unevenness it is the most important consideration. It increases if the draft increases, if the roller setting is increased, or if the twist in a roving is increased and is most prominent when fibers are most entangled, that is, when a card sliver is drafted. Thus drafts are kept small on simple drafting systems, and roller settings are kept as close as possible consistent with proper drafting (if they are placed too close together, long fibers are gripped at both ends, the rollers slip, and crackers are formed in the yarn). Twist should be no more than is necessary to allow rovings to wind on and off roving packages without damage. The waves produced when card slivers are drafted are particularly prominent, but if groups of these are subject to repeated drafts, as happens in draw frames and gill boxes, the waves become less pronounced as the fibers become straighter and more parallel.

The thicker the strand being drafted, the greater the amplitude of the variation produced in absolute units, but expressed as a percentage of the average thickness, the smaller it will be, for example, if drawframe sliver is drawn, it will have a greater amplitude than that of a yarn drawn from a roving, but expressed as a percentage of the mean thickness of the product, it will be less. Thus as material passes through the various stages of a drawing set, the percentage variation of the products in process increases. This cannot be avoided.

The wavelength of the drafting wave is not as important as the amplitude. As far as the spinner is concerned he would say that it determines the distance that is noted between adjacent thick places in yarns. If draft or roller setting is increased, the average wavelength is increased, but these are always chosen to given minimum amplitude. Under such conditions, the average wavelength is rather more than twice the mean fiber length. Thus for a 1-in. staple cotton, the mean wavelength would be about 2-2½ in., and in a worsted yarn with a mean fiber length of, say, 2½ in., the mean wave length would be about 5-6 in.

If this movement of floating fibers in bunches can be restricted, the drafting wave should be suppressed, or at least decreased in amplitude. This should enable more uniform products to be made or—and this is the most attractive course—higher drafts can be used without incurring undue irregularity. This means that fewer drafting operations can be employed.

Such devices as gill boxes have always been used in worsted, flax, and jute spinning in which the motion of short fibers is controlled by beds of fine steel pins penetrating slivers in the drafting zone. Carriers and tumblers were used for the same purpose in drafting worsted rovings. Modern drafting systems employ single and double apron drafting systems and other specialized roller systems for the same purpose for both short and long staple fibers. In these systems, the fibers in the drafting zone are pressed between, or against, other surfaces in such a way that, whereas the long fibers can be withdrawn by the delivery rollers, the floating fibers tend to be restrained and prevented from moving forward prematurely. Although they are a great improvement, they never completely get rid of drafting waves.

In the simple case of two pairs of drafting rollers, the material in the drafting zone tapers in linear denisty toward the delivery rollers. Twist entering the zone with the material fed runs to the thinnest place, which is just behind these delivery rollers, and therefore tends to bind the floating fibers to those being removed by them. This accentuates the drafting wave. If the twist can be prevented from moving forward, it will tend instead to bind the floating fibers to the slowly moving fibers gripped by the feed rollers and so diminish the unwanted effect. It is probable that fiber control devices such as aprons when processing twisted rovings owe at least some of their effectiveness to the fact that they prevent the twist running forward.

Regular Periodic Variations in Linear Density

The type of variation now to be discussed can be quite disastrous when it arises, but it can be avoided because it is caused by defective machinery; it has a wave form of regular periodicity and, in most cases, a regular amplitude.

Regular periodic thickness variations in yarns are caused mainly by two factors:

1. oscillation of roller nip
2. variations in roller speed

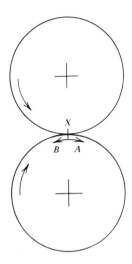

Figure 7.2. Oscillating Roller Nip.

Figure 7.2 shows a pair of delivery rollers with roller nip at X. In some circumstances a mechanical defect can cause the position of the nip to oscillate, as shown by the arrows, between the limiting positions A and B. As the nip moves from A and B it is moving back into the drafting zone, gathering a bunch of fiber ends, and a thick place emerges; when it reverses and moves back to B it is retreating from the fibers, and a thin place follows. This action usually happens once per revolution of the roller, giving a periodic variation whose wavelength is equal to that of the roller circumference; the amplitude depends on the amount of the nip oscillation, which is much exaggerated in Figure 7.2 for the purpose of illustration.

There are two causes of nip oscillation:

1. eccentric rollers
2. varying compressibility of top rollers

If the top roller is eccentric in its bearings, its line of contact with the bottom roller rocks backward and forward on the bottom one as it rotates. Eccentricity of the top roller does not cause any variation in surface speed, since it is driven by contact with the bottom roller through the strand being drafted. If, however, the bottom roller is eccentric, there is, in addition to nip oscillation, a variation in roller speed.

The top roller has a resilient surface so that in contact with the hard steel surface of the bottom roller it can effectively grip the fibers and draft them. If this roller surface varies in compressibility around its circumference, under the pressure applied, it will flatten out more when a soft spot is in contact with the bottom roller. This will broaden the nip each time the soft place comes round.

Different types of top rollers vary in their liability to produce such defects according to their mode of construction and the material from which they are made.

The causes of roller speed variations are:

1. eccentric bottom rollers
2. eccentrically mounted gear wheels
3. inaccurately cut gears and faulty gears
4. roller vibration

When a bottom roller is eccentric, its center of rotation does not coincide with its true center. If the distance between them is e, and if r is the radius of the roller, the effective radii at the nip varies between $(r + e)$ and $(r - e)$, that is, there is a periodic variation in radius of amplitude e and wavelength $2\pi r$. The surface speed is proportional to the radius, and this causes a variation in draft. If the eccentric roller is a front-delivery roller, this produces a periodic variation in linear density of wave length $2\pi r$ and percentage amplitude $100e/r$. If the eccentric roller is not a front roller, the wavelength is the circumference of the eccentric roller multiplied by the draft between it and the front roller.

An eccentrically mounted gear wheel produces a variation in speed similar to that caused by an eccentric bottom roller. The wavelength in the material is the length delivered during one revolution of the faulty gear, which may be any wheel in a train of gears. A gear may be eccentrically mounted because it is inaccurately bored or because the shaft to which it is attached is bent.

Speed variations can also be caused by slow variations in tooth pitch around the circumference of a gear, by meshing gears too deeply, by gears having thick or high teeth, and gears that cannot be meshed properly because they are loaded with too much lint and hard grease.

According to Thomason (9), no draft gears used in cotton processing should have an eccentricity of more than 5/1000 in. unless they have over 100 teeth.

The amplitude of a periodic variation caused by a faulty gear is the same all the way along the frame, whereas that caused by bottom roller eccentricity is likely to be different at different places along the roller, because such eccentricity is usually caused by bending or strain on the roller.

According to Foster (2), roller vibrations can occur in cotton speed frames and consist of jerky movements of back and middle bottom rollers giving rise to fairly rapid vibrations superimposed on the steady rotation of the rollers. It is caused by a stick-slip effect in a worn bearing at the driving end of the bottom back rollers.* These rollers extend the full length of the machine, and

*The middle roller is usually driven from the back roller.

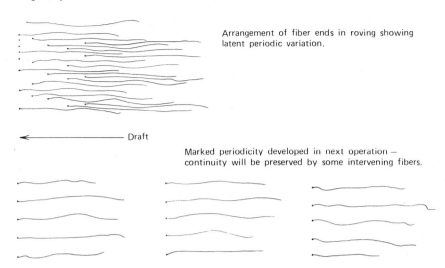

Arrangement of fiber ends in roving showing
latent periodic variation.

◄───────────────── Draft

Marked periodicity developed in next operation –
continuity will be preserved by some intervening fibers.

Figure 7.3. (*a*) Arrangement of fiber ends in roving showing latent periodic variation, (*b*) marked periodicity developed in next operation–continuity will be preserved by some intervening fibers.

since the top rollers are heavily weighted, considerable force is required to drive the bottom ones; this causes wear of the bearings. On applying the driving force instead of the roller rotating, the shaft bends, and the roller does not begin to turn until sufficient force has been built up to overcome the static friction in the worn bearing. The roller then accelerates, but since kinetic friction is less than static friction, the frictional force decreases and the roller slows down again. Instead of rotating steadily, the back and middle rollers move in jerks, and their motion is a series of rapid stick-slip vibrations of 20-30/sec superimposed on a steady rotation.

With a vibration of this frequency, groups of fibers are alternately held back and released by the middle rollers, and groups of fiber ends are spaced periodically in the roving every 1/20th sec or thereabouts. At usual front-roller speeds, the length of roving produced in this time is between 0.3 and 0.7 in., and successive groups of fibers overlap so closely that no appreciable irregularity is apparent in the roving as a whole. However, if this roving is drafted again at the spinning frame, the groups of fiber ends are taken in turn and are drawn out, and the yarn may contain a very prominent regular periodic variation in linear density (see Figure 7.3).[†]

[†]In practice the roving is reversed before spinning. Because the fibers are variable in length, the other ends of the fibers in the groups are more dispersed, and the amplitude of the variation is reduced. The same concealed periodic effect can arise in the spinning of flax in which the length of fiber can be many times the circumference of a faulty roller that arranges fiber ends in a periodic way.

As Foster has stated, if the frequency of vibration is f per second and the speed of the front rollers at the faulty roving frame is v inches per second, the wavelength of the latent periodicity in the roving is v/f inches and in the next product Dv/f if D is the draft at the following machine.

Roller vibration is likely to occur in any long slowly rotating roller; it is less likely at high speeds. To prevent it, worn bearings and worn shafts should be corrected, and friction should be reduced and kept at a minimum by keeping bearings clean and in good alignment. Long, thin, overhanging shafts should be avoided to avoid flexibility in driven members.

Sometimes roller vibration is set up for the reasons mentioned in the previous section—gear teeth set too deeply in mesh or gear teeth rubbing against an obstruction. In these cases, the wavelength is the length of roving delivered during one tooth movement of the gear.

It is probable that unwelcome periodicities of this kind are more prevalent in cotton spinning than in worsted spinning. This would be expected because rollers on worsted machinery are of larger diameter than on cotton machinery. Thus any eccentricities in rollers would be expected to be proportionally less significant. Wavelengths of periodicities will also be correspondingly greater.

The control of quality in yarns leads to the discovery of such periodic defects by the use of testing instruments and should lead to a search for their origin in the sequence of yarn-making machinery. As the above analysis shows, possible sources are many, and Thomason (9) provides an extremely useful and comprehensive discussion of each source in cotton pickers, cards, drawing, roving, and spinning frames.

Fortuitous Events Causing Unevenness

Irregularities under this heading can be of a wide variety of types. In the Zellweger, Uster publications, (10-12), these have been divided into two categories: (1) thick places, thin places, and neps that occur relatively frequently and are called "imperfections" and (2) a variety of faults that are relatively "seldom occurring" consisting of piecings, fly, slubs, crackers, long thick places, long thin places, knots, snarls, and loops.

Imperfections

Short, small, thick, or thin places may sometimes be caused by excessive amplitude of the drafting wave at the spinning machine. Their origin is in the drafting elements, most likely of the spinning machine. Too high a production rate in carding can also increase their frequency.

Neps are often caused in carding, and in the case of cotton, immature or dead fibers are often responsible. Combing, if used, should remove them.

It should be possible, therefore, to control the prevalence of these imperfections by attention to processing procedures. Their importance in particular cases is determined by the appearance of the fabrics made from the yarn. One must equate the level of care and control exercized in the processes with the standards required in the cloth. Such faults could be removed by yarn clearers, but their frequency prohibits this, and the resultant knots would usually be just as objectionable as the faults they replace. Remedies therefore lie in attention to the mechanical details of processing.

Seldom occurring faults

These are usually longer and of greater diameter than the imperfections. They are usually removed by yarn clearers and replaced by knots. However, the fewer of these faults there are, the fewer stoppages of winding machines occur; the fewer knots, the more efficient the process and the better the cloth.

Piecings are caused by end breakages in spinning and have the appearance of two ends of yarn corkscrewing around each other with, sometimes, a loose end spun in. The fewer ends down in spinning the fewer piecings.

Fly in the spinning and preparation rooms can collect on machinery and be blown or fall on creeled rovings or running yarns. The bunches of fibers are so short that they do not draft. Roller clearers and clearer aprons on roving frames build up collections of short fibers that may become detached and be spun in. Materials that ride on yarns are loosely attached by a few fibers that are bound in by twist; if bunches of fibers become incorporated at an early stage, they will form a slub—a thick place with little twist. Damaged pins in gill boxes can accumulate fibers and give rise to slubs. Fibers may collect on winding frames and be dislodged to form fly on yarns.

Crackers at spinning machines are caused by too close a setting on the drafting rollers, or too little pressure on the rollers. This causes the material to stop drafting momentarily. During this short time, throughput is held up, and a lot of twist accumulates at the front roller. These places look like short piecings with excessive twist.

Thick ends, spinners doubles, and long thin places are caused by (1) silver and roving piecings being made too thick or too thin and being drafted out in subsequent operations, (2) missing or extra doublings during doubling (particularly in worsted spinning), (3) at a drawframe producing more than one sliver through one pair of rollers, faulty subdivision produces one sliver too thick, the other too thin, (4) faults at a ring frame caused by double creeling or collection of an adjacent roving before drafting or the lashing in of yarn from an adjacent spindle.

Knots are caused when ends break during spinning, twisting, and winding and are obviously reduced in number the more efficient these processes are. They are also formed when a yarn package on a creel runs out and the next has to be tied in; therefore, the larger the creel package the fewer the knots. If delivery and supply packages can be adjusted in size to run out at the same time, some knots can be avoided, just as piecings can be in drawing if creeled packages of rovings are of such a size as to form an integral number of delivery packages.

Snarls usually occur in high-twist yarns if the tension is released; if the tension in such a yarn is released during winding a loop may be formed that will not pull out as the twist becomes entangled at the neck of the loop.

Shortage or the lack of training of supervisory and cleaning staff can increase the number of these faults; they can only be avoided by greater care and attention. They can be cleared from yarns using yarn clearers on winding machines but can only be replaced by knots.

The remedial action to be taken is determined by the type of fault and type of cloth. In worsted cloths, those which can easily be picked off in burling are best removed in that way. The long faults that have to be mended, which is a costly process, are better replaced by knots.

Unevenness in Condenser-Spun Yarn

Since the condenser system used in the woolen and cotton waste systems is quite different from the extensive roller drafting procedures used in cotton and worsted systems, it is pertinent to ask how the ideas discussed in the preceding sections apply in such circumstances. The same amount of research work has not been done in these areas, but a number of facts are well substantiated.

The same argument regarding the optimum distribution of fibers in a random manner is equally applicable to a woolen or cotton waste card as to worsted and cotton cards; therefore, one would expect no woolen yarn to have a coefficient of variation less than that given by the formulae in 7.1.4. All measurements are in accordance with this expectation.

Since there is no roller drafting, one cannot expect to find drafting waves. Spinning uses a type of stretch drafting, traditionally it is thought that mule spinning reduces some of the variations in linear density present in the roving, and this has been confirmed for small drafts by Angus and Martindale (13), but a contrary opinion is expressed by workers at WIRA (14). If there is any decrease in irregularity in spinning it is, however, not a very large one. Perhaps it depends on the irregularity of the roving itself; if it is fairly uniform to start with, it may be less affected than if it contains substantial irregularities. The extent of improvement might be less with ring-spun than with mule-spun yarn, since the free length that is stretched is much less in the former case.

Regular periodic variations are found in woolen yarns (14) because of the following:

1. the joints between successive layers or bands of intermediate feed slivers fed to the finisher card
2. irregular speed of a feed-lattice at the finisher card
3. tape doffer with faulty card clothing

4. a finisher doffer operating in a worn bearing that causes the doffer-cylinder setting to vary

It is also known that if workers, instead of being driven smoothly, rotate in a spasmodic manner, they deposit an intermittent feed of fibers on the cylinder via the stripper, which produces regular periodic variations in linear density in the output. Faulty card clothing on finisher doffers has produced a similar effect per revolution, and this may be a local fault that only affects one or two neighbouring rovings at the condenser.

Unlike the periodic defects produced in roller drafting that may have a wavelength of a few inches corresponding to the circumferences of relatively small diameter rollers, the defects in woolen yarns are of the order of a few to many yards in length, corresponding to the large roller sizes. It is also to be noted that faults in rollers are more important the nearer these are to the condenser end of the card; unevenness because of faulty parts at an earlier stage of processing are considerably reduced by the subsequent smoothing effects of workers, doffers, and intermediate feeds.

The fortuitous faults in cotton and worsted yarns are by their very nature likely to be absent from woolen yarns or much less frequent in their incidence. Faults caused by incomplete or inefficient opening and cleaning at the card are, however, more likely to be present, since if they are present in the carded roving they have little chance of being removed during the single spinning operation.

The types of unevenness that have been discussed so far are concerned with variation along strands. In the case of condenser spinning, there is, however, a type of variation that is very important whose characteristics are quite different from anything encountered in the processes employing roller drafting. This is a variation in average linear density between the many ends of roving produced by the condenser (15).

The condensers of woolen and cotton waste cards usually produce simultaneously upwards of 100 rovings, depending on the width of the machine and the width of the tape. Ideally, each roving would have exactly the same linear density, but this is never the case Figure 7.4 is a graph showing a typical example of the percentage differences in count (linear denisty) from end to end *across* the face of the condenser of a woolen card. It shows deviations from the mean value, and it will be seen that some ends are as much as 7% heavier than the mean, and some 6% lighter. Something can sometimes be done to reduce these differences, but in woolen carding, a range of ± 5% can be regarded as good performance.

To obtain such graphs, a length of at least 50 yd of roving should be collected simultaneously from each end and weighed. If this is done a number of times, it will be found that, although the mean count may vary somewhat as carding proceeds, the pattern of variation expressed in graphical form as in Figure 7.4 is

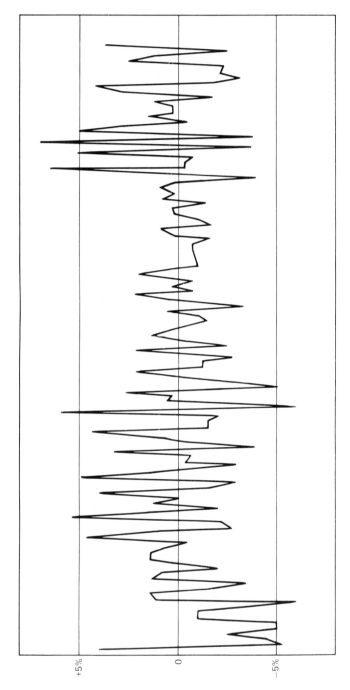

Figure 7.4. Variation of count of ends across condenser of a woolen card.

234

persistent. This observation, together with the fact that the pattern of variation on another machine is different, indicates that the reasons for these count differences reside in the particular machine.

It has been found that the pattern of variation across a woolen card is unaffected by such things as change of production rates, change of roving count, change of type of feed rollers to finisher card, fettling, that is, cleaning the finisher card, varying the amount of oil applied to the blend. Differences arise over a period of time in a mill, and changing the condenser tapes produces a completely new pattern. This suggests that the major cause of the pattern is the way in which the web is divided by the tapes. The differences arising over a period of time are caused by changes in the tapes brought about by use; for example, some may be stretched more than others because of the accumulation of fibers around tensioning rollers.

This theory is substantiated by the fact that, if the tension of one bank of tapes is increased by moving the tape tensioning roller, the rovings produced by that bank, which are directed to one condenser bobbin, become heavier, and those on other bobbins become correspondingly lighter. Some condensers in use can be maladjusted in this way, and as a result there are unnecessarily large differences in count variation across the condenser. This can be checked by filling all bobbins simultaneously with the same length of roving and weighing them before spinning (making due allowance for bobbin weights). Differences in weight indicate differences in average linear density of roving per bobbin, which can be reduced by changing the tensions of banks of tapes by degrees until such differences are eventually eliminated.

This in turn suggests that the remaining differences are caused by tension differences between adjacent tapes. This has been verified by measuring the tensions of the tapes in a series tape condenser and comparing these with roving count variations; there is a very close correspondence—tight tapes produce heavy rovings. This suggests that, when the web enters the dividing rollers, tight tapes bind the fibers more securely to this surface of the rollers than do the slack ones. Thus a fiber that is gripped by adjacent tapes of different tensions will tend to be taken by the tight one and robbed from the slacker tapes in a consistent way. If this is a correct explanation, one would expect that, if the length of the fibers being carded were substantially increased, they would then extend over more tapes. When the web is divided this would complicate and alter the manner of the distribution of fibers; with longer fibers the effect of tight tapes might spread to further ends, and robbing would be more extensive. Changing the blend does affect the pattern of variation, and the range of count variation between ends does increase as the fiber length increases.

Thus in most cases the greater part of this type of variation is generated by the condenser tapes. If, however, the material on the card is not uniformly spread across the width of the machine because of intermediate feeds not being properly set, or if air currents blow the material about, one may find certain sections producing heavy or light ends. Sometimes side-ends are very light. Keeping a sharp point on the card clothing and the avoidance of strong draughts across the face of the card that blow material inwards or outwards minimize this effect.

THE MEASUREMENT OF VARIATION IN LINEAR DENSITY

Methods for Measuring Unevenness

Before the advent of instruments for measuring the unevenness of strands the only method available was, in the case of tops, slivers, and rovings, to cut them up into unit lengths and weigh these. In the case of yarns, these could be wound on black boards, and this presentation of evenly spaced threads against a contrasting background was studied.

The cutting and weighing method is extremely slow and tedious; it was used in early research work, but is now used as a means of calibrating unevenness-measuring instruments that measure the linear density of a strand by indirect means.

The examination of yarn wound on black boards is an entirely subjective method and does not provide an absolute measure of unevenness nor does it lend itself easily to the need to keep records. It was mostly used by cotton spinners who would make comparisons between yarns in this way or compare a yarn with a standard. It is still used in this way to grade cotton yarns for appearance (16); yarn specimens wound on black cards are compared with photographs of standard grades. As the specification for this standard method explains, judgment of the yarn appearance takes into account unevenness, fuzziness, neppiness, and visible foreign matter. If such a method is used to judge evenness, it is doubtful whether an observer can altogether divorce evenness from these other aspects of irregularity. A similar comparison can be made by examining knitted webs made from the yarns under investigation and has been used in the assessment of worsted yarns (17). In methods of this kind, many precautions must be taken: sufficient yarn must be taken from a number of cops to form a representative sample, they must be of the same count; and the specimens must be prepared and mounted under identical conditions. To reduce the effect of personal errors resulting from the subjective nature of the test, a number of observers should be used and statistical methods of rank correlation adopted to assess the results. Having done all this rather tedious and time-consuming work, the result is not directly concerned with yarn unevenness as defined above, but with irregularity of apparent yarn diameter or thickness.

Many indirect instrumental methods have been used to measure unevenness. Photoelectric cells have been used to measure thickness variations in yarns that have been traversed across a slit passing a beam of light to the cell (18); the problem of the importance of fibers protruding from the body of the yarn has been a complicating factor. More recently (19) another instrument uses the resonant frequency of a length of yarn as a means of establishing its linear density over a short length of yarn. The relation between linear density, m, length, l, tension, T, and resonant frequency, ν, is given by

$$\nu = \frac{1}{2l} \sqrt{\frac{T}{m}}$$

In the instrument described, a 5-cm length of yarn is used and under a fixed tension is caused to resonate. As the yarn passes through the resonance, oscillation is controlled automatically and recorded continuously.

Although many of these methods are interesting, they have little to commend them in favor of instruments that have already been commercially adopted and widely accepted in industrial and research laboratories. These are of two types: the compression type of unevenness tester that is mechanical in operation and the electronic capacitance type. Both operate continuously and relatively quickly and at any instant give a reading that is directly proportional to the linear density of the strand element that is within the instrument's sensing unit at that instant.

Compression-Type Tester

The best known of these are the Saco-Lowell Roving and Sliver Tester, the Pacific Evenness Tester and the WIRA (Wool Industries Research Association) Levelness Tester. They all operate in the same way and consist of a pair of cylinders or wheels as in Figure 7.5. The lower wheel W_1 has a groove of rectangular section cut in its periphery and rotates in fixed bearings; the upper wheel W_2 has a tongue of rectangular section around its circumference, and this fits almost exactly into the groove in the lower wheel. A sliver or roving is placed in the groove of W_1 into which it is compressed by the tongue of W_2 carried on a pivoted beam and adequately loaded. As W_1 is rotated, the sliver or roving is drawn through and as it varies in linear density the upper wheel rides up and down on it. The beam carrying W_2 therefore oxcillates, and these oscillations are magnified by mechanical linkages that drive a pen on a moving paper tape, thus recording the variations.

The accuracy of this instrument in recording unevenness depends on two factors. (1) the sliver or roving must be thick enough to fill the width of the groove in W_1 and (2) this having been ensured, the applied load P must be high enough to quickly compress the material to a steady value of x on entering the groove. Wheels with wide grooves are used for card slivers or worsted tops, since finer rovings would not fill such grooves, a range of wheels with narrower

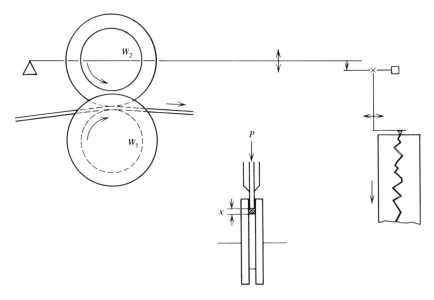

Figure 7.5. Principle of compression-type unevenness tester.

grooves are provided for progressively lighter rovings. Three or four sets of wheels cover the whole practical range of thicknesses.

Such instruments are calibrated by measuring slivers and rovings at various points and correlating the readings with the weights of short lengths cut from the same points. If the load P is big enough, correlation coefficients are quite high (e.g., 0.93), and readings are directly proportional to weight per unit length. These are therefore reliable instruments for measuring unevenness of slivers and rovings; the small amount of twist in rovings is not great enough to affect the compression of the roving in the groove so that it always fills it.

Such an instrument is, however, no good for yarn for two reasons. First, grooves for yarn would need to be so fine that it is doubtful whether they and the associated tongues could be made with sufficient accuracy; high magnifications would need to be used, since yarns are so fine. Equally important, yarn twist varies considerably and whereas the thick, soft places could well be compressed into the rectangular shape of the groove, the thinner places, being much more tightly twisted, would not be squeezed into a rectangular shape. A linear relation between weight per unit length and instrument reading would no longer obtain, and the record would no longer be a true record of unevenness.

Capacitance levelness tester have many advantages over these mechanical compression type testers, but the latter are accurate and can be very useful. There is another reason why they merit inclusion here—they are the basis for many of the modern autoleveling systems otherwise called servodrafters, in which the linear density of slivers leaving a carding machine, or entering a draw-frame or gill box, is monitored by such a pair of wheels and the information regarding unevenness, instead of being recorded, is fed backward or forward to drafting mechanisms in which drafts are adjusted to eliminate or reduce the variations in linear density.

Capacitance-Type Unevenness Testers

There are a number of these testers, of which the Uster Evenness Tester is the most widely used. A strand of textile material is passed at a constant speed between the two parallel plates of a capacitor. As the strand varies in linear density and displaces the air normally present, this produces a variation in capacity that is proportional to the mass of textile material between the plates. This capacitor is included in circuits that transform the capacity variations into variations in an electric current. This varying current can be used to operate a pen recorder that plots a record of the unevenness on a paper chart. It can also be fed into computers that calculate and display statistics characterizing the unevenness.

The basic requirement of this type of unevenness tester is that the output of the measuring circuit be directly proportional to the linear density of that part of the strand within its capacitor, that is, the relationship between capacitance and the mass of fiber between the plates must be linear. This is so if the ratio of fiber to air in the capacitor (the filling factor) is not too high—less than about 0.3.

Under these conditions the capacitance is proportional to the relative permitivity* of the dielectric between the capacitor plates, and this depends on the relative permitivity of the fibers and that of any moisture they contain. Different fibers have different values of relative permitivity, but these differences are not as great as the differences between values for fiber and water. The effect of these differences is decreased by increasing the frequency of the applied voltage, and for this reason the Uster Unevenness Tester uses a frequency of nearly 30 Mc/s.

It is nevertheless advisable that the strand to be tested should have uniform moisture regain throughout its length, and it is therefore necessary that the atmospheric conditions in which the strand is stored should be stable for some time immediately before testing—tightly wound packages should be conditioned in this way for 48 hr and strands in loose form for 24 hr (20). Providing that conditions are *stable* in this way, it is not necessary to precondition strands in the testing room, since this can waste unnecessary time, but the specimens must be tested *immediately* on being brought there; otherwise moisture changes may begin to occur and uniformity of moisture distribution be upset.

Fiber blending should also be reasonably uniform to avoid spurious errors.

*Relative permitivity, known previously as dielectric constant or specific inductive capacity.

Uniformity of fiber distribution in a yarn has been investigated with the capacitance evenness tester. If a bicomponent yarn is tested and one component can be removed by a solvent, then if the blend is not uniform the unevenness will be greater when this component is removed.

All capacitance-type evenness testers produce a chart displaying unevenness, and instrument sensitivity can be adjusted to change the amplitude of the irregularities on the chart. A number of speeds at which strands can be processed are usually available and with similar provision for varying the speed of the paper chart, the length of yarn recorded on an inch of paper can be varied over wide limits depending on whether one is particularly interested in very short- or very long-term variations.

The Uster Evenness Tester has the most developed ancilliary recording equipment, consisting of chart recorder, integrator, imperfection counter, and spectrograph. The integrator computes percentage mean deviation, U%, or coefficient of variation, V%, of linear density. The integrator has a timing device incorporated that allows the operator to record the unevenness over any period of time up to 5 min.

The imperfection indicator simultaneously counts the thin places, thick places, and neps in a yarn and is connected to the evenness tester through which the yarn is passing. Thin places in a time determined by a preset time switch are recorded on a counter, and a sensitivity switch can be set so that all thin places either 30%, 40%, 50%, *or* 60% *below* the average value of linear density can be counted. A thick place is counted if a control limit is overstepped; this limit is adjustable at 35%, 50%, 70%, or 100% *above* the average value. Neps may be counted on a basis of a nep 1 mm long having an average cross section four times that of the yarn in the least sensitive setting; other sensitivities are based on neps with cross sections 2.8 X, 2 X, and 1.4 X the mean cross section. The imperfection indicator has a built-in average value corrector; in the case of count variation, the relative sensitivity remains constant.

As discussed in pages 247-251, the unevenness records can be considered to contain a continuous spectrum of wavelengths. The Uster Spectrograph is a 35 channel analyzer that is connected to the Evenness Tester and analyzes its output in such a way that all wavelengths from 0.5 in. to 20 yd are determined and registered on a graph known as a "Spectrogram." On this graph, the wavelengths are plotted as abscissae on a logarithmic scale and the amplitudes as ordinates. The length of time for which data is accumulated and computed can be anything up to 10 min, determined by a preset time switch. When the test is completed, the operator presses a button and the spectrogram is registered automatically on a paper chart.

A later development has been to employ the capacitor-type sensing unit to count the fortuitous type of yarn unevenness discussed in pages 230-236. There

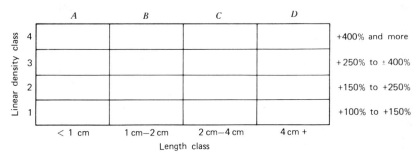

Figure 7.6. Classimat grades.

have been a number of systems of this kind, but the one most widely developed and used is the Uster "Classimat" system (21).

For the purposes of this system, a series of "Classimat" grades was arbitrarily established, as shown in Figure 7.6. The yarn "faults" are divided into 16 grades based on length and linear density; as such a fault passes through the sensing capacitor, its length and linear density are monitored and classified according to this grading system.

A "Classimat" installation consists of six sensing units that are applied to six yarns being wound on a winding machine. These units are each connected to the same classifying instrument; thus signals from all yarns are fed into the same counters. This instrument has a display of 16 counters arranged as in Figure 7.7. They are so connected that A1 counts all "faults" shorter than 1 cm whose linear density is more than double the average linear density of the yarn; A2 also counts all shorter than 1 cm but only those greater than the mean linear density by more than 150%, A3 counts all shorter than 1 cm but greater than mean linear density by more than 250%, etc. Similarly *all* class B faults (1-2 cm in length) more than double mean linear density are recorded at B1, those of class B greater than +150% at B2 etc.; that is, within each length group the classifying instrument provides cumulative frequency values. To obtain actual numbers in groups 1, 2, 3, and 4, the number in 2 has to be subtracted from 1 and the number in 3 subtracted from 2, etc.

With such an installation on a winding machine, large quantities of yarn can be examined in a nonwasteful way since the wound yarn need not be withdrawn from production. Quality control standards can be established. Following this, yarn-clearing devices working with similar sensing units can be set to pass faults in certain grades and clear all those of greater length or thickness. The grades to be rejected can be established objectively on the basis of customer requirements and economy of operation.

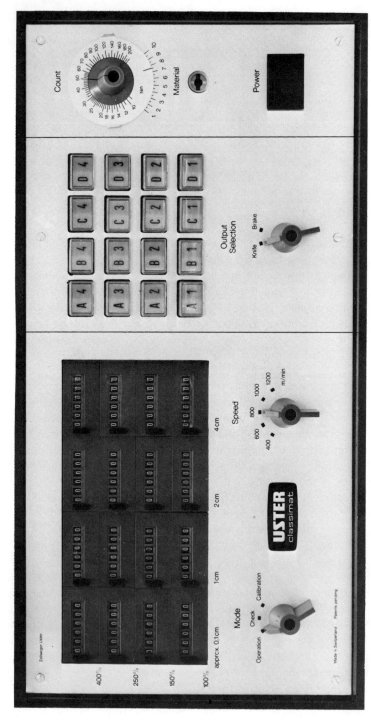

Figure 7.7. Uster Classimat Classifying Instrument (Courtesy Zellweger Ltd., Uster).

242

THE ANALYSIS AND INTERPRETATION OF UNEVENNESS DATA

Index of Irregularity of a Yarn

Equation 3 on page 223 postulates a minimum coefficient of variation, V_r, for a yarn, the unevenness of the ideal yarn in which fibers are assembled in random order. If unevenness caused by drafting waves is always present, the measured coefficient of variation of an actual yarn, V_m, will be greater than V_r. The ratio $\frac{V_m}{V_r} = I$ is called the "index of irregularity" and should therefore always be greater than unity.*

Whereas V_m is a measure of the unevenness of a yarn, I is a measure of the degree to which a yarn approximates to the ideal yarn and is therefore a measure of machinery performance; the less fiber control there is in drafting the greater I will be, and conversely, the greater this control, the nearer I will approach to unity.

Zellweger Ltd. (Uster) at one time gave the data shown in Table 7.1 as typical of yarns made by different systems. This data shows (1) that cotton yarns contain a greater nonrandom component of variation than worsted yarns, that is, drafting waves are more important in cotton yarns and (2) that this is more noticable the lower the quality of cotton.

Table 7.1. Typical Values of Index of Irregularity for Yarns

Yarn		Even	Average	Uneven
Cotton Carded	60 tex	2.6	3.4	4.5
	30 tex	2.2	2.8	3.4
	20 tex	2.0	2.5	3.0
Cotton Combed	20 tex	1.6	2.0	2.5
	10 tex	1.5	1.7	2.0
	5 tex	1.2	1.4	1.6
Worsted		1.2	1.35	1.5
Woolen	200 tex	1.7	2.1	2.8
	100 tex	1.5	1.9	2.5
	50 tex	1.4	1.7	2.2

These data have since been replaced by a series of graphs showing "experience values for yarn irregularity." A typical member of this series for combed cotton

---

*If the distribution of unevenness is normal, the percentage mean deviation, U, measured by some instruments, is proportional to the coefficient of variation, V, and I can be represented as $\frac{Um}{Ur}$.

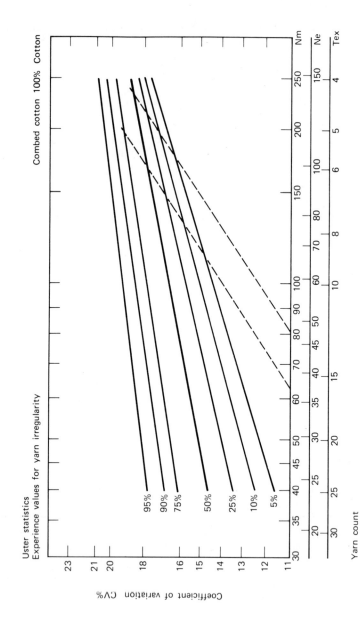

Figure 7.8. Experience values for unevenness values of combed cotton yarns (Courtesy Zellweger, Uster).

yarns is shown in Figure 7.8 and shows CV% plotted against count; the line labelled 50% shows an average value, the value of CV% for a full range of counts below which values for 50% of manufactured combed cotton yarns fall, and similarly for the other percentage values indicated. The dotted lines indicate the irregularity of ideal yarns made from cotton. A spinner can therefore compare the unevenness of his yarns against the "experience values" or against the values for an ideal yarn; the former comparison is a measure of the quality of his yarn as compared with others, the latter of the performance of his machinery. Similar graphs are published as Uster Statistics for other yarns such as carded cotton, polyester/cotton, and polyester/wool. In the case of all-wool worsted yarns, the Uster graphs have been superseded by a series of ten graphs published by the International Wool Secretariat on behalf of the International Wool Textile Organisation. Figure 7.9 is the first of this series and shows expected levelness values (CV% or U%) for average yarns made from wool fibers of nine different mean diameters into a range of counts; these values were obtained from a survey of worsted yarns produced in the principal yarn-

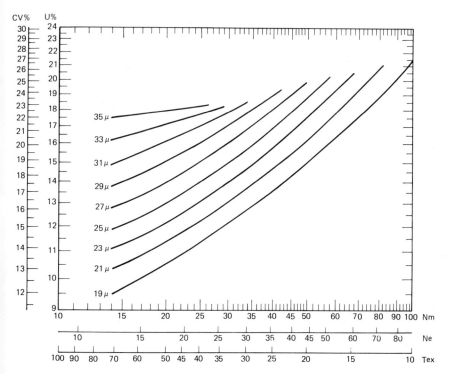

Figure 7.9. I.W.T.O. experimental unevenness values for average wool worsted yarns (Courtesy International Wool Secretariat).

producing countries of the world (35). For each of the fiber qualities shown in Figure 7.9 there is given another chart showing a series of seven curves, and a curve for the ideal yarn, similar to those of Figure 7.8, thus any yarn whose levelness has been measured can be related to the world-wide performance of comparable material, and it can be determined where it lies between the best 5% of production (below the 5% line) down to the worst 5% (above the 95% line).

The incidence of regular periodic irregularities in a yarn would increase I in proportion to the seriousness of the defect, but the fortuitous type of fault mentioned in pages 230-236 would not, since the measured coefficient of variation V_m would not be significantly affected by the incidence of thick or thin places occurring infrequently; a length of yarn might well be tested without encountering one.

K Values of Slivers and Rovings-the Theory of "Grappes"

The use of an index similar to the index of irregularity for slivers and rovings was first employed by Huberty; it was indicated by K. Huberty's K differed from I in that it did not allow for variation in linear density of the fibers making up the sliver.

Huberty found that values of K were high for slivers and decreased as drawing proceeded; when the similar index I is calculated, the same is true. Table 7.2 shows values of I calculated from data given by Foster (2) relating to the drawing of Texas cotton, and Table 7.3 gives analogous figures for wool processed on the worsted system (22).

Table 7.2. Values of Index of Irregularity for Texas cotton strands (Foster)

Texas Cotton	Count	n	V_m	V_r	I
Third drawframe sliver	0.12	25,200	4.8	0.63	7.6
Slubbing	0.6	5,050	7.9	1.48	5.3
Intermediate	1.5	2,000	9.0	2.37	3.8
Roving	4.0	758	13.9	3.85	3.6
Yarn	30.0	101	22.7	10.53	2.2

Table 7.3. Values of Index of Irregularity for worsted strands (Wool Sci. Rev.)

Index of Irregularity	I
Sliver from rectilinear comb	18
Finished top	8
Sliver (final stage)	1.60
Yarn	1.20

These figures seem to show that drawframe sliver or top is far from the random fiber arrangement that has been postulated, but that drafting and doubling reduce the discrepancy.

Vroomen and Montfort (6) have described experiments in which a number of slivers were drafted together (doublings = E), the draft employed (D) being equal to the number of doublings ($D = E$). The sliver emerging was therefore of the same linear density as any one of those entering, and the unevenness of each was measured; there was a decrease of unevenness on drafting. This process of drafting and doubling was then repeated, and it was found, in the case of both cotton and wool, that after three or four operations V did not continue to decrease, but became stabilized, as on a plateau (palier), as did K (or I). They state that if fibers draft as individuals, it might be expected that the arrangement of fibers would become more nearly random as drafting and doubling proceed and K (and I) would continue to decrease. Vroomen and Montfort have therefore proposed that fibers do not draft as individuals, but in clusters—*grappes*—and that it is these grappes that are arranged at random in slivers. They found on repeating their experiments with a lighter sliver that the same stabilization of K occurred, but at a lower value than before, and explained this by suggesting that the grappes now contained a smaller number of fibers. The yarn, by this theory, is regarded as the limiting case of the process in which the grappe is approaching single fiber size, and the theory is approaching the theory of the ideal yarn based on the random arrangement of single fibers.

In favor of this view of the drafting behaviour of fibers, it is pointed out that in the case of a wool top there are 1000 leading fiber ends per millimeter, and with a doubling of eight tops there will be 8000 fiber ends per millimeter entering the first gill box; the authors regard it as unlikely that the drafting rollers can handle each as an independent unit. At a later stage of drawing, at the finisher, it is stated there will be only 300 leading ends per millimeter of feed.

This theory would seem to explain many experimental facts; as yet it has not altogether been completely assimilated. A question that might arise is whether the grappes that have been visualized as being withdrawn by the drafting rollers might be identified with the thick places of what has previously been referred to as the drafting wave.

Short-Term and Long-Term Variations

As explained on page 226, each operation of roller drafting introduces variations in linear density with an average wavelength of about 2½ times the mean fiber length. In the case of cotton this might therefore be about 2½ in., and if the draft at the next machine were 6 these waves would be drawn out to about 15 in. At the same time, this machine would introduce drafting waves of its own, also about 2½ in. long, and further stages of drawing would draw both sets of waves to longer lengths and introduce new sets of their own. Moreover, those introduced in strands at early stages would be modified by being overlayed by others in strands with which they were doubled. A single yarn is therefore a most complex spectrum of wave-like variations of different lengths and amplitudes arising from each drafting operation, together with variations caused by random fiber arrangements and, perhaps, mechanical defects. The variations in any strand range in length from a few inches, corresponding to those introduced at the last machine, to differences in linear denisty of long lengths corresponding to waves introduced at the very early stages of drafting, or to unevenness in card sliver originating at an even earlier stage of processing. With the introduction of autoleveler devices in early stages of processing, the latter should be of diminishing importance; it has been pointed out, however, that these mechanisms have little effect on the internal evenness of lengths of sliver of the order of 2 meters or less; therefore, they cannot be expected to have a beneficial effect on lengths of yarn less than 100 meters (22).

One can then picture a yarn, or any other strand, as being composed of variations of all lengths: short-term variations of a few inches in length, and long-term count variations that are manifest if long lengths of yarn, say 120 yd, are reeled and weighed. There is no definition of what lengths qualify for the description short- or long-term, and the expression medium-term is sometimes used to cover intermediate lengths. A.S.T.M. standard D1425-67 suggests that lengths less than 5 cm are usually the basis for short-term unevenness; although this might be appropriate for short staples spun on the cotton system, it would hardly seem appropriate for longer fibers, for example, short-term variation produced in worsted spinning with a 4-in. fiber would have a wavelength of about 8-10 in. It has been suggested (23) that in the case of woolen yarns, short-term should refer to wavelengths up to 6 or 8 ft, medium-term up to 100 yd, and long-term up to the length of the roving on a condenser bobbin. These terms are likely to remain qualitative rather than to have specifically defined lengths allocated to them; what is important to the student is that he should recognize the wide and continuous spectrum of wavelengths in yarn unevenness.

Variance Length curves

If one divides a strand into lengths of l yd there will be unevenness "within" these lengths, and if the lengths themselves are weighed they will be found to differ in linear density amongst themselves, that is, there is unevenness "between" the lengths. The sum of these two will be the total unevenness of the strand. If l is increased (say it is doubled), variations " within" increases, because opportunity is provided for longer wave variations to come within the longer length and add to the variations already there, and variation "between" decreases because the linear density of the doubled length averages out the differences between the two original lengths, l.

If $CV(l)$ is the coefficient of variation *within* a length l of a strand, and $CB(l)$ is the coefficient of variation *between* lengths l, $(CV(l))^2$, written as $V(l)$, is

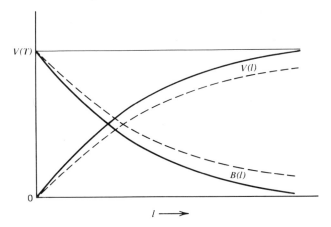

Figure 7.10. Variance–length curves.

called the normalized variance within lengths l, and $(CB(l))^2$, written as $B(l)$, is called the normalized variance between lengths l. The laws of statistics state that the normalized variance of the whole length of material $V(T)$ is given by

$$V(T) = V(l) + B(l)$$

that is, one does not add the coefficient of variations but their squares to get the total variance of the strand.

These two cases are shown in Figure 7.10 by the full lines. The curves of $V(l)$ and $B(l)$ plotted against l are known as variance-length curves for the strand, and the total variance $V(T)$ is shown as a straight line. Since $V(l) + B(l) = V(T)$, one is a mirror image of the other.

If strands differ in the proportions of short-term and long-term irregularity they contain, the variance length curves will reflect this. In Figure 7.10 we have already considered the case of one yarn; if a second strand with the same total variance contained a greater proportion of long-term variation, this would mean that $B(l)$ would be greater for high values of l and would have a $B(l)$ curve shown by the broken line.

[In this case, since by hypothesis $V(T)$ is the same in both cases, the second strand must contain less short-term variation $V(l)$. This is shown by constructing the $V(l)$ curve, shown by the broken line, as a mirror image of the broken $B(l)$ curve.].

It will be apparent therefore that the unevenness of a yarn cannot be adequately expressed by a single value of coefficient of variation of linear density relating to unevenness within an arbitrarily chosen length, which is usually what is provided by an instrument; *the complete picture of yarn unevenness depends on the relative importance of variations of a wide variety of lengths.* It is this that determines the appearance of cloth made from the yarn, and comparisons between yarns are only valid on the basis of such complete information.

Unfortunately, to obtain all the information required, even with the use of electronic instruments, is a very laborious task, and variance-length curves have never been widely applied as an industrial tool for quality control purposes. Methods have recently been described (24) whereby instrumental unevenness records have been recorded directly on punched tapes. Processing these in a suitably programmed computer can produce variance-length curves very rapidly. Recording of irregularity data on magnetic tape would enable the fastest computers to be used, and there seems no reason why the labor of preparing variance-length curves should be any problem in the future if such recording facilities are used.

Long-Term Unevenness

Long-term unevenness may be checked by reeling a number of lengths of yarn from each spinning frame and observing the count variation between the individuals in the sample taken. Each mill may have its own standards based on its own sampling procedures and experience, but little information regarding such values has been published. Excessive variation in count is undesirable, but it is not possible to say what is good and what is bad without consideration of many factors; for example, one would expect more count variation in woolen yarns than in worsted yarns for the reason given in Section 7.18.

G. M. Bornet, (25, 26) has done world-wide surveys of count variation in single yarns covering over 200 mills and over 1000 yarns of all kinds. His unit of length for count determination was about 100 m, and in his largest survey 48 bobbins were tested for each yarn (4 from each of 12 different spinning frames). As a result, he has suggested the values shown in Table 7.4 as a table of ratings for overall count variation for yarns produced by different systems.

Table 7.4. Count Variation in Yarns—Coefficient of Variation (Bornet)

	Cotton Systems	Worsted System	Condenser Yarns Including Woolen	Filament
Very regular	Below 2.0	Below 1.8	Below 2.5	Below 0.5
Regular	2.0 -2.75	1.8-2.4	2.5-4.0	0.5-1.0
Medium	2.75-3.5	2.4-3.0	4.0-5.5	1.0-1.5
Irregular	3.5 -4.25	3.0-3.6	5.5-7.0	1.5-2.0
Very irregular	Above 4.25	Above 3.6	Above 7.0	Above 2.0

Bornet's findings also include the following.

No relation was observed between count variation and either yarn count or fiber (except in the case of jute).

The spread of count variation within a spinning system is strikingly greater than the difference in general level between the systems (this is shown in Table 7.4).

When count variation is high, the major part is not caused by short-term unevenness at earlier processes of spinning (an unavoidable source of variation) but by deficiencies in machinery, in supervision, or in quality control methods, all of which can be remedied. Deficiencies in quality control allow variations to creep in because of count differences arising between shifts, days, frames; differences in draft between deliveries, machines, or times of spinning; periodic faults; malfunctioning of autoleveling devices.

The Measurement of the Wavelength of Periodic Variations

The variance-length curves in Figure 7.10 show the kind of curves obtained from a typical yarn. This curve is made up of components caused by the different elements that cause unevenness.

Breny (27) and Olerup (28) have calculated the shape of the $B(l)$ curve for an ideal strand in which fibers are arranged at random. This is shown in Figure 7.11a for fibers of mean length 40 mm; if the effect of drafting waves are added, the line will lie somewhat above this, depending on the amplitude of the waves. Figure 7.11b shows a $B(l)$ curve for a regular sinusoidal variation of wavelength 20 cm, and Figure 7.11c shows it added to the random effect. It will be seen that it gives rise to ripples on the curve that are not very prominent. This was confirmed in a practical case by Dyson and Schofield (26). If more than one regular periodicity were present in a yarn, it would be very difficult to resolve them and measure their wavelengths.

The variance-length curve is, therefore, not a very suitable method for measuring wavelengths of periodic variations in strands.

Autocorrelation

Another method of analyzing unevenness data is the method of autocorrelation, which produces diagrams known as correlation periodograms or correlograms. In this method, the linear density at points along the unevenness record are correlated with the linear density of other points distance x from them. If $x = 0$, the correlation is between like points and correlation coefficient $r = +1:0$. As x increases, the correlation decreases, and the course of the correlogram depends on the type of variation present.

Figure 7.12a shows the correlogram for a strand in which fibers of mean length 40 mm are arranged at random. For values of x less than the fiber length, there is some correlation, since some fibers are common to strand sections at both

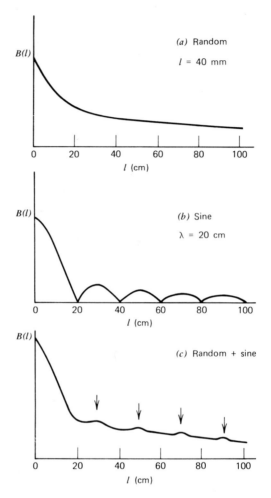

Figure 7.11. Variance length curves (courtesy Zellweger Ltd., Uster).

points, but beyond the length of the longest fiber, the linear density of sections is quite random and $r = 0$. The correlogram for a regular sinusoidal variation, Figure 7.12b, varies from $r = +1.0$ at $x = 0$, λ, 2λ, etc., to $r = -1.0$ at $x = \lambda/2$, $3\lambda/2$, $5\lambda/2$, etc. Figure 7.12c shows a correlogram for a combination of random and sinusoidal variations. The effect of random variation plus drafting waves is similar, but the peaks on the correlogram do not occur at regular intervals, are sometimes irregular in shape, and tend to be damped out. They are difficult to interpret (8).

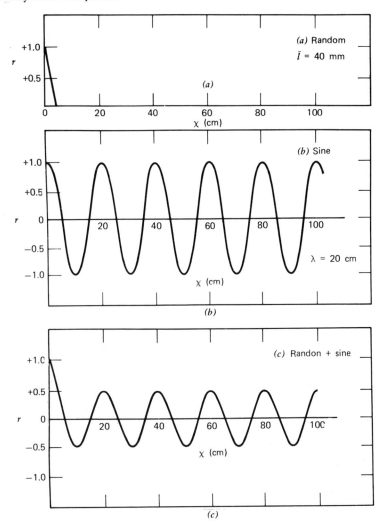

Figure 7.12. Correlograms (courtesy Zellweger Ltd., Uster).

This method has not been used much because no commercial instruments are available to perform the analysis; the computation is extensive and time consuming; with the use of magnetic data recording and the use of a digital computer this can now be overcome.

The Spectrogram

The Uster Spectrograph is a frequency analyzer that produces a record of the

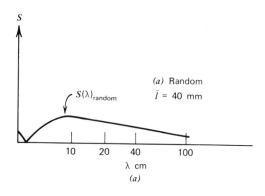

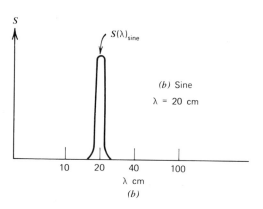

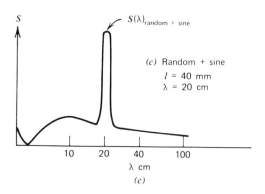

Figure 7.13. Spectrograms (courtesy Zellweger Ltd., Uster).

wavelengths present in the unevenness data, and the ordinate corresponding to a given wavelength is a measure of the relative importance of that particular wavelength in the data analyzed.

Figure 7.13*a* shows the spectrogram for a strand in which fibers of mean length 40 mm are arranged at random; it shows a flat peak at rather more than twice the mean fiber length. Figure 7.13*b* shows the periodogram for a sinusoidal variation with wavelength of 20 cm. Figure 7.13*c* shows a combination of the two. The height of the "chimney" is dependent on the amplitude of the periodic variation and is a measure of its importance relative to the random variation indicated by the flatter curve. For a given wavelength, only one peak, or chimney, appears, and this makes it easy to resolve and identify different wavelengths if more than one regular periodicity is present. For this reason, the spectrogram has been adopted in industry as the most reliable and convenient way of investigating and locating the machinery defects that give rise to periodic variations in the linear density of textile strands.

As stated earlier, unevenness caused by random fiber arrangement is always to be expected; thus a curve such as that in Figure 7.13*a* is to be found as the basis of all spectrograms. In the Zellweger "Uster" publications (29), this is referred to as the ideal spectrum, and it can be calculated. For a strand with fibers of uniform length, the curve is given by the formula:

$$S(\log \lambda) = k \frac{\sin \frac{\pi l}{\lambda}}{\sqrt{\frac{\pi l}{\lambda}}}$$

Where $S(\log \lambda)$ is the amplitude of the spectrum corresponding to wavelength λ
            ploted as abscissa on a logarithmic scale

$k = \dfrac{1}{\sqrt{\pi n}}$ where n is average number of fibers in strand cross section

l = fiber length
λ = wavelength

The maximum of the ideal spectrum lies at about 2.3-2.7 the mean fiber length, the lower value corresponding to more uniform staples.

In the practical case, another component caused by drafting waves is superimposed. This is shown in Figure 7.14, in which the lower curve is the ideal spectrum; the hatched area is that caused by unevenness produced by drafting waves. The greater the effect of this source of variation, the greater this superimposed amount and the higher the spectrogram. Thus a yarn with a higher index of irregularity will have a higher spectrogram. It follows from the data given in Table 7.1 that carded cotton yarns should have higher spectrograms

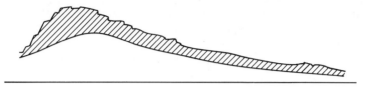

Figure 7.14. Spectrogram for yarn showing, lower curve, spectrogram for the ideal yarn.

than combed cotton yarns, and worsted yarns have flatter periodograms than cotton yarns. This is found in practice, confirming that drafting waves are less significant in worsted spinning than in cotton spinning.

Since the drafting wave has an average wavelength 2-2½ times the mean fiber length, its contribution has a maximum amplitude at this point of the spectrogram scale. This is not a very different position from that caused by random fiber distribution, and as the wavelength of the drafting wave varies a lot it also produces a similar flat maximum. When the two are superposed they more or less coincide. However, if other rather flat peaks appear at higher wavelengths in a yarn spectrogram, these are interpreted as the effects of the drafting wave from previous processes.

Chimneys protruding above the smooth course of the spectrogram indicate regular periodic variations in the strand, and their importance is proportional to the ratio of the height of the chimney to the height of the underlying curve at that point. Their wavelength is read off on the wavelength scale, and processing details such as drafts, roller diameters, etc. may enable the mechanical cause of the unwanted period to be deduced.

The way in which the shape of the spectrogram changes from cotton card sliver to yarn provides an interesting picture of drafting (30). In the card sliver the spectrogram is of almost the same height for all lengths with one or two small chimneys (Figure 7.15). This shows that, in addition to random fiber distribution, variations appear at nearly every wavelength up to 20 yd. The variations produced by the picker are reproduced.

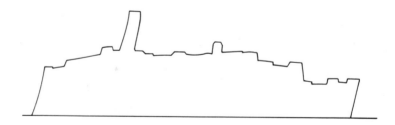

Figure 7.15. Spectrogram for cotton card sliver (Locher).

As drafting proceeds, the amplitude caused by the longer waves begins to disappear, the effect of these being reduced by doublings, and as the strand gets finer the spectrogram becomes more and more like the ideal spectrum. This is in accordance with the data given in Table 7.2.

It has also been stated (29) that in the roving stage the peak that appears is not at 2.5-3 times the mean fiber length but at 3-4 times. The reason adduced for this is that the fibers are not yet drafting as individuals, but as bunches "of which the entire length is somewhat greater than the length of the single fibers." This would be in accordance with the theory of grappes outlined earlier.

EVENNESS IN RELATION TO CLOTH APPEARANCE

The appearance of knitted and woven fabrics can depend on many yarn properties such as hairiness, neppiness, and irregularity of color, and twist or linear density. This section is concerned solely with the effects of unevenness, and these are dependent not only on the character and magnitude of this unevenness, but also on cloth factors such as weave structure, cover, compactness (e.g., degree of felting in the case of woolen cloths) and, if only some cops are affected by abnormal unevenness, by the number of shuttles used in weaving. The following discussion must therefore be regarded only as providing guide lines that may help in assessing particular situations; a yarn that may give rise to an objectionable appearance in one cloth may be quite acceptable in another of different character.

The Effect of Nonperiodic Variations

If the short-term unevenness is distributed uniformly throughout a yarn and is free from periodic machine faults, when this yarn is used for weft knitting, or as weft in a woven fabric, the appearance of the fabric will depend on the magnitude of the variation (31). In a range of similar fabrics made from yarns of increasing unevenness, as the coefficient of variation increases, the fabrics become more and more cloudy; eventually the thick and thin places become prominent and the appearance is described as "flaky." At a certain point on this scale of increasing unevenness, the fabrics would be downgraded as seconds. The yarns used in the experiments referred to were cotton yarns, but there is no reason to doubt that similar results would be obtained with other yarns.

In some yarns there can be irregular unevenness (32), that is, some sections have greater unevenness than others. This results in a patchiness in cloth appearance that may be more objectionable than overall unevenness of greater average value. Thus it is again to be noted that one cannot rely only on instrumentally determined values of percentage variation to form criteria of yarn quality; it is always valuable to examine the evenness charts as well.

Medium- and long-term variations give rise to streakiness in woven and knitted fabrics, the length of streaks being commensurate with the lengths of the thick parts of the yarn. These "faults" may arise from the causes dealt with in an earlier section. Long streaks of this kind may now become apparent in the warp direction. If the long-term variations are so long that they appear as count differences between cops—as can more readily happen in condenser yarns—the effect in the cloth can be to produce weft bars of different weight per unit area that may be visible to the eye.

Effect of Periodic Variations

Periodic variations in linear denisty of yarn can produce diamond patterns in cloth (31-33). It will be clear in what follows that similar regular periodic variations in color or twist could produce similar defects; in fact, according to Foster (32), most faults of this kind in worsted fabrics are caused by periodic twist variations caused by faulty spindle drives during spinning or twisting, or by twist redistribution by traverse mechanisms in winding processes. In woolen and worsted yarns, periodic variations can also be caused by uneven moisture content of yarns on packages; if yarn on the outside of a cop is moister than that inside, then, on winding, the yarn at these points may be stretched more than that inside and a periodic variation in linear density will occur equal in length to the chase of the yarn on the nose of the cop. Cloth faults caused by a variety of effects of this kind have been dealt with by Poole, Haley, and Turner (35).

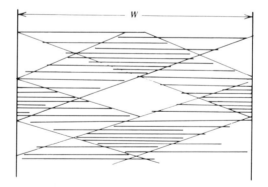

Figure 7.16. Diamond pattern in cloth when $\lambda = W + x$.

Figure 7.16 shows the kind of diamond pattern that appears if the length of the period, λ, is *slightly greater than* the pick length, W, that is, the actual length of the yarn in the pick. To indicate the periodic effect, half the length of a period in this diagram is shown by a line, and the thinner half is omitted; this

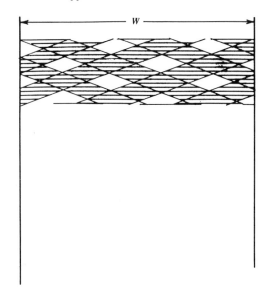

Figure 7.17. Diamond pattern when $\lambda = \dfrac{2W}{5} + x$.

exaggerates the effect. In this case, the cloth width at any point contains two diamond repeats, one complete at the center and half a diamond at each side.

Figure 7.17 shows the pattern produced when λ is slightly greater than $\dfrac{2W}{5}$; there are now five diamond repeats across the cloth width. A similar result would be obtained if λ were *slightly less than* $\dfrac{2W}{5}$.

Thus if R^* is the number of diamond repeats in the width

$$R = \frac{2(W \pm x)}{\lambda} \qquad \text{where } x \text{ is a small amount.}$$

As x increases the steepness of the diagonals becomes less; a small change in cloth width also alters the ratio, and patterns may appear and disappear along a length of cloth.

According to Catling (33), the prominence of the diamond barring increases as the amplitude of the periodic fault increases, and, for a given amplitude, is more noticable the higher the pattern ratio.

In yarns made on cotton spinning frames, periodic effects can arise with wavelengths that are quite short by comparison with the pick length in a woven cloth,

*Foster calls R the pattern ratio; Catling calls $\dfrac{R}{2}$ the pattern ratio.

or the course length in a weft knitted fabric. These can produce quite objectionable patterns, and in knitted fabrics the diagonal lines of the diamonds change direction from time to time, presumably as the pattern ratio varies, and a Moiré effect may be seen (31). It is stated in the same article that periodic effects become objectionable when the corresponding chimney on the Uster spectrogram is half the height of the underlying spectrogram at that point.

When $\lambda = 2W$, $R = 1$. There is now only one diamond in the width, and this looks like a zig-zag bar in the cloth. As the wavelength increases still further, the diamond form becomes less recognizable, but the bars across the cloth may still be seen. Such long wavelengths are to be found in woolen yarns; a faulty doffer may have a circumference of about 100 in. and, if a spinning draft of 1.6 is used, a fault of about 160 in. may arise in the yarn. This is about three times the pick length in a woolen cloth, and bars will be seen in the cloth.

REFERENCES

1. Martindale, J. G., *J. Text. Inst.,* 1945, **36**, T35.
2. Foster, G. A. R., *Manual of Modern Cotton Spinning Vol. 4, Pt. 1.* The Textile Institute, Manchester, England, 1958.
3. van den Abele, A. M., *J. Text. Inst.,* 1951, **42**, P162.
4. Taylor, D. S., *J. Text. Inst.,* 1955, **46**, T284.
5. Breny, H., *Ann. Sc. Text. Belges,* 1962, No. 2.
6. Vroomen, F. and F. Montfort, *Studies in Modern Yarn Production.* The Textile Institute, Manchester, England, 1968.
7. Vroomen, F., *Wool Sci. Rev.,* 1970, **39**, 18; ibid, 1970, **40**, 43.
8. Foster, G. A. R. and J. G. Martindale, *Shirley Institute Memoirs—Series A,* 1941, 125.
9. Thomason, W. A., *Uster News Bulletin,* 1962, No. 3.
10. *Uster News Bulletin,* 1963, No. 4.
11. *Uster News Bulletin,* 1965, No. 6.
12. *Uster News Bulletin,* 1965, No. 7.
13. Angus, J. and J. G. Martindale, *Text. Res. J.,* **26**, 698, 1956.
14. *Woollen Carding,* WIRA, p. 3, 1968, Leeds, England.
15. *Wool Research*—Vol. 4, *Carding,* WIRA, Leeds, England.
16. A.S.T.M., Standard Method of Test, D2255-64. Part 24, Philadelphia, Pa.
17. *Wool Research*—Vol. 6, *Drawing and Spinning,* WIRA, Leeds England.
18. Chamberlain, N. H., *J. Text. Inst.,* 1944, **35**, T61.
19. Downes, J. G. and B. G. Leary, *Text. Rex. J.,* 1958, **28**, 497.
20. A.S.T.M. Standard Method of Test, D 1425-67, Part 24, Philadelphia Pa.
21. *Uster News Bulletin,* 1968, No. 11.
22. *Wool Sci. Rev.,* **25**, 32, 1964, International Wool Secretariate, London, England.
23. *Woollen Carding,* WIRA, p. 1, 1968, Leeds, England.

24. Dyson, E. and B. Schofield, *J. Text. Inst.*, 1968, **59**, 529.

25. Bornet, G. M., *Text. Res. J.*, 1966, **36**, 506.

26. Bornet, G. M., *Text. Res. J.*, 1966, **36**, 519.

27. Breny, H., *J. Text. Inst.*, 1953, **44**, P. 1.

28. Olerup, H., *J. Text. Inst.*, 1952, **43**, P. 290.

29. *Manual for the Uster Spectrograph*, Pt. II, 133,900-IE, Zellweger Ltd, Uster, 1959.

30. Locher, H., *The Spectrograph "Uster"*, 133,700, Zellweger Ltd, Uster, 1954.

31. *Uster News Bulletin*, No. 15, 1971.

32. Foster, R., *J. Text. Inst.*, 1952, **43**, P. 742.

33. Catling, H., *J. Text. Inst.*, 1958, **49**, T232.

34. Poole, E.J.D., C. Haley, and J. Turner, *Illustrated Practical Problems*, WIRA, 1964, Leeds, England.

35. Grignet, J. and P. Goddard, *Wool Sci. Rev.*, 1973, **44**, 17.

SUGGESTED READING

Manual of Cotton Spinning, Vol. 4, Pt. 1, by G. A. R. Foster. Butterworth and the Textile Institute, Manchester, England.

Wool Research, Vol. 4, *Carding*, Ch. 8. WIRA, Leeds, England.

Wool Research, Vol. 3, *Testing and Control*, Chs. 4 and 9, WIRA, Leeds, England.

Regularity of Tops, Slivers and Yarns, Wool Sci. Rev., Nos. 24, 25, 26, International Wool Secretariat, London, England.

8

Yarn Structure in Relation to the Aesthetic and Tactile Qualities of Apparel Fabrics

INTRODUCTION

Among the more important factors in the design of apparel fabrics are the *look*, the *feel*, the *performance*, and the *cost*. Of course, other considerations may be more important from time to time, but there is no question that the visual and tactile qualities are always fundamental to fabric design. The look of a fabric depends mainly on the interaction of color and surface texture. The feel of a fabric depends on several physiological-psychological complexities that are not yet fully understood.

In addition to the look and feel, aesthetics are also concerned with the sound and smell of fabrics. In general, the sound referred to is the frictional sound of an apparel fabric sliding over itself or over another surface. Usually, a fabric must have a crisp, smooth surface to generate a characteristic sound. A fabric with a soft, napped, or hairy surface is less capable of producing a distinctive swishing or rustling sound. The smell of a fabirc is usually the aroma generated by a particular finish. Once the fabric is converted into apparel, the aroma is caused essentially by laundering, dry cleaning, or permanent soiling.

The purpose of this chapter is to discuss the role of yarn structure in the look and feel of fabrics designed for apparel.

VISUAL AESTHETICS

It is easily observed that the contribution of yarn structure to the visual aesthetic qualities of a fabric is transmitted mainly, if not completely, through the surface geometry of the constituent yarns. The components of yarn surface geometry are the several macroscopic topological features that make up the characteristic grain of the yarn. Examples of these features are the fiber orientation, fiber packing density, yarn twist ridges, yarn imperfections, yarn hairiness, fiber blend effects, yarn flatness or nodular projection, etc. These yarn surface features combine to communicate usually a certain degree of texture, softness, luster, smoothness, and bulkiness (or body) in apparel fabrics.

Texture, softness, and bulkiness in a fabric are aided by three basic yarn structural features. These features are a low degree of fiber orientation with respect to the yarn axis, a great amount of fiber protrusion from the yarn surface, and a low fiber-packing density in the yarn structure. Conversely, luster and smoothness in a fabric are aided by a high degree of fiber orientation with respect to the yarn axis, little or no fiber protrusion from the yarn surface, and a high fiber-packing density in the yarn structure. Twist plays a major role in relation to yarn surface features and fabric aesthetics. For example, using the same size yarn, a smaller amount of twist provides for a softer and bulkier look in a fabric than does a greater amount of twist. Moreover, a variety of effects can be produced in a fabric through the use of balanced and unbalanced twist constructions in plied yarns. The unbalanced twist constructions provide unusual nodular effects in yarn and fabric. Also, overtwisting is used for various novelty and rough-surface effects in fabrics.

Many of the fabric aesthetics attributed to yarn structural features are magnified, masked, or modified to some degree in fabric finishing. In fully brushed or semibrushed fabrics, the textural features of the constituent yarns tend to be masked or are masked completely. In clear-finished fabrics, the textural features of the constituent yarns tend to be emphasized.

ROLE OF YARN STRUCTURE IN VISUAL AESTHETICS

The three basic yarn structures of commercial importance that provide for a distinctly different look in apparel fabrics are the spun, filament, and textured filament.

The spun yarns provide a unique combination of structural features that contribute a constantly varying or a kind of natural textured appearance to fabrics that is not possible with any other yarn structure. This unique combination of structural features is: fiber nonlinearity on the surface of the warp because of a high level of twist; variation in yarn linear denisty and apparent diameter; imperfections in yarn appearance because of neps, slubs, slugs, and so on; fiber blend effects that tend toward a heather rather than solid shade, if desired; hairiness caused by fiber ends protruding from the surface of the yarn; and, excluding the surface fuzz zone, a high fiber-packing density in the core of the yarn. Although stretch and high-bulk filament yarns are also capable of producing a textured look in fabrics, the look is much more uniform and lofty (bulky) and not as random, grainy, and hairy as the look produced by spun yarn.

All the features that contribute to spun yarn texture also combine to provide visually subtle surface roughness, a fair amount of softness, good bulk or covering power, and low luster in apparel fabrics. Some major differences do exist among spun yarn structures. Combed yarns have more luster and smoothness than carded yarns because of improved fiber orientation, relatively less protruding surface fiber, a relatively higher degree of twist, and, in the case of cottons, singeing of the surface to remove hairiness. Conversely, carded yarns tend to project a greater sense of softness, warmth, and bulkiness than combed yarns. Combed yarns tend to have relatively fewer yarn imperfections than carded yarn from the same stock. This is caused mainly by the extra effort in combed systems to remove short fiber, which is considered to be a major source of yarn imperfections. A large number of yarn imperfections need not always be considered visually displeasing, however. The denim look is a good example of a design that incorporates yarn imperfections, practically to a maximum extent, into an apparel fabric that is known for its casual look. Conversely, by conventional design, a minimum of yarn imperfections is tolerated in more formal apparel.

Filament yarns also provide a unique combination of structural features that contribute a fabric appearance not possible with spun or textured yarns. The appearance of filament fabrics is practically textureless or very fine in texture as compared with the appearance of fabrics composed of spun or textured yarns. The yarn structural features responsible for the fine texture look of filament fabrics are: maximum fiber orientation with respect to the yarn axis, an almost flawless uniformity of yarn structure, and a potentially high filament-packing density. The uniformity of linear density and of apparent diameter in filament yarns is the result of the manufactured regularity in the number, linear density, and cross-sectional shape of composite filaments along the length of the yarn structure. The high degree of filament linearity and packing efficiency tend to emphasize visually this uniformity of filament yarn structure.

This unique combination of structural features in filament yarns also provides the ultimate in luster, sheer, and smoothness in apparel fabrics. On the other hand, filament yarns are not efficient for the provision of bulkiness, covering power, or a soft look in fabrics. No protruding fiber ends or filament loops are normally associated with filament yarns. Moreover, with filament yarns, the opportunity for intimate fiber blending and associated heather effects common with spun yarns does not exist. However, the luster of filament yarn structures can be toned down through the use of delusterants to semidull or dull level. Also, filament yarns can be irregularly twisted to provide a fine slubly texture or nonsmooth look in apparel fabric.

Textured filament yarn structures combine some features of the spun yarns with some features of the filament yarns. The unique combination of structural features found in fabrics made from textured filament yarns are: a high degree of fiber nonlinearity becuase of texturing of filaments, a very uniform yarn structure in terms of linear density and apparent diameter, and a very low fiber-packing density throughout the yarn structure. These features combine to provide a highly textured appearance, a great amount of bulk and covering power, and a soft look in fabrics. The textured appearance is somewhat more intense but much more uniform than the grainy texture provided by spun yarns. Also, the nonlinear filament orientation plus the protrusions of filament loops from the surface can lead to the appearance of lower luster and a matted surface in fabrics made from textured yarns.

Many techniques exist for the stretch and high-bulk texturing of filament yarns. Accordingly, structural differences do exist among the variously textured yarns. These differences are not as great, however, as the differences among spun, filament, and textured filament yarns. The major difference is that high-bulk filament yarns have a low filament-packing density on the surface and higher filament-packing density in the yarn core. The packing density of filaments in stretch yarns tends to be more uniform throughout the yarn cross section.

TACTILE AESTHETICS

The appeal of apparel fabrics depends on the interactions of the visual effect of fabric texture and its feel in the hand of the consumer. It is not easy to separate the visual from the tactile effects, because most individuals are accustomed to seeing and feeling a fabric simultaneously. Indeed, many visual and tactile effects do seem to go together.

The tactile aesthetics of apparel fabrics are commonly referred to as the fabric hand or handle. The hand of a fabric is defined as the total response or combination of impressions that arises when a fabric is touched, squeezed, rubbed, or handled by any other manner. In subjective panel testing, several descriptive terms are used to characterize the feel of a fabric (e.g., soft, crisp, firm; hard, harsh, boardy; dead or mushy, live or wiry; cool, warm, dry, etc.). Several fabric properties are measured in an effort to objectively characterize the many aspects of fabric hand. These properties are: fabric bending stiffness, compressibility, resilience, extensibility, bulk retention, friction, drape, shearing stiffness, hairiness, density, etc. Because fabric hand is such a major factor in consumer appeal and clothing comfort, many techniques have been proposed for the complete and objective assessment of fabric hand. No reliable technique is available, however, that takes into consideration *all* of the physiological-psychological factors that are associated with the feel of a fabric. For example, little work has been done in measuring apparent surface temperature, surface roughness, real surface contact, and other topological properties of fabrics.

As in the case of visual aesthetics, the contribution of yarn structure to the tactile qualities of fabrics is also transmitted through the surface geometry of the constituent yarns. Exceptions to this tendency, once again, can be found in the case of heavily napped, brushed, or felt-finished fabrics and in the case of apparel fabrics that have been given heavy coatings or chemical treatments.

However, the tactile qualities of a fabric are also dependent on the compressive behavior of the fabric. The dimensional stability of the cross section of the constituent yarns plays a major role in fabric compression. Thus the behavior of yarn cross section during fabric compression is quite fundamental to the hand and comfort of apparel fabrics.

ROLE OF YARN STRUCTURE IN TACTILE AESTHETIC

The role of yarn structure in tactile aesthetics of fabrics is somewhat dominated by yarn twist. For example, fabrics composed of yarns with higher levels of twist are known to have higher bending stiffness, less compressibility, less fiber mobility, lower surface friction, less bulkiness, and less potential contact with a contiguous surface than similar fabrics composed of yarns with less twist. Increased yarn twist leads to greater internal (fiber-to-fiber) friction within the yarn structure; the constituent fibers or filaments tend to bend as a group rather than individually, thereby increasing the bending stiffness of the yarn. The increased fiber entanglement and internal friction caused by yarn twist also provide for a more dimensionally stable yarn structure that does not deform as much under compressive loads. On the yarn surface, the segment fiber length between points of entanglement is reduced with increased yarn twist. This effect severely restricts fiber mobility and the chance of snagging of fibers or filaments. Yarn twist creates lower surface friction and potential surface contact because of less yarn flattening under low levels of compressive loading. Increased twist tends to reduce softness, covering power, and bulkiness, in general, and hairiness in the case of spun yarns.

Fiber linearity and fiber-packing density in yarn structures are also important to the tactile qualities of a fabric, when not masked by twist. In untextured filament yarns, the fiber linearity and packing density are quite high. Consesequently, the yarn leads to a smooth, uncompressible feel in fabric. With similar yarns that have been textured, the low packing denisty of filaments and the non-linear protruding filament loops produce a soft, compressible but resilient (spring back) feel in fabric.

COMFORT IN APPAREL FABRICS

A great deal has been written on the subject of comfort in apparel and apparel fabrics. Unfortunately, most of the reported work has been subjective, and conclusions have been based on rather questionable samples, measurements, and interpretations of interactions. This suggests that all the fundamental parameters of comfort have not been clearly defined. However, the authors wish to recommend to the reader a recently published source that is basic for the understanding of comfort in apparel. This work is a joint effort: *History, Physiology and Hygiene* by Renbourn and *The Biophysics of Clothing Materials* by Rees. Several of the following statements concerning comfort in apparel fabrics have been selected from the conclusion found in this fine reference.

According to Rees, the thickness of apparel fabrics strongly affects certain comfort factors such as warmth or the amount of heat insulation and the ability of the body to dissipate its insensible perspiration and heat. In general, the greater the fabric thickness and bulk, the greater the thermal insulation. Regardless of fiber content, warmth or thermal insulation is dependent on the entrapment of air over a wide climatic range of wind, temperature, and humidity conditions.

It is also true that certain materials are better insulators or conductors of heat than others. For example, a flat sheet of metal feels cooler than a flat sheet of wood. However, a smooth or polished wood feels cooler than rough wood at first touch. With textile structures, the rate of heat transfer depends much more on the real area of contact between skin and fabric surface than on the type of fiber used. In general, napped fabrics feel warmer than clear-finished fabrics. Similarly, fabrics with rough surfaces feel warmer than fabrics with smooth surfaces.

The dissipation of perspiration and body heat is also influenced by the porosity and wicking characteristics of apparel fabrics. Pore size affects air resistance and water repellency. Usually, the larger the pore size, the lower the resistance to air flow and the lower the water repellency. Wicking efficiency is affected by fabric geometry and by the geometry of fibers in yarn and fabric. Usually, hairiness and random fiber arrangement lead to slow rates of wicking and surface wetting.

Rees concludes that the overall comfort of an apparel fabric depends on the proper *combination of values* for pore size, air permeability, water vapor permeability, thermal insulation, surface contact with skin, and several other fabric properties. Also, it is even more important that the particular combination of values for these various properties be maintained under a range of fabric compressions and tensions that would be encountered during actual usage. Drastic changes in comfort can occur whenever the combination of values of various fabric properties is altered. In considering all these remarks, it should be kept in mind that the effect of finishing, coatings, or laminar construction could change substantially the values of particular comfort factors. Furthermore, the loss of any finishes in subsequent usage, laundering, dry cleaning, etc., can also have a substantial effect on values for these comfort factors.

APPAREL FABRICS UNDER CONDITIONS OF ACTUAL USE

One fabric or several layers of various fabrics may be worn as an apparel ensemble. The components of garments, when being worn, are usually subjected to very light pressures and are generally in the uncompressed state. However, exceptions to this generality are found in the following examples: (1) soles of stockings when a person is standing, walking, or running, (2) parts of garments underneath a person sitting or lying down, and (3) parts of garments that are loaded (especially in the vicinity of the knees, elbows, and derriére) when a person is working or exercising. Furthermore, although the entire apparel ensemble being worn is not under great pressure, subtle or very lght pressures are distributed throughout, and localized areas can encounter heavy pressures. As Rees concludes, it is important, therefore, that comfort parameters be measured under conditions that relate to actual use.

Whenever an apparel fabric is compressed or whenever localized pressure occurs, several important comfort factors are affected. Values for many of the comfort parameters are changed because of deformation of the constituent yarn structure and the consequential effects on the internal structure of the fabric. For example, as the fabric structure is compressed, the fabric thickness and bulkiness are reduced because the structure of the constituent yarns deform and flatten out. As the yarn structure flattens, changes will occur in the size and shape of the intersticies (fabric pores), in the fiber orientation at the yarn surface, and in the real area of contact between fabric and skin or any other contiguous surface. Thus changes can occur in values for air permeability, water vapor permeability, rate of surface wetting and wicking; water repellency, thermal insulation, cool or warm feel, and other important comfort parameters. Moreover, the combination of values for these comfort parameters could be quite different for apparel fabrics in a compressed, as opposed to an uncompressed, condition.

Because fabric thickness or bulkiness is related to so many comfort factors, it has been found quite practical to use fabric compressibility and resilience as indicators of changes in many of the comfort parameters mentioned in the previous paragraph. This means that fabric compressibility and resilience can be used in predicting the comfort of apparel fabrics, in general. Compressibility is the proportional reduction in the thickness of a material under prescribed conditions of increased pressure or compressive loading. Resilience is the degree to which a material recovers from compressive deformation. Obviously, if an apparel fabric has slight change in bulk or thickness with low rates of compressive loading, its combination of values for important comfort factors will not change substantially. Such a fabric should provide uniformity of comfort under a wide range of conditions. If, on the other hand, an apparel fabric shows a substantial change in bulkiness or thickness with low rates of compressive loading and very poor (quick) recovery from compressive loading, one can expect some discomfort during actual use of the fabric.

ROLE OF YARN STRUCTURE IN FABRIC COMPRESSION

The behavior of yarn structure and, more specifically, yarn cross-sectional shape during fabric compression is quite complex. The specific manner in which the yarn behaves depends on the interaction of such variables as constituent fibers, yarn structural features, fabric structural features, effects of fabric finishing, and the nature of the fabric deformation. For example, similar yarns composed of different fibers could deform to the same degree or differently, depending on the specific fabric structure or finishing processes. The same yarn would deform more in loose woven or knitted fabrics than in tightly woven or knitted fabrics. Moreover, the same yarn in fabrics with long yarn floats between interlacings or interloopings will deform more than in fabrics with short yarn floats. If a fabric finish tends to reduce fiber segment mobility, the constituent yarns deform less under compressive loads than in the case of finishes that increase fiber segment mobility. Although each variable (fiber, yarn, fabric, finish) is quite important, it is the interaction of all of them together that really determines how a yarn deforms under fabric compression. In spite of the risk of oversimplifications, generalities can be stated concerning the effect of yarn structure alone on the deformation of yarn during fabric compression.

From any commercially available man-made filament, it is possible to produce a variety of yarn structures (such as spun, filament, twisted filament, high bulk, and stretch of similar linear density. If these yarns were woven or knitted into a standard structure, a variety of fabric bulkiness or thickness would be expected. Also, under the same conditions of fabric compression, a variety of yarn deformations would be expected. To illustrate this point, yarn structures of similar linear density and fiber composition are ranked for relative bulkiness in Table 8.1. Referring to Table 8.1, under relaxed conditions (no tension, no compression), the highly textured stretch and high-bulk yarns would produce fabrics with the greatest bulk or thickness. Spun yarns would produce fabrics with the next level of thickess, followed by fabrics made from twisted filament yarn. The least bulk or fabric thickness would be found in samples made from twistless filament yarns.

Table 8.1. Relative Bulk and Surface Contact of Yarn Structures Under Relaxed and Compressed Conditions

Yarn Structure	Relative Bulk		Relative Contact	
	Relaxed	Compressed	Relaxed	Compressed
Spun	3	1	3	5
Untwisted filament	5	5	1	1
Twisted filament	4	3	2	3
High-bulk filament	1	2	5	4
Stretch filament	1	4	5	2

Note: 1 is the highest in rank, 5 is the lowest.

Under compressive loads, however, fabrics made from spun yarns would provide the greatest bulk for insulative and dissipative purposes. The spun yarn, with its hard core of fiber geometric entanglement in the center of the yarn and an outer fuzz zone of protruding fiber ends, would maintain to a considerable degree the structural integrity of the yarn cross-sectional shape under low rates of compressive loading. The high-bulk yarn would flatten out more than a comparable spun yarn because of the tendency of the projecting filament loops to bend and comform to the pressure. The looping filaments and the random, twisted filament entanglement in high-bulk yarns would probably provide more bulk than the twisted filament yarn. However, the twisted filament yarn can withstand more compression without flattening out than either the stretch or untwisted filament structures. The bulkiness normally associated with stretch textured filament yarn cannot be maintained even under the slightest compression because of the lack of filament entanglement in the structure.

Rankings according to relative contact of yarn structures with contiguous surfaces under relaxed and compressed conditions are also listed in Table 8.1. Relative contact is the estimated proportion of fibers or filaments in the cross section of a yarn that makes actual contact with a surface contiguous to the yarn. It can be seen that the rankings of yarn structures for relative contact are somewhat the reverse of rankings for relative bulk. Under relaxed conditions, untwisted filament yarns tend to establish the greatest relative contact with contiguous surfaces. This is due to a high degree of fiber linearity with respect to the yarn axis and to the absence of twist or other entanglement that would restrict the flattening out of the yarn cross section via filament spreading. The twisted filament yarn structure offers the next level of relative contact with slightly reduced fiber orientation with respect to the yarn axis and less spreading out of the filaments caused by the restrictions imposed by twist. A great amount of twist and hairiness in a spun yarn leads to less relative contact with contiguous surfaces than the untextured filament structures. Under relaxed conditions, the high-bulk and stretch-textured yarn structures are under the negligible pressure of their own weight only. Consequently, under relaxed conditions, with the high degree of fiber nonlinearity and the low degree of fiber packing, relatively little contact is made between the textured yarns and a contiguous surface.

When these various yarns are subjected to compression, however, the rankings of the structures change considerably with respect to relative contact. Once again because of its superior ability to resist compression, the spun yarn does not flatten out to as great an extent as the other yarn structures. This is because of the complex fiber geometry and entanglement, the high degree of twist, and the substantial fiber packing found in spun yarns. Consequently, under conditions of yarn or fabric compression, spun yarns provide for the least relative contact with contiguous surfaces. The stretch yarn cross section deforms quite easily under compression and flattens out to provide for substantial contact almost to the same extent as untextured filament yarns. In the case of high-bulk textured yarns, the filament loopiness and yarn cross section are also easily depressed under compressive loading but not to the same extent as stretch-textured structures.

A practical technique for assessing the relative contact potential of yarn structures is the measure of yarn friction under various initial tensions. It is well established that, under properly controlled conditions, yarn friction test data can reflect the relative degree of contact between the constituent fibers in yarn and the frictional surfaces over which the yarn specimen passes. For example, a filament yarn with no twist has much greater friction than the same yarn with only a few turns per inch of twist under the same test conditions. The filament yarn with no twist flattens out and its constituent filaments make greater proportional contact with the frictional surface than those filaments in the twisted structure. Twist also restricts the tendency of the yarn to flatten out, even under high tension or compression. Furthermore, the twisted yarn structure tends to pass over a frictional pin surface riding on filament segments associated with the summit of the yarn twist ridges rather than on all of the filaments that appear to be in contact with the frictional pin surface. Minimal initial tensions during friction tests on yarn specimens are analogous to relaxed conditions for yarn and fabric structures. Substantially higher initial tensions on yarn specimens are analogous to compressed conditions in yarn and fabric structures.

   In conclusion, the role of yarn structure in the aesthetic and tactile qualities of apparel fabrics is quite prominent. Many of the factors that affect comfort in apparel fabrics are dependent on the ability of the constituent yarns to resist compressive deformation. Regardless of fiber content, spun yarn structures appear to provide a superior resistance to compression in apparel fabrics. This is because of the unique structural geometry of spun yarns.

SUGGESTED READING

Materials and Clothing in Health and Disease: History, Physiology and Hygiene—Medical and Psychological Aspects, E. T. Renbourn; *The Biophysics of Clothing Materials,* W. H. Rees, H. K. Lewis, London, 1972.

Friction in Textiles, H. G. Howel, K. W. Mieszkis, and D. Tabor, Butterworths Scientific Publications, London, 1959.

Clothing Comfort and Function, Lyman Fourt and Norman R. S. Hollies, Marcel Dekker, New York, 1970.

Yarn Structure in Relation to Luster, L. Fourt, R. M. Holwarth, M. B. Rutherford, and P. Streicher, *Text. Res. J.,* 1954, 163.

The Subjective Assessment of the Roughness of Fabrics, H. C. W. Stockbridge and K. W. L. Kenchington, *J. Text. Inst.* JTI 1957, T26.

A Method of Measuring Yarn Softness and Its Use to Show the Effect of Single and Ply Twist on the Softness of 31/2 Cotton Yarns, E. L. Skau, E. Honold, and W. A. Boudreau, *Text. Res. J.,* 1958, 206.

A Study of the Comparative Comfort Properties of Nylon Bulk-Yarn Knitted Fabrics with Those Prepared from Cotton, Wool and Continuous-Filament Man-Made Fibre Yarns, E. R. Kaswell, L. Barish, and C. A. Lamond, *J. Text. Inst.,* 1961, P508.

Factors Affecting the Thickness and Compressibility of Worsted-Spun Yarns, W. J. Onions, E. Oxtoby, and P. P. Townsend, *J. Text. Inst.,* 1967, 296.

The Hairiness of Cotton Yarns—An Improvement Over the Existing Microscopic Technique, B. C. Goswami, *Text. Res. J.,* 1969, 31.

Yarn Hairiness: A Survey of Recent Literature and a Description of A New Instrument for Measuring Yarn Hairiness, A. Barella and A. Viaplana, *J. Text. Inst.,* 1970, 438.

The Nature of the Hairiness of Open-End Spun Yarns, A. Barella, *J. Text. Inst.,* 1971, 702.

9

Conversion of Staple Fibers into Yarn

FUNDAMENTAL PRINCIPLES OF PROCESSING

General Considerations

Staple fibers have a very wide range of physical properties, most of which have some effect on the methods used in processing them into yarn. Before processing, flax fibers may be 3 ft long, whereas some cottons are less than half an inch; supplies of man-made fibers are free from extraneous physical or chemical impurities, whereas bales of cotton or wool contain many types of impurity from which the fibers must be separated. Questions of fiber shape and fiber organization also have some influence on processing methods; most staple fibers are discrete, individual entities, but this is not true of jute, in which adjacent long fibers are interconnected by branches that run from one fiber to another and produce a lattice structure that is broken up during processing. As another example, cotton is a relatively straight fiber, whereas wool is both crimped and curly, this situation is further complicated by the fact that cotton is a relatively inelastic fiber that, once straightened out, tends to stay in a straightened form, whereas wool fibers, being very elastic, make the most of any opportunity during processing to return to their previous curly and crimped state. Such differences in staple fibers have influenced machine design in diverse ways; thus cotton spinning machinery looks quite different from that required to make a jute yarn, and the differences between the woolen and worsted processes seem to be even greater. These differences in the machinery used in different spinning industries are sufficiently great to be confusing, and to conceal the fact that the *fundamental* principles of yarn making are the same for all staple fibers.

It might also be mentioned here that to be commercially acceptable a fiber must be tough enough to withstand the stresses and strains of the yarn-making processes. Many naturally occurring fibers fail on this score and have not therefore been admitted as textile fibers; the possession of adequate tensile properties to withstand the rigors of yarn-making processes was one of the first requirements with which man-made fibers had to comply.

Fundamental Operations

The process of yarn making consists of converting a stock of fibers into a yarn whose composition, color, count, and twist are specified. Irrespective of differences in machinery used for different fibers, the operations that are fundamental to the process are:

1. opening, cleaning, and mixing
2. sliver formation
3. attenuation of the sliver (i.e., drafting)
4. twist insertion to prevent fiber slippage in rovings and yarn
5. packaging of the yarn.

The spinner, as well as complying with the specifications, must ensure that the yarn is sufficiently even and strong to be commercially acceptable, and also that it is produced as economically as possible.

In modern processes, acceptable evenness requires (1) that automatic devices (autolevelers, servo-drafters) be used in the early stages of drafting to ensure effective control of the linear density of slivers in process, so that long-term variation of linear density of the yarn is minimized, and (2) that procedures are followed that avoid the unnecessary introduction of excessive irregularity during drafting.

Economic production implies a minimum amount of machinery since this involves a smaller capital outlay, less floor space and therefore smaller overhead expenses, and a smaller labor force. A reduction in the number of operations may at some point have a detrimental effect on yarn quality; this possibility has been greatly reduced by new machinery that has been perfected in recent years, but the practical outcome is still a sequence of machines in which the economy of production must be balanced against the quality of the product.

Opening, Cleaning, and Mixing

The early stages of opening, cleaning, and mixing usually occur together, the detailed procedures being determined by the state in which the fiber stock is received in the mill. Cotton is received in bales that have been hydraulically pressed to a density of 35-40 lb per cubic foot. In addition to cotton fibers these bales contain impurities—particles of leaf, stalk, seed, sand, and dust, all of which must be removed before the fibers are formed into a yarn. Since the bale is so highly compressed, machinery must be employed that can open up the contents of the bale to such an extent that these impurities can be extracted. Thus in the case of cotton processing, opening is a necessary preliminary to cleaning.

In the case of new wool (as distinct from reprocessed wool), the fibers are always accompanied by appreciable quantities of wool grease as well as vegetable matter, sand, and dirt. Since wool is also press-packed, the grease prevents opening by simple mechanical methods, and it must be removed before opening of the fibers can take place. In this case, then, an initial cleaning is a necessary preliminary to opening, after which the particulate types of impurity are more accessible and can be separated by mechanical means.

In the case of fiber stocks of man-made staple fibers, these do not contain any solid impurities; cleaning is unnecessary. Man-made fibers are originally produced as continuous filaments, and groups of these, called a tow, are cut into the desired lengths to form staple. At the cutting point, the fiber ends sometimes become stuck together; unless they are carefully opened, groups of fibers may tend to go through the yarn process together, producing slubs. Man-made staple fibers are also press-packed for economy of transport; therefore, on both these counts, opening is necessary.

The irregularity of yarn is discussed in Chapter 7 and the importance of yarn evenness is demonstrated. In order that a yarn be as uniform as possible in appearance and in tensile properties, it is necessary that fibers of differing lengths, fineness, color, etc., should be mixed and distributed through the final product as uniformly as possible. In the case of man-made fibers, these variations in properties may not be as great as in the case of natural fibers, but they do exist; therefore, good mixing is still essential. When different types of fiber are blended together, the need for uniform mixing is of paramount importance; even so, fibers that are appreciably dissimilar in dimensions cannot be blended together because, in such circumstances, optimum machinery settings cannot be provided for dissimilar components, and the presence of such dissimilar components might well be prejudicial to the properties of the yarn.

Mixing cannot be very effective unless the fibers are well separated from one another; in fact, one could say that perfect mixing cannot be achieved unless each fiber is completely separated from its neighbor at some point in the process. Complete separation can be achieved easily in substances like sand or gravel, in which the particles are separate from one another, but fibers in the early stages of processing are much entangled and must therefore be opened as thoroughly as possible to achieve good mixing. This initial entanglement introduces the possibility of fiber breakage during mechanical opening.

The first three fundamental operations in making yarn from staple fiber are therefore inseparable; thus a particular machine specifically designed to carry out one of them usually carries out the other two operations to some degree. For example, a beater used in cotton opening is primarily designed to open and clean, but in passing through such a device some fiber mixing inevitably occurs; in wool scouring, the process is designed primarily for cleaning, but considerable opening takes place, as well as some mixing. The student should bear in mind that, whatever fibers are being processed, these are the three objects of all the operations prior to forming a sliver. Machinery used for the purpose should be judged on the effectiveness and economy with which it achieves these operations. It should also achieve them with the minimum of damage to the fibers in transit.

Sliver Formation

The next essential practical requirement in processing is to form a sliver. There is no basic, fundamental reason why this should be so; if, after a thorough opening, cleaning, and mixing, fibers could be continuously extracted in sequence from the bulk (as is done by a hand-spinner) and assembled in a sufficiently uniform strand, at the correct rate to constitute a yarn of the desired count, the need for sliver formation and the subsequent processes of attenuation would disappear. This, however, is not possible at the present time.

The sliver is usually formed on a carding machine, and carding is an important operation in most yarn making sequences; cotton, cotton waste, worsted, semi-worsted, woolen, jute, and flax tow systems all employ carding machines when using either natural or man-made fibers. On the other hand, the carding processes cannot be said to be an essential feature of the processing of fibers into yarn, because flax is converted into linen yarn without carding, as are long wools. The reason in both cases is that they are long fibers, and it is desired to preserve the fiber length. If they were processed on carding machines, these long fibers would need extremely large workers and strippers; otherwise they would be stretched and broken by becoming wrapped around these rollers. The sliver is therefore formed in another way: bunches of fibers are fed manually to gill boxes so that the fibers are as far as possible aligned in one direction, the continuity of the feed is maintained, and after the gilling operation they issue as a continuous sliver. Short staple yarns have also been made from cotton without carding by opening fibers with a beater mechanism and assembling them into a sliver in an air stream; the resultant yarns, however, were not always good enough to make this a viable process.

To form a sliver, the feed of fibers must be separated almost to single fiber formation on the card so that they can be reassembled side by side. This very complete fiber separation also enables the carding machine to perform a very efficient cleaning operation.

All this must be achieved with a minimum of fiber breakage, since fiber length is an attribute that makes for more uniform, smoother, and stronger yarns, but there are interesting exceptions to this golden rule—the case of bast fibers. Jute fibers are 10 -15 ft long and are interconnected by fibrous strands that start in one fiber and branch off and become part of an adjacent fiber; this repeated occurrence forms a lattice-like fiber structure. The carding process breaks down these long fibers so that they can be processed subsequently on drawing machinery suitable for processing fibers about 10 in. long and also, by tearing apart the branched structures, produces finer fibers. Jute is the cheapest of textile fibers so its conversion costs cannot be high; its physical properties are such that it is easily broken, and the fact that the card does this is used as a convenient basis for subsequent processing. Flax, although not carded, is also shortened in length during drawing in gill boxes.

Fibers issue from the carding machine as a fine web. This web is gathered together and passes through a funnel from which it issues in rope-like form as a sliver which then must be reduced in thickness in subsequent processes to that of the required yarn. There are two exceptions to this—woolen spinning and cotton waste spinning. In these systems (known as condenser spinning systems), the carded web is continuously divided into strips about a centimeter wide—the exact width depending on the count of yarn to be spun— and each of these is rubbed into an untwisted roll of fibers called a slubbing or roving. This splitting and rubbing is done by a machine known as a condenser. The rovings are then attenuated by about 50% to attain the required linear density of the yarn; spinning, therefore, immediately follows carding in these cases.

Sliver Attenuation—Drafting

Except in the woolen and cotton waste condenser systems of spinning, card sliver has many thousands of fibers in its cross section, whereas most single yarns that are required will have about 100 or fewer. The sliver must therefore be attenuated, or drawn finer, and this is done by passing it through drafting rollers. The simplest arrangement would consist of two pairs of rollers through which the sliver is passed. The first pair, the feed rollers, have a surface speed of V_f, and the second pair, the delivery rollers, a surface speed, V_d. Thus the fibers emerge at a faster speed than the speed of entry, and the sliver thickness is reduced in proportion; the reduction equals the ratio of roller speeds, $\dfrac{V_d}{V_f}$, which is called the draft.

This simple arrangement requires that the distance separating the roller nips must be comparable with the length of the longest fiber in the sliver. It is explained in Chapter 7 that this drafting system generates irregularities in the emerging product and that the greater the distance between the rollers, the greater will be these irregularities. The roller setting is therefore kept to a minimum, which is usually rather less than the length of the longest fiber; if it is made much shorter, too many fibers are gripped at both ends, drafting is impeded, and the roller surfaces are damaged.

In a drafting system, the feed rollers grip a relatively large group of fibers that is being fed forward at a slow speed. The delivery rollers grip a smaller group of fibers that is being drawn forward at a greater speed. These two groups of fibers form two interpenetrating tufts that are being separated from one another in a continuous way as the drafting proceeds. As this separation proceeds, the fibers of each tuft are drawn through the fibers of the other, and this straightens them to some extent and makes them more parallel to each other. If the fibers are not already parallel, fiber parallelization is therefore a by-product of drafting.

The fibers issuing from the card in web form are in a fairly entangled state; such an entanglement is in fact necessary to enable the web to hang together, and to be transported without breaking. When it is gathered together and compressed into a sliver, its cohesion is probably increased. Card slivers therefore have sufficient interfiber cohesion to give them adequate strength for handling, for packaging in various forms, and for withdrawing from these packages for further processing. As the fibers become more parallel during drafting, this cohesion is reduced, and slivers must be handled with more care; otherwise, the fibers will slip, and unwanted, irregular stretching of the sliver will occur. This will introduce unnecessary unevenness and perhaps excessive end breakages at subsequent stages. The slivers may have to be assisted from cans and supported at later machines to avoid these consequences.

As sliver becomes attenuated, its cohesion becomes so slight that no amount of care will avoid damage during processing unless additional strength is imparted. In many cases this is done by twisting the sliver, which may now be called a slubbing or roving*; in some worsted processes, the additional cohesion is provided by rubbing the roving between rubbing aprons. This twist, or rub, imparted to a roving must be sufficient for it to be wound on to a package, and unwound from that package, without being damaged, that is, without being stretched irregularly. In neither case must the twisting or rubbing increase the cohesion to such an extent that the fibers will be prevented from drafting easily in the next drafting zone.

Mixing occurs during drawing because a number of slivers produced at different times at one stage of processing are doubled together and fed simultaneously, side by side, to the next machine. In many modern drawing sequences, this mixing, together with increased fiber parallelization, is the chief purpose of at least the first drawing operation, because in such cases no overall attenuation is achieved; the sliver emerging from the first drawing stage may even be heavier than any of those entering.

---

*The distinction between slubbings and rovings is not clearly defined; slubbings are usually the first intermediate products made with twist and have a high linear density; products of later drawing machines are called rovings.

Drafting therefore results in attenuation of the sliver, parallelization of fibers, and mixing caused by the attendant doublings. All these features make some contribution to the properties of the final yarn. As fibers are drawn parallel, this becomes an inherent characteristic of a cotton spun yarn, or a worsted spun yarn, in contrast to condenser spun yarns which are produced direct from a strip of card web with no roller drafting. A yarn produced by roller drafting tends to be lustrous, lean (to use a worsted term), and to produce cloths that are themselves lustrous and clean finished. In condenser spun yarns, the fibers are still in an entangled state and cannot be so closely packed together, although they may have been straightened out a very little during processing on the condenser and in spinning. These yarns therefore have a different structure and a smaller bulk density, that is, for the same linear density they have a rather greater diameter than roller-drafted yarns.

Formation of the Yarn

When the strand of fibers has been reduced to the specified linear density (count), it must be given sufficient strength to enable it to be fabricated into a cloth of some kind. This has been traditionally accomplished by twisting it; the amount of twist inserted is determined by the count, the fiber properties (mainly length and diameter), and the purpose for which the yarn is to be used. Warp yarns, where maximum strength and elasticity are required in the weaving process, have more twist inserted in them than do weft yarns from the same fiber, since the prime requirement for a weft yarn is covering power which is greater for a more softly twisted bulky yarn; nor is the weft yarn subjected to as much mechanical treatment—abrasion and repeated stretching—during weaving. Hosiery yarns are spun with even less twist to attain maximum bulk consistent with adequate strength for processing.

The required amount of twist per unit length of yarn having been specified, the length of yarn that can be delivered by a spinning machine in a given time (the production rate) is then determined by the maximum number of revolutions the yarn rotating mechanism can make in that time. Mechanical limitations to spindle speed have thus become an obstacle to the achievement of higher production rates in spinning, and the past few years have seen the introduction of new methods of yarn twisting to surmount this barrier, namely, open-end or break spinning, and self-twist spinning. Other developments propose the abandonment of the use of twist altogether as a means of imparting strength to a yarn; this leads to twistless spinning in which the necessary interfiber support is provided by adhesives.

The longer the fiber the less the amount of twist required for a given count of yarn. For example, the *lowest* twist multiplier $\dfrac{\text{Twist}}{\sqrt{\text{count}}}$ used for cotton yarn is for Egyptian weft and is about 3.2, based on cotton counts. Converting to worsted counts, this multiplier becomes 2.60, which is even greater than the value of 2.15 normally used for a *hard twisted* Botany warp yarn (1) which is made from longer fibers than Egyptian cotton.

Packaging of Yarn

The yarn produced must be wound on some form of package to enable it to be stored and transported in a tidy and convenient way. This necessary provision has no effect on yarn structure; therefore, no particular attention is paid to it here. We merely note that the forms of packages on which yarn is produced at the spinning machine are determined by the type of spinning process used. The size of the package in use determines the frequency of doffing and, on ring-spinning machines, the maximum spindle speed that can be employed; it also governs the number of knots that must be tied in the yarn in later operations. Such considerations are concerned with productive efficiency rather than yarn quality.

Combing

The discussion of the fundamental operations of yarn making has omitted any consideration of combing. This process is not a fundamental necessity in making a yarn; it is a means of achieving, or enhancing, some desirable yarn properties.

As explained, drafting results in some degree of fiber parallelization, and when twist is inserted, this enables adjacent fibers in a yarn to bind together more intimately over longer lengths and to maximize the strength it is possible to develop in the yarn. It also enhances luster.

In spite of this the potentialities of some fibers are not fully realized. Fibers arrange themselves in such a way that their ends, being under minimum tension when the yarn is formed, tend to migrate to the surface and protrude. Thus for a given count of yarn, the shorter the mean fiber length, or the greater proportion of short fibers, the more fiber ends there are, and the hairier and less lustrous the yarn will be. Other things being equal, such a yarn will also tend to be weaker.

For this reason, some slivers are combed, and in this process a proportion of the short fibers (those up to a selected length) are removed. This consists in broad terms of taking the sliver to pieces in successive tufts, combing both ends of the tuft with metal combs, and reassembling the combed tufts into a new sliver. As well as removing short fibers, this fine-combing process provides a very effective means of performing a final cleaning operation by removing remaining particles of impurity, neps, and any remaining fiber entanglements or adhesions, and of more effectively straightening fibers and parallelizing them. The amount of fiber removed is correlated with the length of the longest fibers extracted, and this is determined by machine settings.

The considerations governing combing, and the degree of combing employed, are largely economic. Until recently, it could have been said that all worsted-type yarns were combed. These long and comparatively expensive fibers are used for making yarns that are themselves relatively costly and can bear the cost of combing to enhance their quality. In recent years, however, a semiworsted, or half-worsted, process has been developed whose distinctive characteristic is that combing is omitted. It is used mostly for relatively long man-made fibers that do not contain a lot of short fibers, and for the production of carpet yarns in which the qualities of yarn luster, strength, and fiber parallelization are of secondary importance.

Only the best quality cottons (long and fine) are combed. This brings out the full luster of the fibers by getting maximum parallelization and, by removing short fibers, realizes the full potentialities of the fine, long fibers for fine spinning; this makes possible the production of fine, lustrous, quality fabrics that can well bear the cost of the combing process. Short and medium-length cottons are not combed; by definition they can only produce relatively coarse yarns and cloths. The process would not be economic and with shorter cottons is not mechanically feasible. When man-made fibers are processed in the worsted process, they are combed after carding; in the cotton process they are not combed.

In the processing of line flax into linen yarn, an analogous process known as hackling is always used before drawing. Large bunches of fibers (the fibers are almost 3 ft long) are combed at both ends before being assembled into a sliver which is drawn into a yarn. Again the product is of such a high quality that it can command a price to pay the costs of combing. The short fiber (tow) removed in this process is still a few inches long and is carded, drawn, and spun into flax-tow yarn in a process that also includes combing on a rectilinear comb.

It is a prerequisite of any combing process that the fibers to be combed should be disentangled and reasonably parallel. The combing of a sliver in which the fibers are entangled has the same result as the combing of matted locks of one's own hair, that is, fiber breakage and the production of excessive amounts of waste fiber.

THE MAJOR STAPLE YARN PROCESSES

Basic Systems

The major part of the world's staple yarn is produced by one or other of the following systems:

carded cotton spinning   worsted spinning   line flax spinning

combed cotton spinning   woolen spinning   flax tow spinning
cotton waste spinning   jute spinning

It is assumed that the student is familiar with, or can see and study, machine details; the intention here is to consider what is happening to fibers passing through the machines and the effect of this on subsequent yarn-making processes and resultant yarn properties.

Cotton Spinning

Table 9.1 shows the sequence of operations carried out in the carded and combed cotton systems.

Table 9.1. Cotton Spinning Systems

Carded Cotton		Combed Cotton	
Bale opening or digesting	O, C, M	Bale opening or digesting	O, C, M
Opening and cleaning	O, C, M	Opening and cleaning	O, C, M
Picking (scutching)	O, C, M	Picking (scutching)	O, C, M
Carding	O, C, M, S	Carding	O, C, M, S
Drawing 1	D, M, P	Drawing	D, M, P
Drawing 2	D, M, P	Lap formation	D, M, P
Roving	D, T, (M)	Combing	O, C, M, P, S
Spinning	D, T	Drawing 1	D, M
		Drawing 2	D, M
		Roving	D, T, (M)
		Spinning	D, T, (M)

O = opening, C = cleaning, M = mixing, S = sliver formation, D = drafting (attenuation), P = fiber parallelization, T = twist insertion, (M) = mixing occurs if more than one end is drawn together at these processes.

Opening and Cleaning

In modern cotton systems, fibers pass automatically through all processes up to and including picking—known as scutching in Great Britain—flow-regulating devices being employed in the hopper-feeding mechanisms placed at various points in the sequence of opening and cleaning machines. Mixing takes place in a fortuitous and random way in the hopper-feeders, which feed different sections, and in the air streams transferring fibers from one section to another. The use of a number of blending hopper-bale-openers, each feeding the same feed lattice and each being supplied from a large number of bales representative of the blend in process, ensures that cotton from a large number of bales is being mixed at any one time.

Opening is done mechanically by assembling the fibers in a thick wad or lap and feeding them forward by feed rollers to a beater revolving at high speed. Beaters vary in type, but the principle of operation is that they strike the material from the feed rollers in small tufts. This releases particles of impurity, which either fall out or are thrown out by centrifugal force through a series of closely spaced grid bars surrounding the beater. This grid prevents the majority of fibers from falling out, and they are carried in an air stream to the next machine, where they may be reassembled for a repetition of the process; opening and cleaning is done in stages, with the tufts being broken up into smaller and smaller units.

In order that beaters reduce the feed to small tufts, the feed rollers must grip the fibrous feed very securely; this serves to consolidate the material into a more solid mass and embed the trash more securely in it. This defeats the object of the process and also breaks up the brittle vegetable material into smaller particles, making it even more difficult to remove.

Recent opening machines avoid compressing the partly opened material. One recent machine, the step opener, has a series of six beaters arranged successively on an inclined plane at an angle of 45° (Figure 9.1). The material is fed in at the bottom, is taken across grid bars by the first beater, and encounters the second beater. As the motions of the two beaters are opposed, the material is beaten and agitated between them, dirt falls out through the grid, and the fibers pass upward to the succeeding beaters. Such a machine is said to remove 35-45% of the impurities. A similar action takes place in the older types of conical beaters in which the material is treated between beaters in a conical arrangement of grid bars and an upward air stream carries away opened cotton that has become bouyant. In another modern machine (Figure 9.2) rollers feed material *in a thin layer;* this is gently opened by a revolving beater and picked up by an air stream moving at high velocity into a tapering channel that is then turned sharply through an angle. At the outside of the bend is an opening through which the heavier particles are spat while the more bouyant fibers turn the corner with the air stream which is removed by a fan.

In the latter machine it is clear that some heavier unopened particles of material may be thrown out with the trash, and this is generally true of all cleaning devices; speeds, settings of critical parts, positioning of parts, and speeds of air streams have to be adjusted to strike a compromise between the amount of trash removed and the amount of useful fiber lost at each stage. At no point can trash be removed without forfeiting some fiber.

The better quality cottons contain less trash and therefore need less cleaning. This is convenient, since the fibers are longer and finer and would suffer from severe treatment; this is also true of man-made fibers. Trends in pickers have therefore been towards types of beater with gentler actions, for example, the Kirschner beater, for use with the finer, longer fibers.

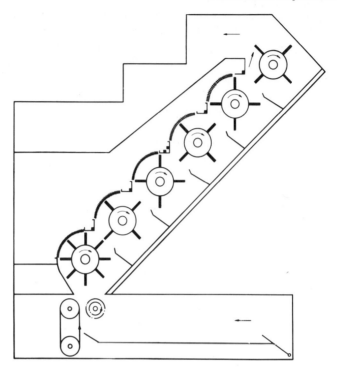

Figure 9.1. The step opener (courtesy Platt International Ltd.).

Modern trends are toward more automation, and the first process, in which the bales were formerly torn apart by hand so that slabs of fiber were presented to the bale openers, is now being mechanized by a type of machine on, or in, which the bale is placed and shredded into tufts that are fed into the opening line. The smallest units handled are therefore bales. At the other end of the opening process, there is a trend towards the elimination of lap formation and the feeding of the card directly by a specialized hopper known as a chute feed. Apart from cutting out manual handling, this also avoids any difficulties in lap formation; on the other hand, it eliminates a quality-control point. The picker (scutcher) made a lap of a given length that was weighed, and only those which were produced within certain weight tolerances were passed forward; the others were returned to the mixing at the beginning of the opening line. This ensured a relatively even feed to the cards with some assurance of constant average linear density of card sliver. With the adoption of chute feeds this possibility is now absent and the monitoring of card sliver with autoleveling devices at the first drawframe is essential.

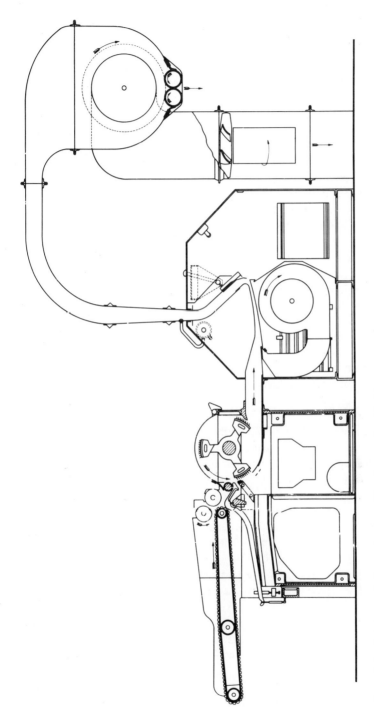

Figure 9.2. "Air stream" cleaner (courtesy Platt International Ltd.).

Carding

The taker-in region of the cotton card is designed to open the material to a still more finely divided state and to remove more trash. Following transfer of material from taker-in to cylinder, the interaction of cylinder and flats provides the last opportunity for the removal of impurities, and, since it is desirable that these working points should not be overloaded, it is important that the cleaning by the taker-in be as efficient as possible. The speed of the taker-in relative to that of the cylinder and its setting relative to the parts adjacent to it should be such as to give maximum trash removal with a minimum loss of good fiber, together with a uniform well-opened web with a minimum nep content. One cannot necessarily achieve all these things at the same time, and therefore compromises must be accepted.

Before the recent radical increase in taker-in speeds, and the redesign of cleaning devices below the taker-in by some manufacturers of high-production cards, it was found (2) that, with a normal cylinder speed of 168 rev/min, the best all around taker-in speed was between 420 and 600 rev/min, the higher speed gave better cleaning (at the expense of rather higher fiber loss) and a minimum amount of nep. If the taker-in speed approached the surface speed of the cylinder, nep content increased, as did the amount of good fiber lost; the web also became rather cloudy. Mote knives at the usual settings recommended by machine makers usually remove a maximum amount of trash with a minimum amount of good fiber. In general, the undergrid under a taker-in, or any other separating edge, removes more trash the closer the setting, but, at the same time, more useful fiber.

In the region of the taker-in, fine fibers, and in particular long synthetic fibers, require more gentle treatment, and the shape of the teeth of the taker-in clothing is modified to take account of this; the angle of the leading edge of the tooth is varied according to the severity of the action required.

The material is stripped from the taker-in by the cylinder and passes on between cylinder and flats, then under the front plate to the doffer. At the doffer, fibers are removed and eventually enter the card sliver. We consider what happens to fibers at three points—between cylinder and flats, at the front plate, and between cylinder and doffer.

Figures 9.3 and 9.4 show the two main actions that take place on carding machines They show two closely spaced, relatively moving surfaces clothed with card clothing. In Figure 9.3, surface A represents the cylinder with fibers partly embedded in its clothing; surface B is initially free from fibers and may be moving to left or right. The motion is usually to the right, and, in the case of the cotton card, surface B represents the flats moving at speeds of anything from 2 to 8 in./min, whereas the speed of surface A (the cylinder) will be at least 2000 ft/min. In practice, the surface of the flats is tilted so that the setting between the surfaces at X is 0.030 in. and at Y 0.010 in. With these close settings, protruding fibers, neps, and impurities tend to be caught on B where the particles

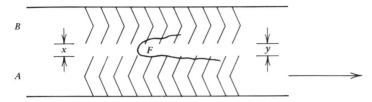

Figure 9.3. Carding action.

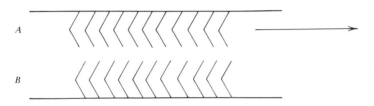

Figure 9.4. Stripping action.

become impaled, and the fibers are partially disentangled and separated from one another, some being retained on surface B, whereas the rest are swept on by A. This is termed a carding action and takes place between the cylinder and doffer as well as between cylinder and flats.

In Figure 9.4, surface B is relatively slow moving, with fibers embedded in its surface, and the teeth of surface A moving past it at as shown will, if set close to B, engage the fibers on it and draw them away from this surface, whose teeth are inclined in such a direction as to assist rather than impede this action. Surface B is therefore stripped of its load of fibers, which is carried away by A; this is the way in which the taker-in is cleared by the cylinder of the cotton card.

The fibers and trash collected by the flats are slowly transported to the front of the card, and as the flats are in turn removed from the cylinder each is stripped of its accumulation, which is entirely removed from the carding process.* If the speed of the flats is increased by a few inches per minute, the carding action is not impaired, but as the area of clean flat surface coming into action every minute is increased, more flat strip is accumulated and removed.

The detailed observation of individual fibers as they pass through a card is virtually impossible; there may be 100 million in a cotton card at any instant, and the behavior of individuals between flats and cylinder cannot be detected visually. DeBarr and Watson (3) employed an occasional radioactive tracer fiber in material being carded and followed its passage by Geiger counters placed at

*This is the only type of card in which this occurs; all other spinning systems employ roller cards on which the material picked up by the slowly moving surface is later returned to the stock in process. It seems to be an essential requirement of the cotton process that neps and trash be removed; this is presumably why the flat card has evolved in this industry.

various points around the card. They found that a large proportion of fibers is retarded with respect to the cylinder by being caught on a flat for varying, but very short, lengths of time, and a very small percentage of these—consistent with the amount of flat strip being made—are not released at all but remain there. According to Charnley (4) "approximately two-thirds of the fiber material in the flat strips is similar in quality to that in the lap and is received by the flats immediately they position themselves on the bends at the back of the card...and takes no further part in the carding action." It is only the remaining third that is concerned with the interaction between flats and cylinder, it is rather shorter in length and contains the nep and trash. This interpretation of the intersurface action is based on the observation that long fibers are at the base of the wire of the flats.

The removal of flat strip contitutes a loss of valuable fiber material, say up to 3%. Its amount can be controlled by varying the setting of the front plate to the cylinder surface; a close setting reduces the amount, a wider setting increases it. A close setting will therefore reduce the loss of good spinnable fiber, but it will also result in less trash and nep being removed—again the optimum setting is the result of a compromise between these two considerations. According to Morton and Cheng (5), the reason for the decrease in the amount of flat strip as the lip of the front plate is moved nearer the surface of the cylinder is that this intensifies the current of air being carried under the plate by the quickly moving cylinder. Some of this air comes down between the flats, and, as its velocity increases with a closer plate setting, it tends to draw more of the fibers into contact with the cylinder wire, which peels off more of the strip. It is not thought that the air currents themselves remove the fiber, but only that they assist the cylinder to rob fiber from the flats.

The relative motion of cylinder and doffer has been stated in Figure 9.3 in which B represents the surface of the doffer. Some of the fiber is transferred to the doffer surface B and is removed. De Barr and Watson (3), using the radioactive tracer-fiber technique, showed that in general only about 20% of the fibers are removed the first time they reach the doffer. On the average, a fiber makes 15-20 circuits of the cylinder before being removed; some fibers make more than 60 circuits.* As the number of circuits of the cylinder made by a fiber increases, the probability of its eventual transfer to the doffer decreases, but even after 200 circuits there is still a possibility of removal, although it is not known what determines this. This action shows that some amount of mixing is performed by a cotton card as fibers fed in are filtered out by the doffer after different numbers of revolutions of the cylinder. However, in view of the high speed of the cylinder, the effect cannot be very great; if an average fiber is held up for 16 revolutions of the cylinder, this only represents a delay of 6 sec if the

---

*These findings refer to experiments on a card clothed with wire card clothing and not the metallic clothing used on modern cards.

cylinder speed is 160 rev/min. Even a fiber making 60 circuits of the cylinder would emerge only about 22 sec after another fiber that entered the machine with it but that was removed by the doffer at their first encounter.

The mixing power of the cotton card has also been demonstrated by Butorovich (6) and Borzunov (7). The former varied the rate of feed to a card in a periodic manner and showed that the card smoothed out the variations to some degree. Borzunov arranged for a sudden (step) change in the feed; if fibers passed through and were immediately removed by the doffer there would be a similar step change in the output; in practice, as anyone who has started or stopped a card knows, the sliver thickness gradually builds up on starting, that is, when increasing the rate of feed, and gradually falls on stopping. From the rate at which the output responded to a step change in input, Borzunov found, for different cards at different settings, that the average time for a fiber to pass through ranged from 1.6 to 7.2 sec. An earlier experimenter, De Swann (8), showed that if a small amount of fiber was fed to a card with no doffing comb operating, it distributed itself over cylinder and doffer in a definite ratio, and if the cylinder was stripped and the machine run with no doffing comb, some of the material on the doffer would be transferred back to the swift.

Recent developments in carding that have increased production rates three- or four-fold have been based on the use of rigid metallic card clothing instead of flexible wire clothing. It is also necessary, when processing at these high rates, to reduce the load of fiber on the cylinder and to minimize the amount of trash going forward to the cylinder and flats. To achieve the first, the surface speeds of cylinders have been increased to something like 4000 ft/min, and for the second reason the taker-in surface speeds have been increased to about 2400 ft/min and every effort made to improve cleaning arrangements at the taker-in. Since doffer speeds have been correspondingly increased to cope with the increased output, doffing combs have been replaced by doffer stripping rollers common to the woolen industry, and web crushing rollers of the type used in the woolen industry have been imported into the cotton industry to crush the impurities in card webs.

Fiber Hooks

An important consideration concerns the configuration of the fibers in the sliver after it leaves the card. Morton and Summers (9) and Morton and Yen (10) were the first to investigate this by introducing a small number of black tracer fibers into a lap feed and studying their shape in the card sliver. They divided the fibers according to shape into five groups, as shown in Figure 9.5. Group 5 fibers were those so rolled up and entangled that they could not be assigned to any of the other groups.

Morton and Summers, working with viscose staple fiber, and Morton and Yen, working with Egyptian cotton, obtained the analyses of fiber shapes in card

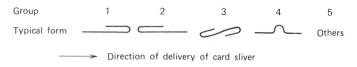

Figure 9.5. Shape and orientation of fibers from card.

sliver shown in Table 9.2. In both cases not only did they find that the majority
of fibers were hooked as in group 2, but they also found that the hooks of group
2 were larger than those of group 1.

Table 9.2. Percentage of Fibers in Fiber Groups Shown in Figure 9.5 (Morton et al.)

Group	1	2	3	4	5
Viscose staple	16.2	47.8	13.2	20.8	2
Egyptian cotton	10	42	8	25	15

   This effect is now well established—it is accepted that the majority of hooks in
card sliver are on the trailing ends of fibers as they leave the card and the
minority are on the leading ends. They have therefore been designated as majority hooks and minority hooks, and their removal in later processes is an important consideration in the technology of cotton spinning.
   Morton and his co-workers suggested that the reason why the trailing hooks
are in the majority is that, at the point of transfer from cylinder to doffer, part
of a fiber becomes attached to the doffer and the cylinder then sweeps the rest
of it past the point of attachment so that the tail of the fiber emerges first. This
is indicated by fiber F in Figure 9.3. Ghosh and Bhadhuri (11) have more recently stated that the taker-in, the flats, and the stripping of the doffer have no
effect on hook formation, which supports the view that the source of their
production is at the point of cylinder-doffer interaction; changing the ratio of
the speeds of these two parts affects the number of hooks formed. According to
Fiori, Simpson, and De Luca (16), increased production rates can result in a
decrease in the number of majority hooks and an increase in the number of
minority hooks. With medium and short cottons, the latter effect is not very
marked; thus in these circumstances, increased carding production can be
beneficial in decreasing the total number of hooks.
   The presence of hooked fibers in slivers, or other strands in process, reduces
the effective fiber length in the strand, and properties that benefit from length
of constituent fibers thereby suffer. For example, if hooks persist into the yarn
the yarn will be weaker, and there will consequently be more breakages in spinning and subsequent winding operations.
   Between each operation of processing the strand is reversed so that fiber hooks
are presented alternately head first and tail first to the feed rollers of successive

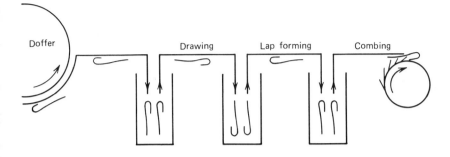

Figure 9.6. Reversals of fiber hooks prior to combing.

machines. Figure 9.6 shows the orientation of majority hooks proceeding from card to combing machine. These preparatory processes involve drafting and so disentangle the fibers and remove some hooks, but there are still many hooks left at the combing machine, and the majority are presented head first to the comb cylinder which readily straightens them out. If the preparatory processes contained one more or one fewer process than this, the majority of the fibers would be presented tail first for combing; when released by the nips they would have hooked trailing ends, would be treated as shorter fibers, and many would be removed as waste. Thus two intervening processes, leading to three reversals between carding and combing, give less waste than one or three such processes, straightens the majority of fibers, and ensures that the maximum proportion of spinnable fibers is retained in the combed sliver.

Table 9.2 shows the proportion of different groups of fibers found by Morton and Yen (10) in Egyptian cotton card sliver. Table 9.3 shows the proportion in the ribbon lap before combing and in the combed sliver in the same experiments. It will be seen that, after combing, 90% of the fibers have no hooks, and the number of fibers so entangled as to be unclassifiable has been reduced to 3%. It should be noted that the majority hooks in group 2 are pointing in the same way in card sliver and ribbon lap (see Figure 9.6) so that the percentage has been reduced from 42 to 25, but the direction of fiber ends has been reversed by combing, so the comparable figure in combed sliver is 5%; there are now no fibers with hooks on both ends.

Table 9.3. Percentage of Fibers in Various Fiber Groups as Drawing and Combing Proceeds (Morton and Yen)

Group	1	2	3	4	5
Card sliver	10	42	8	25	15
Ribbon lap	11	25	5	52	7
Combed sliver	5	2	0	90	3

Morton and Nield (12) found that by interposing another drawing operation between carding and combing the amount of comber waste produced was increased from 13.1 to 16.8%, but if the extra process was accompanied by an additional reversal of sliver at some point in the process, so that the majority hooks were still presented head first at the comb, the waste produced was only 11.45%. Thus an additional drawing process prior to combing does give rise to more parallel fibers and fewer hooks, but this is more than offset by the presentation of majority hooks in the wrong direction unless this is artificially overcome by a specially arranged sliver reversal. This indicates that the lap normally presented for combing is far from being in an ideal condition for combing and results in more waste being extracted than otherwise need be. This illustrates the importance of the preparatory processes before combing.

The presence of hooks on fibers in the drawing and spinning processes might well be of importance with modern tendencies towards higher drafts, higher speeds, decreased number of doublings, and fewer operations. Given a sufficient number of drawing operations, the fiber parallelization effected might remove most of the hooks, but with the reduction in number of operations some might persist to the spinning stage and affect spinning performance. De Luca and Fiori (13-17) have carried out a long series of experiments to examine these effects, and their results may be summarized as follows:

1. Hooks of both types are most easily removed when drafted in a trailing direction. Minority hooks are more easily removed than majority hooks, probably because they are shorter. Increasing the draft generally reduces fiber hooks. Increasing the weight of sliver in drawing assists in hook removal.
2. Evenness of roving and yarn is improved when majority hooks are drafted in a trailing direction.
3. There are more ends down the greater the number of hooks entering spinning, and the yarn is weaker. Fiber hooks have a more critical effect on these factors the finer the yarn.
4. Combing and postcombing operations so effectively reduce fibers hooks that the above effects are negligible for combed yarns.

These workers have therefore stated that the best results are obtained with carded yarns if the majority hooks are trailing in the first two drawing operations, leading at roving, and trailing at spinning. Such procedures require reversing the creels at first and second drawing. They advocate that whenever possible majority hooks should trail in the spinning process. As state earlier with regard to carding, higher rates of production lead to fewer majority hooks with a resultant improvement in spinning, but if the cylinder begins to overload, hook formation increases.

Drawing and Spinning

The fundamental importance of yarn evenness is stressed in Chapter 7. Medium and long-term unevenness have been reduced by the invention of evenness control systems applied to drawframes. These, known as autolevelers or servo-drafting systems, measure the slivers entering the drawframe and regulate the draft proportionally, thus the mean linear density of the emerging sliver remains constant. In the great majority of cases, the ingoing slivers are passed through a pair of tongued and grooved rollers similar to those used in compression-type evenness testers (Figure 7.5); the linear density is measured and stored in a memory device carried in a rotating cylinder and, when a particular section reaches the point in the drafting zone at which it is reduced in thickness, the information regarding its linear density is recovered and the draft is varied accordingly. This is usually done by varying the speed of the feed roller so that the machine production rate (determined by delivery roller speed) is unaffected. Other types of sensing units utilize capacitance, photoelectric, and pneumatic principles, but the mechanical compression system is used most. Such draft control systems eliminate the need for a lot of doublings, which were formerly used to reduce long-term variations, and help to shorten the drawing process.

Short-term unevenness in yarns is caused by drafting waves generated in spinning and roving machinery and has resulted in great attention to the placing of restraints on the motion of floating fibers in the drafting zones of these machines. This has produced hundreds of inventions that can be divided into (1) double-apron drafting systems, in which the drafting fibers are gently gripped between two bands traveling at the surface speed of the back rollers; thus floating fibers are held back until they are almost into the nip of the delivery rollers, whereas long fibers being drafted are allowed to slip through, (2) single-apron drafting systems, in which a similar band is used to support the strand and carries various arrangement of relatively light rollers that are used to restrain floating fibers, or (3) specialized roller systems that attempt to perform the same functions. Double-apron systems now dominate the scene; they are used to produce commercially acceptable yarns with higher drafts, and this has again enabled the drawing process to be shortened. Old machinery worked on drafts of the order of 6 to 8; modern controlled drafting systems use drafts of the order of 30.

Worsted Spinning

Worsted systems show more complexity and variety than cotton systems. This is because wool and animal fibers have a much wider range of physical properties than cotton fibers, and optimum conditions of processing require a wider range of methods of treatment.

Wool is received as fleeces in press-packed bales and is first sorted into different grades based on characteristics of length, fineness, soundness of staple, and color. Only sound wools of good color and suitable length are used in the worsted process, the others being used for woolen yarn. This is an extensive subject well treated by Von Bergen (18); it is essentially a process of fiber selection in which a little dirt is liberated fortuitously as a result of the handling that the wool receives.

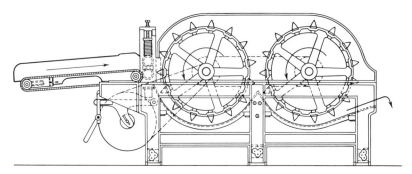

Figure 9.7. Double cylinder wool opener.

Blends of wools are then made in which wools with different selected characteristics are mixed to provide the top or yarn characteristics ultimately required.
The price of the product is also an important consideration.

The blend is next cleaned and opened, and the first step may be dusting or
shaking. This is done in a double-cylinder wool opener, as shown in Figure 9.7*.
By beating the wool between revolving spiked beaters, some loose dust, sand,
and dirt can be shaken out. Clearly this is only worth doing if there are worthwhile quantities of loose material to be removed, but the removal of significant
quantities of such material assists in the economical use of scouring agents in the
next process, and by opening the wool to some extent, the process facilitates the
penetration of the liquor into the locks of wool. Some wools are obtained from
the pelts of dead sheep and the process may leave lime on the fibers; if this
enters the scouring bowl, insoluble lime soaps (which are difficult to remove) are
deposited on the fibers, and a preliminary dusting or shaking before scouring
reduces this difficultly.

Wool contains grease, suint, sand, dirt, and vegetable matter. Wool grease varies
with type of wool, but is largely a mixture of higher fatty alcohols and fatty
acids and has a melting point of 40-45°C. It may be emulsified by alkaline solutions at temperatures above the melting point, but wool fiber can be damaged by
hot alkaline solutions; therefore the temperature of scouring liquors using soap
and alkali is restricted to 55°C. Suint, which is exuded from the sweat glands of
the sheep, consists mainly of potassium salts of various fatty acids and is soluble
in water; this aqueous solution assists in the emulsification of grease. Soft water
must be used or a deposit of lime soap will be formed on the wool. A neutral
scouring process using nonionic detergents can be used instead of soap and
alkali, in which case a higher temperature (60-70°C) can be used, which improves
emulsification of the grease. Such detergents do not yellow the wool as alkali
tends to do, nor are the conditions so conducive to the felting of the fibers in
process.

*Figure 9, Von Bergen's *Wool Handbook,* Vol. 2, Pt. 1, p. 20.

The wool processed by either of the above methods is fed to a series of scouring bowls by a hopper feeder that provides a uniform flow of wool at a correct rate to ensure a uniform and adequate treatment. Fine wools, which have a higher grease content, require more treatment than medium and coarse wools, but three or four bowls with decreasing detergent concentration remove the greater part of the impurity. The last bowl provides a rinse to reduce the alkalinity of the wool. In the first bowl, the wool grease is emulsified and partially saponified, the dirt and grease droplets are suspended in the soap solution and are removed with it when the wool is squeezed between the delivery rollers placed at the end of the bowl; the liquor removed is returned to the bowl, and the wool passes to the next one for another similar cleaning operation.

This long-established method of cleaning wool (and other animal hairs) has depended on the cheap availability of water and the easy disposal of grease-laden effluents. Changing circumstances in both these respects call attention to the alternative possibility of degreasing by organic solvents. The main considerations in comparing the two systems are economic, but attention must be paid to the fact that organic solvents may be toxic. The ultimate choice of process will depend on the relative cost of effluent treatment and disposal on the one hand, and the cost of solvent and of solvent recovery processes on the other, as well as the overriding consideration of possible legislation. At the moment, most wool is processed in aqueous solution.

When a wool fiber is immersed in water, its two components (the orthocortex and the paracortex) swell at different rates; the fibers then begin to bend and twist because different swelling forces develop in the two parts, and this fiber movement promotes felting. In addition, if fibers are propelled through the scouring liquors by forks, they also become entangled and felted to an extent, depending on the particular type of mechanism. Very dirty wool requires vigorous agitation during scouring, and this increases fiber entanglement.

After the scoured wool is dried, it is carded and anything that has led to felting or fiber entanglement before carding increases fiber breakage and nep formation during carding and produces a greater proportion of short fiber in the carded sliver. This results in more short fiber—noil—being extracted by the comb and less spinnable fiber—top—being produced; in fact, the ratio of top to noil—the tear—is often used by research workers as a measure of fiber breakage in carding. Felting of fibers during scouring is therefore a most undesirable feature, and modern developments in scouring techniques have sought to reduce it. The use of nonionic detergents that reduce alkalinity and solvent degreasing methods that avoid fiber swelling have been mentioned; the other method is to avoid the mechanical propulsion of freely floating fibers through scouring liquors. In this method, which has many variants, the wool may be carried on an open-meshed porous belt and be sprayed with hot aqueous detergent solutions, or with organic solvents. In one system the wool is firmly held between such a porous conveyor belt and a revolving perforated drum while it is conveyed through the scouring liquor and sprayed from both sides by pressure jets.

Table 9.4. Worsted Systems

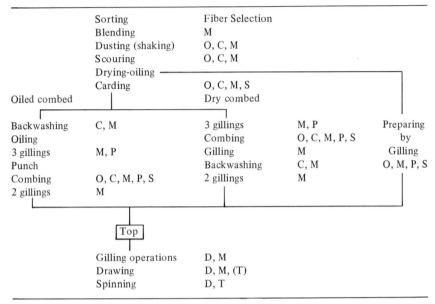

	Sorting	Fiber Selection		
	Blending	M		
	Dusting (shaking)	O, C, M		
	Scouring	O, C, M		
	Drying-oiling ──			
	Carding	O, C, M, S		
Oiled combed		Dry combed		
Backwashing	C, M	3 gillings	M, P	Preparing
Oiling		Combing	O, C, M, P, S	by
3 gillings	M, P	Gilling	M	Gilling
Punch		Backwashing	C, M	O, M, P, S
Combing	O, C, M, P, S	2 gillings	M	
2 gillings	M			

Top

Gilling operations	D, M
Drawing	D, M, (T)
Spinning	D, T

(T) Depending on process.

After scouring and drying, the wool is ready for the spinning process, which takes place in two stages—top-making and spinning. The worsted spinning industry is horizontally organized into these two sections, but some firms carry out the complete process. All wool is processed in its natural color (undyed) up to the end of the first stage.

For the majority of wools, the first operation is carding. However, when wool contains about 7% or more of vegetable matter, this cannot be completely removed by carding and it is removed by carbonizing; wool that has been treated in this way is rarely processed on the worsted system—it goes to the woolen process. Another exception concerns long fibered wools like luster wools, mohair, and alpaca, which are made into fabrics that utilize the lustrous character of these fibers to the full. Carding such long fibers between and around the rollers of a worsted card causes excessive fiber breakage; thus many short fibers and hairy yarn would be produced. To avoid this loss of lustrous quality, these materials are *prepared* for the comb by a series of gilling operations. A fiber length greater than about 7 in. represents the point at which the preparing process is adopted, but wools of this order of length might be carded if required for purposes where yarn luster is of no importance, such as in the case of hosiery yarns (19).

After carding, the wool is prepared for combing. In one system known as the English or Bradford system, 3½% of oil is applied before combing; in the other no oil is applied, apart from ½% before carding. This latter process has been called the French or Continental process, and in recent years a so-called American system, also using dry combed wool, has been developed. Modern trends in spinning technology have modified all these systems and have reduced them to two basic systems—oil-combed and dry-combed. These terms are used here. The oil-combed system traditionally used the longer carding wools (4-7 in.), and the dry-combed used shorter fibers; modern dry-combed systems are now able to process the longer fibers as well as being well adapted to spin shorter staples. The reason for the two systems is historical, but the processing of the longer fibers in oil increases fiber cohesion and, together with other features of the oil process such as the drafting of twisted strands, produces lean, smooth yarns traditionally much favored in the British wool textile industry for the manufacture of clear-finished worsted fabrics. Dry-combed systems produce fuller yarns.

Carding

Worsted carding takes place on roller cards fitted with workers, strippers, and fancy rollers. The principles of operation in performing opening, cleaning, and fiber mixing are the same as for woolen cards.

The card usually consists of two cylinders and their associated doffers, known respectively as the breaker and finisher sections, with a hopper feed and a licker-in or small breast to open the fiber material and present it at a uniform rate, and in a partially opened state, to the breaker cylinder. Burr beaters are set close to some of the rollers, for example, strippers; these are metal rollers with protruding blades set parallel to the carding roller and revolving at high speed; they strike and remove any burr or other piece of vegetable matter riding on the surface of the clothing because it is too large to penetrate the wire. When the wool has passed through the breaker section, the web may be treated by other devices, for example, crushing rollers. In this opened state, impurity is accessible for treatment and, having been treated, is more readily removed in the finisher section.

About ½% of oil, or the same amount in emulsion with water, is applied before carding; it minimizes fly and the liberation of dust in the card room and lubricates the fibers. This latter is often regarded as significant, but whether it is so is doubtful, since oil content does not affect fiber breakage. Too much oil increases nep formation and with modern metallic clothing makes the cylinder difficult to clear; the wool forms a lap around the cylinder and results in the sliver delivery breaking down.

It is also advisable not to dry the wool too much before carding since dry fibers are less elastic and form more neps (20). According to Townend and

Spiegel (21) the fewest neps are formed when the regain lies between 30 and 50%. Although it is inadvisable to store wools at such regains before carding because this renders wool liable to mildew, there is no harm in feeding relatively moist wool to a carding machine, since it very quickly reaches moisture equilibrium with the surrounding atmosphere, and this might even be beneficial in helping to avoid the formation of static electricity that can cause the web to break down. Neps are also liable to form if settings between cylinders and workers are too wide, if the card wire is not sharp, and if production rates are so high that the cylinder is overloaded (increasing cylinder and doffer speeds will reduce this). In spite of these factors affecting nep formation, it is generally recognized that much nep is due to its incipient formation during scouring where local fiber entanglements occur. Most fiber breakage in carding is also caused by this, as has been emphasized; given normal card settings fiber breakage is not affected by running conditions and seems to be concentrated in the earliest stages of opening on the card where the wool enters the machine and is subjected to the tearing action of the card teeth while still being gripped by the feed rollers.

Across each card cylinder between the last pair of workers and strippers and the doffer is the fancy roller. The card wire of this roller has long teeth and acts rather as a wire brush; it is set to penetrate the teeth of the cylinder slightly, and, since the surface speed of the fancy is greater than that of the cylinder, it sweeps through the cylinder teeth and raises the fibers to the surface so that they are more easily captured and removed by the doffer. It is interesting to contrast the need for a fancy roller on a worsted card and the lack of it on a cotton card, and to consider the different fiber characteristics and the different mechanical constructions of the two cards in this region just before the doffer.

In spite of the raising action of the fancy, fibers are not necessarily removed by the doffer at their first encounter. The probability of removal is possibly not very different from that on the cotton card, and this results similarly in a circulation of fibers around the cylinders, giving a mixing action; it also helps to overload the cylinder. Any action, such as increasing doffer speed, or increasing the efficiency of the fancy, that helps to improve the efficiency of fiber removal will reduce the load of fiber on the cylinder and should reduce nep formation.

Fibers in worsted card sliver are hooked, and again majority hooks trail (22). Less research has been done on fiber configuration in worsted card slivers than in the case of cotton, and although Belin and Taylor (22) report that, in one case examined, the majority hooks were longer than the minority hooks, in another case they were the same. It is not known how the fancy action or the fiber properties might affect this; wool fibers tend to return to their original shape when released from the constraints placed on them in card clothing.

Preparation for Combing

During carding, fibers have been well opened and small traces of wool grease, and dust, which escaped scouring because the wool was in lock form, are now more easily accessible. In the oil-combed process, the card sliver is therefore backwashed, that is, it is given a mild scouring treatment. A large number of slivers, say 60, are passed side by side through a wash bowl containing a soap or detergent solution, through a second one giving a rinse, through a drier, and then into a gil box in which they are drafted. Before entry to the gill box the oil is added so that it begins to be spread uniformly over the fibers from this point. In the dry-combed process, backwashing occurs after combing at the point shown in Table 9.4. According to Von Bergen (23), the producer of dry-combed tops "is primarily concerned with the straightening and crimp removal of individual wool fibers, as their natural crimp and tendency to curl may assume positions which would impede the French spinning process ...... the main object (of back-washing) is a thorough wetting of the individual fibers to make them pliable and adaptable for stretching when dried under tension". In the oil-combed process, combing is carried out on a Noble comb in which the pins used for combing and the bed in which they are fixed are steam heated. The combing of oiled wool containing moisture through *heated* pins considerably enhances the sleek and silky appearance of the fiber arrangement in the top and contributes to its spinning quality.

In accordance with general principles, the gilling processes before combing are introduced to reduce fiber entanglements. The pins of the gill boxes control the movement of short fibers and minimize the development of unevenness in the slivers, as well as assisting in fiber straightening. The Punch in the oil-combed system is a machine that assembly winds slivers for insertion into the creel of the Noble Comb and in itself contributes nothing to the process—except that it introduces a sliver reversal.

It has been shown (22, 24) that increasing the number of gillings between carding and combing progressively reduces the amount of fiber entanglement and the amount of noil removed during combing, but the benefit of additional gillings diminishes, as one would expect, as the operation is continually repeated. There was little practical advantage to be gained by employing more than the normal three preparatory gillings.

The amount of noil removed at the comb depends on whether the sliver is presented with majority hooks trailing or leading. In oil combing, using Noble combs, the least noil is produced with majority hooks leading, which is in accord with the industrial practice of having three preparatory gilling operations and a sliver reversal at the Punch. Dry combing employs the rectilinear comb similar in action to that used in cotton combing, but with machine dimensions increased to accommodate longer fibers; it is unexpected, therefore, to find in this type of worsted combing that the least noil is usually produced when the sliver is fed with majority hooks trailing (22). This is apparently not always true; it depends on the wool, and in some cases with low noil settings less noil is produced with majority ends leading. The suggestion is made that the behavior is determined by the hook-length distribution and would seem to represent another case in which the different properties of cotton and wool fibers—in particular the tendency of the latter to adopt crimped and curled forms when free to do so—result in different behavior in processing. It is suggested by Belin and Taylor (22) that with the most common hook-length distribution and machine settings, if majority ends lead, both ends of a fiber may be gripped by the nippers so that the hook may be broken by the rotating comb. This increases the amount of short fibers and so produces more noil than when the minority ends lead.

Combing and Top-Finishing

The purposes of combing have already been given, and details of the combs that have been mentioned are given by Von Bergen and in other textbooks. Some comparisons between Noble and rectilinear combing have been made by Belin, Taylor, and Walls (25). The gilling operations following combing are designed to improve the evenness of the combed sliver, to produce a top with an acceptable moisture content and linear density that forms a standard product for use by spinners, to blend the output of a number of combs, and to package the top sliver in suitable form for storage, handling, and transport. It is sometimes said that they also continue fiber parallelization, but whether they can achieve anything significant in this regard after that produced in combing has not been proved.

Preparing

The preparing of long fibers is done by a series of six gill boxes. The scoured wool is fed manually to the first box, and, passing through the fallers in the drafting zone, the fibers are partially disentangled and straightened, they are then delivered to a long, endless, revolving apron around which they wrap and on which a sheet of fibers builds up. When this has reached a sufficient thickness, it is broken, removed, and fed to the next machine in which the operation is repeated. The third and subsequent machines collect a sliver in cans, and by using several cans to feed the last three boxes, effective mixing and uniformity of sliver is achieved. Fiber parallelization is improved throughout the process, and the pinning of the fallers becomes finer as the gilling proceeds; later boxes may be intersector gill boxes. If backwashing is required, it may take place before the penultimate gilling; oil is added after backwashing, or, if this is omitted, at this same gilling or the one preceding it.

Top Making from Man-Made Fibers

Man-Made fibers in staple-fiber form can be handled by conventional wool-processing worsted machinery. There is no need for cleaning, and therefore processing commences with carding. Since the material is reasonably uniform in properties, and cleaning is unnecessary, the worsted-type card used can be of the most straightforward kind and is usually clothed with metallic clothing. Back-washing is not necessary, and no oil is added; otherwise, either the Noble or rectilinear comb can be used, and the top finishing processes are necessary for the reasons already given.

The use of man-made fibers is worsted spun yarns is growing, and therefore the use of the Noble comb, which is essentially used for oil-combed wools, is declining, accentuated by the fact that knitting, especially of man-made fiber yarns, is growing at the expense of weaving, and the Noble comb is not necessary for knitting yarns.

The number of worsted combs as a whole is declining for the same reason—the use of man-made fibers in worsted spun yarns. It is an unnecessarily expensive procedure to cut up tows containing continuous filaments of man-made fibers, card them so that they become entangled, and put them through this long top-making process to rearrange the fibers in parallel form. Machines known as tow-to-top convertors have been developed to break or cut the filaments in the tow to the required staple length, to shuffle the fibers so that the continuity of the sliver is maintained, and so form the top. This process also avoids the wasteful production of unwanted short fibers that are removed as noil in the older process.

It might be noted that, where worsted spun yarns are made from blends of wool and man-made fibers, or blends of man-made fibers alone, the tops for each type are made separately and blending takes place later (see Chapter 10).

Top Dyeing

Tops are always made from fibers in their natural color, but worsted yarns may be spun white or colored. In the latter case, tops are dyed, and this causes the fibers to adhere together. To loosen up this slightly matted state, they are recombed; the recombing process is followed by two finisher-gilling operations producing colored tops. The blending of colored tops is also mentioned in Chapter 10.

Drawing and Spinning

The principles of worsted drawing and spinning are the same as for cotton spinning, but the machinery is somewhat different. The longer fiber length means that drafting zones must be longer, and this always provided room for fiber control devices to be inserted between the drafting rollers, which was not possible in early cotton drafting systems. Thus the first machines in the process, the counterparts of the cotton drawframes, had faller bars in the drafting zone, and the bed of pins so formed controlled the movement of floating fibers in the drafting of slivers. When the sliver became so fine that it could be damaged in handling between processes, it had to be given cohesion. In the oil-combed system this was done by inserting twist after drafting, in the continental dry-combed system by rubbing the strands between reciprocating rubbing leathers.

Once twist had been inserted into the oil-combed system, gill boxes could no longer be employed, since the pins could not easily penetrate the twisted strand without being damaged. The method of fiber control traditionally adopted was to employ two or three sets of carriers and tumblers between the drafting rollers; these ran at progressively increasing speeds from back to front and consisted of a driven steel bottom roller—the carrier—on which the drafting strand rested, and a light wooden cylindrical roller—the tumbler—that sat on the strand above the carrier. Although the slight pressure of the tumbler on the strand might have restrained floating fibers from moving forward, the fact that the twist was prevented from running towards the thinnest section near the front rollers undoubtedly made the greatest contribution to uniform drafting; the twist, being held back, anchored the floating fibers more securely to the slowly moving fibers being fed forwards. This method of fiber control was adopted on all open drawing and spinning machines in the Bradford oil-combed system, and only small drafts of the order of 6-10 could be employed; up to eight stages of drawing were needed.

The continental dry-combed system employed a rotating roller covered with pins, called a porcupine, to excercise some fiber control; it was placed just behind the front drafting rollers so that the fibers were drawn through its pins. The maximum draft that could be applied was 4.5, probably because of poor short fiber control provided by such a device, and up to ten operations of drawing were necessary to perform the attenuation from top to yarn.

Modern Drawing and Spinning Processes

The improvement of gill boxes resulted in the development of the modern high-speed intersecting gill box or pin drafter. This has improved fiber control, increased production rates, and allows heavier sliver weights to be processed. The fitting of autoleveling mechanisms to pin drafters has removed the need for continual doubling to achieve evenness; doublings are now used in these machines to achieve mixing. At the spinning machine, the successful development of double-apron drafting for long staples has enabled drafts of the order of 30 to be employed. Single-apron drafting is sometimes used, consisting of a lower apron above which three or four tumbler rollers control the floating fibers by virtue of the pressure they exert on the strand being drafted.

The modern European worsted system consists of three pin-drafting stages—the first of which is fitted with sliver autoleveling—followed by a balling head finisher with a roving rubbing mechanism producing a twistless roving, and a high draft ring-spinning frame. The modern American system replaces the balling head finisher, with a flyer roving frame producing a roving with minimum twist. Apart from this difference the two systems are practically identical; the last two stages of roving and spinning usually use double-apron drafting, but it may be single apron. The fact that they are equally able to spin man-made fibers or dry-combed wool, or blends of these, and because process development is being increasingly influenced by the greater use of man-made fibers, have resulted in these systems being widely adopted.

If the roving frame is omitted and coarse yarn is spun from the product of the third pin drafter, this is known as *sliver to yarn spinning.*

If the system is preceded by a worsted card only and card sliver is fed to the first pin drafter, this is the *semiworsted or half-worsted* system.

Apron systems cannot be used for drafting oil-combed wool. The Ambler Superdraft (ASD) unit was designed for drafting twisted, worsted rovings with the normal 3½% of oil, although its use is not limited to such material, and man-made fibers can be processed equally well. With this device ultrahigh drafts, that is, superdrafts, can be applied; drafts of 50-150 are normal, and up to 500-600 are possible with fine fibers, including man-made fibers. Controlled drafting is achieved by the inclusion of two devices in the drafting zone just behind the delivery rollers; these consist of a pair of small tension rollers and a "flume." The tension rollers are only about 0.3 in. in diameter; the lower one has flanges 0.075 in. apart and accommodates the top one, which has fine flutes on its surface. Both rollers are driven, and the top one rests on the roving in the groove, their surface speed is about 12% greater than that of the feed rollers. The flume is a narrow funnel-shaped guide of rectangular cross section, 0.75 in. long, tapering in width from 0.075 to 0.065 in. It points into the nip of the delivery rollers, the distance from which is adjusted to give the most even yarn. The wide range of possible drafts that can be applied enables a complete range of counts to be spun from a roving of standardized linear denisty.

Redesign of the ASD unit has produced the Uniflex controlled drafting system shown in Figure 9.8. The flume C is now integral with the housing of the top tension roller, and the drafting zone includes a pair of carrier rollers E and F, the bottom one of which is driven; these rollers enable twistless rovings to be drafted also; when twisted rovings are being processed, the top carrier F is removed.

The Uniflex system can apply high drafts (14-40) or superdrafts (greater than 40), it can handle twistless or twisted rovings, oil-combed or dry-combed wools, 100% man-made fibers, or blends with wool. Table 9.5 shows the two variations of the system employed for the drawing and spinning of oil-combed or dry-combed materials. The Ambler Draft unit referred to is incorporated in the

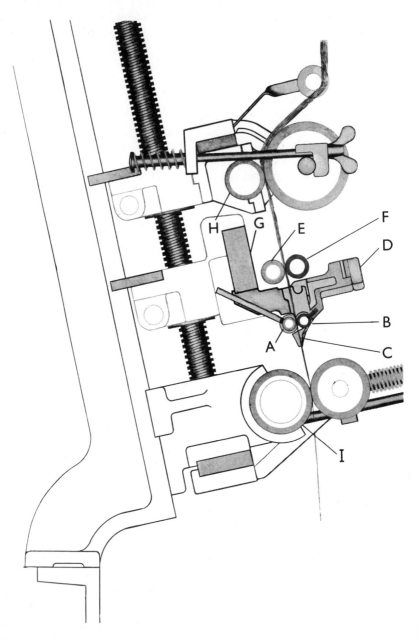

Figure 9.8. Uniflex spinning (courtesy of Platt International Ltd.).

304

Table 9.5. The Uniflex System

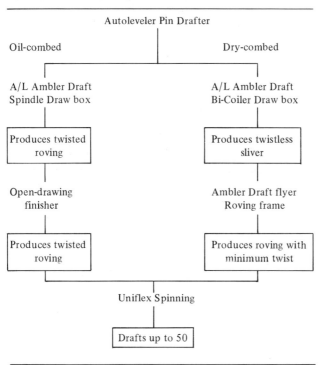

drafting zone of the draw boxes and consists of a flanged bottom tension roller accommodating an upper roller under 12 lb pressure, both positively driven, followed by a pair of plain rollers ½ in. in diameter weighted up to 1¾ lb and also both positively driven, which are placed just behind the delivery roller. The position of this unit is adjusted to produce the most even strand.

The Uniflex drafting unit has been further modified using a tension roller with a wider groove and a flume with a wider channel; thus heavier roving can be drafted. The finisher on the left hand side of Table 9.5 is then omitted, and the spindle draw box produces a roving of standard linear denisty of 140 dms/40 yd from which a range of counts up to 1/24s can be spun using drafts up to 200 on the Uniflex spinning frame. This is known as the *heavy roving system*.

Woolen Yarn-Spinning System

As mentioned in the previous section, wools that, for any reason, are not suitable for worsted processing are used in the woolen system. This is not to say, however, that only inferior materials are employed in this system. Woolen yarns have a character of their own, and many good sound wools are used in this process to produce good quality yarns for blankets, carpets, hand-knitting yarn, and tweed cloth. In addition to new wool, wool recovered from pulled rags is also used; these include knitted rags, woven rags, old rags, new rags (tailors cuttings), white rags, and colored rags of a wide range of qualities. Yarn and cloth wastes from mill processes are also used, as are all kinds of man-made fibers, hairs, silk, and cotton.

Given the assembled blend of any of these fibers, the processes of yarn making consist simply of carding and spinning, but the processes that may be used before blending are extremely varied, and it would be beyond the limits of this section to detail them. For example, rags must be opened by pulling; yarns must be opened by garnetting; to eliminate cotton in the form of seams etc. from wool materials the rags must be carbonized, that is, treated with acid to destroy it; if materials going into the blend are different in color, the blend may have to be dyed to give the appropriate solid shade and the original color may have to be removed before dyeing; different colors may be blended; wools heavy in vegetable matter will be carbonzied before blending. Individual components of the blend coming from different sources may be prepared separately before blending, and the blend may have some further treatment to prepare it for carding. Whatever the procedure, the blend should not contain lumps or particles that may cause damage to the card clothing, nor should they be so difficult to disintegrate that opening is not complete when the particle (e.g., a piece of hard thread) has passed through the carding machine. In modern practice the blend is as well cleaned and mixed as possible before carding; the woolen card is an expensive machine, and although opening, cleaning, and mixing are its important functions, unnecessary demands on it are uneconomic in that they may require an unnecessarily long machine; uneconomic, frequent stoppages for fettling (cleaning) and the possibility of damage to card clothing during difficult or too frequent fettling. Preparation for carding is therefore equally as important as carding itself. Table 9.6 illustrates the Woolen System using new wool.

Table 9.6. The Woolen System

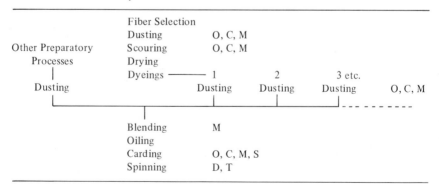

Scouring is the same as in the worsted process, and preliminary dusting may or may not be advisable. The dried wool may be dyed in lots to a number of different shades and the different colors assembled with fibers from other sources. These are weighed in the required proportions in batches or are processed continuously, and each lot is dusted separately or together before blending, the means adopted depending on the process. In modern processes the material is blended in large circular bins in which it is distributed by a rotating spreader; it is then passed through an opening machine, and the whole operation may be repeated to improve the blending. At the end of the process after the maximum amount of dust has been removed, oil is applied; the application of oil after cleaning makes cleaning easier and avoids the entry of sticky dirt to the card.

Since the draft used in spinning is usually not more than 1.5, the strand that is twisted to form the yarn is virtually formed on the card and must therefore be as uniform in count and color as it can be at that point; nothing can be done to improve it during spinning. Given a well-prepared blend, the carding operation is therefore the key operation in the woolen process.

The woolen card usually consists of two sections known as the breaker and finisher parts or, alternatively, the scribbler and carder parts. Each consists, in the most common form, of two cylinders each with its associated workers and strippers, fancy, and doffer. The blend is fed to the machine by a weighing hopper whose function is to feed equal weights of material to the card at equal intervals of time to ensure that the slubbings or rovings emerging at the delivery end are uniform in linear density. Absolute equality of successive weighings by weigh-pans is not possible, but the design of modern hopper feeders is directed towards the improvement of this capability while satisfying a demand for higher production rates.

The condenser at the delivery end of the machine divides the carded web into strips by means of leather tapes, and these strips are rubbed between rubbing leathers to form rubbed slubbings or rovings. These must have sufficient cohesion to enable them to be wound on and off condenser bobbins without being damaged, in particular without being unevenly stretched, since this would cause unnecessary and unwanted unevenness in the yarns spun from them. The fact that tape condensers do not divide the web into strips of equal linear density has already been discussed, and there is no evidence that tapes of any particular configuration are significantly better than others in this respect.

Within the carding machine itself, it will be appreciated, from what was said in Chapter 7, that the various rollers must be set accurately to each other at the selected settings to ensure uniform carding and uniform rovings. This means that all rollers must be accurately cylindrical, which is possible to ± 0.002 in. using suitable cylinder grinding equipment, and must be mounted securely and perfectly squarely, on accurately machined frames mounted on a level floor; this ensures accurately aligned shafts and also has the advantage of ensuring minimum power consumption. It should always be remembered that the basis of good carding practice is the action of well-maintained card clothing mounted on accurate rollers; there may be room for argument as to the most suitable roller speeds, settings, type of clothing, etc., but within reasonable limits these are secondary considerations by comparison with this basic requirement. Recent tendencies have been toward the production of wider machines; the advantage of this is that rate of production is greater for a given provision of peripheral equipment, frames, belts, motors, etc.,–but it is only possible because modern engineering methods allow the necessary accuracy to be attainable with wider rollers.

The action of carding that takes place between cylinder and workers is shown in Figure 9.3, and the same action occurs between cylinder and doffer. The action whereby the stripper strips the worker, and the cylinder strips the stripper, is shown in Figure 9.4. The clearing of rollers by the stripping action is 100% because the teeth of the roller being stripped do not oppose the stripping action. However, in the carding action, the two sets of teeth oppose one another and share the material passing the carding point; this is the essential action on the carding machine and has been the subject of a good deal of research work by Martindale, Townend, and others.

Bearing in mind that the fundamental actions required of the carding machine include opening, cleaning, and mixing, we can consider what takes place at such points. A tuft of material approaching a cylinder-worker carding point is in part arrested by the slowly moving worker, a fraction, p, is attached to the worker; the rest, $1-p$, passes on with the cylinder. The fraction p has been called the collecting power of the worker and values vary from about 0.75 to 0.25 (26). This action opens the material and should release impurities, but since it occurs on the top of the cylinder, there is no means of immediate escape for the released trash; unless it is still free to fall out when it reaches the underside of the machine, it will either be reincorporated into the material passing forward or be pressed into the base of the card clothing, eventually giving rise to the necessity for the machine to be stopped and cleaned. Trash freed at feed points, or where cylinder meets doffer, will fall out more readily. Crushing rollers are often fitted between the breaker and finisher parts to crush particles of impurity when they have become accessible in the tenuous web, and this enables them to be more easily liberated in the finisher part. Cleaning, however, is somewhat fortuitous, and it needs little consideration to appreciate what was said earlier— that the more cleaning effected before carding the better.

Complete opening of fibers is particularly necessary on the woolen card to achieve intimate and uniform association of fibers in the ultimate yarn, because the single process of spinning following carding does little, if anything, to improve it. It has been difficult to establish what is the most favorable mode of material subdivision by worker and cylinder to achieve this result most effectively. Martindale (26) showed that the collecting power of a worker was increased with increased worker speed, no doubt because more clean carding points were presented to the cylinder in a given time; it was also increased, but only slightly, with closer setting of worker to swift and also by grinding worker wire. Collecting power decreased as one passed from worker to worker around a cylinder and increased again when fiber was transferred from one cylinder to the next. Similar results have been reported by Townend (27). These observations do not prove that a high value of collecting power is conducive to good opening, although this is suggested by the fact that closer setting and a keen point on the clothing are both regarded as being advantageous in this respect. It was found, however, that stronger yarn was produced when high values of collecting power were obtained from faster workers on the breaker part. Dircks and Townend (28) also found that increased worker speeds reduced the nep content of a carded web, and also the amount of thread remaining after carding, when worsted and woolen threads were incorporated into the blend being processed. There may therefore be a close connection between opening power and collecting power.

It should be noted that the collection of material by slowly moving workers also produces mixing on the card; that part of the material which is transferred to the worker is delayed by its slow circuit of the worker and is then deposited on material, arriving later when it is transferred back to the cylinder. It then reaches the same carding point again and will be divided a second time, and so on. The action at the cylinder/work point is therefore a series of repeated actions of dividing and doubling. Factors that increase the collecting power will increase this mixing effect except where the result is achieved by increasing worker speed; this decreases the delay on the worker and reduces mixing, since material is not spread over such a long delivery period (26). Increasing the worker speeds on the breaker part of a card is worthwhile if it improves opening because, although reduced mixing might occur, there is plenty of mixing power available in the remainder of the machine. Increasing the worker speeds on the finisher card may, however, reduce yarn evenness (26). Intermediate feeds between parts are provided specifically to improve mixing.

Since the action between cylinder and doffer is the same as that between cylinder and a worker, one would expect the collecting power of the doffer to be affected by similar factors. Martindale (29) has shown this to be so; particularly important is the increased collecting power with increased doffer speed. The fancy also plays its part in this action (29), since it raises material to the cylinder surface preparatory to transfer to the doffer. The behavior of the fancy is more complicated; in most cases increases in fancy speed assist the transfer by increasing the collecting power of the doffer, but if circumstances are such that the cylinder becomes heavily loaded—high feed rates and low doffer speeds—the beneficial effect of high fancy speeds will be reduced, may disappear, and may even be reversed.

The fact that the doffer only removes a proportion of the material approaching on the cylinder is similar to the situation already discussed with respect to the doffer on a cotton card. Wira (30) give values for this fraction ranging from about 0.04 to 0.20, the higher values being obtained with higher surface speeds of the doffer. Thus material is delayed in its passage through the machine in two ways: by being held up on the workers and by escaping transfer to the doffer and recirculation around the cylinder. Wira has defined the *delay factor* for one part of a woolen card as the average length of time it takes for a fiber to be transferred to the doffer after entry to the part. The delay factor for the whole machine is the sum of the delay factors for each cylinder of the machine plus the time fibers spend on the doffers and intermediate feeds. Values up to one minute per cylinder are common.

A little consideration will show, therefore, that high values of delay factor are conducive to good mixing, and since mixing results in the mixing of inequalities in the amounts fed by the hopper as well as the mixing of fiber types, a high value of delay factor also promotes uniformity of linear density of output. On the other hand, if such a high value is obtained by running doffers slowly, reducing their collecting power, the cylinder becomes overloaded and carding suffers, neps increase, and any threads in the blend will not be so well opened. In order that a cylinder can accept a maximum rate of feed of fiber, the doffer should clear the cylinder as effectively as possible, keeping down the load of fibers on the cylinder, that is, it should have a high surface speed.

Other factors affecting the clearance of cylinders in woolen cards are equally important in avoiding overloading and consequent deterioration of the carding action. Since it is the card wire that performs the essential carding action, its properties are clearly vital, and characteristics such as density of card wire, fineness, profile, wire angle, and type of foundation—which affects its rigidity—could all be important factors. Townend and his students have investigated these properties and have studied fiber breakage, nep formation, and ability to reduce hard threads as criteria of performance.

Ashdown and Townend (31) found that denisty of card clothing had no effect on fiber breakage, which is determined more by the state of fiber entanglement on entry to the card; once the initial entanglement is opened up, there is little further breakage. Long fibers suffer more in this respect than short fibers, but the features of the card wire have little to do with it.

According to these authors, long and fine fibers are more prone to nep formation, and they found, when processing fine lambs' wool, that repeated passages through a single cylinder machine increased neps considerably when coarse, open-card clothing was used. Table 9.7 shows quite a dramatic effect; neps per gram were counted after a varying number of passages through a single-part machine clothed with different densities of wire. Up to and including 80/8 wire repeated passages merely manufactured more neps; using 100/10 or finer, repeated passages eliminated them. Other experiments showed that, if these card

Table 9.7. Neps Per Gram Formed by Repeated Passages on a Woolen Card Clothed With Different Densities of Card Wire (Ashdown and Townend)

Number of Passages Through Card	Card Clothing Count/Crown and Number of Points per Square Inch					
	D 64	60/6 144	80/8 256	100/12 400	120/32 576	140/14 784
1	348	413	372	122	143	185
2	590	684	725	35	40	5
3	655	1198	1110	17	3	1

D = Diamond point wire

cylinders were used progressively in sequence, the neps formed on the early, coarse cylinders could not be removed later by the fine ones; carding should start with 100/10 clothing, or at least 80/8, on the first cylinder. Threads were also reduced much more effectively by clothing finer than 80/8. With coarse fibers there was no such effect.

In summarizing his work, Townend (32) states that less neppy webs are produced if more open card clothing is used on cylinders and doffers while retaining conventional pin density on the workers. He also refers to work with Dobson (33) that showed that the angle between the upper part of the card wire and the tangent to the cylinder at its base affects neppiness. If this angle for the cylinder wire is too acute, namely, 57° "then the doffer is unable to capture fibers from the cylinder, the latter becomes too heavily loaded, and this gives rise to excessive nep formation .... if it is too obtuse it is unable to pass the fibers easily and the nep content rises." The best angle for the cylinder wire is 65°; with this angle the neppiness was somewhat affected by the angle of the doffer wire, the best angle for this was also found to be 65°.

The same article also shows the effect of (1) cylinder speed, doffer speed worker speed and fancy speed— in all cases, increased speed produced fewer neps; (2) worker/cylinder and doffer/cylinder setting—in both cases the closer the setting the fewer the neps; (3) grinding of clothing—fewer neps; (4) production rate—the lower the rate the fewer the neps. These experiments were carried out with a variety of fibers and in some cases using rigid metallic clothing.

The spinning machine plays a minor role in the formation of a woolen yarn; it applies a small draft and twists the roving into yarn. It should not damage the material, but on the other hand, there is very little scope for improvement. Yarns are spun on spinning mule or ring frame, and over the years there have been many arguments about which produces the better yarn. There have also been a number of research reports in which it is generally agreed that for most materials there is little or no difference between them, and that differences in

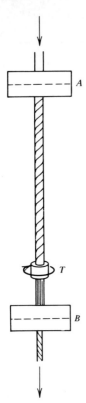

Figure 9.9. Twist distribution in woolen spinning.

the cloth made from the two types of yarn are not noticable. The very inde-
cisiveness of the argument over many years would in itself suggest that the
difference is negligible. On the other hand, less skilled labor is required on ring
frames, and generally the process is cheaper; the percentage of mule spindles has
decreased considerably, and it seems likely that mule spinning will disappear; it
will therefore not be considered here.

In the ring frame, roving is fed from the condenser bobbin and enters the
drafting zone by way of the feed rollers, A in Figure 9.9. The length of the drafting zone is not governed by fiber length; it is about 20 in. In this drafting zone is
placed a false-twist tube, T, through which the roving passes and which imparts a
rotary motion to it, thus twisting it and giving it cohesion. To do this the twist
tube grips the strand gently. This grip is provided by spring jaws, or by centrifugal force generated by the rotation that can be made to close the jaws, and by
friction between the strand and needles, or fingers, placed near the axis of the
tube. In all cases these devices are placed at the exit of the false-twist tube.

In Figure 9.9, the length of roving between A and T receives Z twist; if the rate of rotation of the twist tube is n rev/min and the rate of feed is x in./min, the turns per inch of twist $= n/x$. As x in./min pass through the twister, they carry down n turns of Z twist, which is exactly neutralized by the n turns of S twist imparted per min. to the length of strand below T; thus this length of the strand is twistless. The draft applied by the delivery rollers B would therefore operate on this length, the strand would be irregularly drafted, and poor spinning would result. To avoid this, the yarn-gripping device is placed very close to the nip of the delivery rollers, and the distance TB is effectively zero.

Towend and Jowett (34) investigated the distribution of draft in this system and found that the major part of drafting takes place in the first 4 or 5 in. of the drafting zone where the twist is relatively small; as twist increases, there is little drafting until the strand enters the tube, where some drafting occurs because fibers are drawn through by the delivery rollers.

Cotton Waste Spinning

The system used for the spinning of cotton waste is similar to that used for woolen spinning in that roller cards are used, slubbings are formed on a condenser, and the spinning operation converts them into yarns on ring-frame or mule with a limited draft of the order of 1½-2. As in woolen spinning, the quality of the yarn is largely determined on the card since, although faulty handling in the spinning process can have a detrimental effect on the yarn, the simple drafting of single ends provides no opportunity for remedying any defects arising in carding.

The materials used in this system are soft wastes and hard wastes. The former consists of droppings and fly from opening and carding machines of the cotton spinning process; card cylinder, doffer, and flat strips; comber waste, clearer waste, and mill sweepings. Other soft wastes formed in that process, such as faulty laps, or other lap waste, sliver, and roving wastes, are of clean, better-quality material and are returned to the blend in the cotton mill; only relatively dirty or short-fibered soft wastes are relegated to waste spinning. Hard waste consists of waste yarn from spinning, winding, warping, reeling, doubling, weaving, and knitting processes. No fiber waste is recovered from woven cotton rags, although a small amount from new, white, knitted fabrics might be. Colored yarns are not usually reprocessed; they form part of the large quantity of yarn wastes used for cleaning purposes.

In these materials it is the hard waste that contains the best, that is, the longest fibers, and it will therefore spin finer yarns (7s and 8s cotton counts); the soft wastes contain fibers that have fallen out or have otherwise been rejected from the cotton spinning process and will only spin coarser yarns (1s-5s). For this reason, and with one exception, hard and soft wastes are not mixed. The one exception is the case of comber waste; being the best quality soft waste it is sometimes used in blends in hard waste spinning. Waste obtained from some sources, for example, sweepings, may be a mixture of hard and soft wastes, and these are separated by mechanical means before processing.

The nature of these materials makes the preparatory processes prior to carding most important; on the one hand (hard waste) to disentangle and disintegrate the thread structure, and on the other hand (soft waste) to open and clean what might be material heavily loaded with vegetable matter, sand, and dust. These different needs give rise to the two processes shown in Table 9.8.

Table 9.8. Cotton Waste Spinning Systems

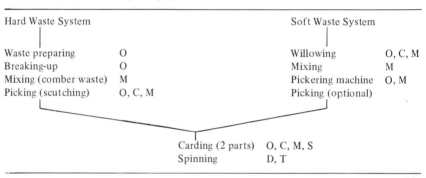

Hard Waste System		Soft Waste System	
Waste preparing	O	Willowing	O, C, M
Breaking-up	O	Mixing	M
Mixing (comber waste)	M	Pickering machine	O, M
Picking (scutching)	O, C, M	Picking (optional)	
		Carding (2 parts)	O, C, M, S
		Spinning	D, T

In the hard waste system, the first machine opens the material by loosening and disentangling waste thread. The breaking-up machine is a powerful opening machine with six strongly pinned cylinders in sequence, each of which strikes the material from feed rollers. After mixing in mixing bins (with comber waste also if used in the blend) the material is further opened by a picker and fed to the card.

In the soft-waste system, cleaning is by a willow to an extent determined by the dirtiness of the waste; this machine is the best cleaner of trashy wastes. After a blending operation, a pickering machine effects further opening and cleaning before carding; it may be followed by a picker.

Carding consists of a two-part machine, which may be hopper or lap fed, with an intermediate feed and condenser. The remarks applying to woolen carding can apply equally to this situation, although very little information has been published relating to research work done on this process. The variations in count from end to end across the condenser are just as marked as in woolen spinning.

Jute and Flax Spinning

The spinning of jute and flax makes use of the same principles as those employed for the fibers so far discussed, but the technology is affected by

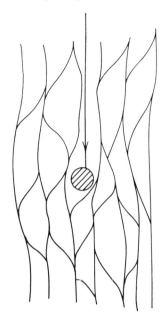

Figure 9.10.   The lattice-like structure of jute being split by a pin.

specialized structural properties of these bast fibers that raise different process-ing problems. In the first place, the fibers are much longer than any of those previously considered; flax fibers range in length from 6 in. to 3 ft—the height of the flax plant, and jute fibers come from the stems of plants, which may reach a height of 15 ft, according to the variety of jute; in the bale they vary from 6 to 14 ft.

The second important structural feature is that the fibers are built up from units known as "ultimates," which are short, spindle-shaped cells of polygonal section that lie side by side, overlap along their lengths, and are cemented together by gums and resins. In the case of flax, these ultimates range from 1 to 7 cm in length and from 7 to 30 microns in thickness (average about 25 microns); the ultimates of jute are 1-6 mm in length (average about 2.5) and from 5 to 25 microns in thickness (average about 18 microns). These ultimates are built up to form fiber bundles that lie just below the cortex or bark, and are arranged in a circle around the woody core of the plant; they run from the root to the tip. In the flax plant there are about 30 such bundles, each containing 50-to 100 ultimates in a cross section. In the jute plant there are more bundles than in flax, each containing about 50 ultimates.

The third distinctive feature is that these bundles are interconnected by fibrous branches a few ultimates thick that migrate from one bundle to another and from a complex network of fibers (see Figure 9.10). This is very marked in the case of jute, in which there are very short lengths between branches, and when it is treated with pins as in carding or gilling, the bundles are split lengthwise into finer fibers and shortened. This reduces the fiber length to a few inches and the number of ultimates per fiber from 50 to about 8. Flax fibers are also much reduced in cross section during processing, and mean fiber length is halved during drawing.

Retting

The extraction of the fiber bundles from the stem of the plant might conceivably be called opening and cleaning; it is accomplished by bacteriological action on the gums fixing the fibers in the plant structure and is called retting. It is carried out by steeping in water, and, after a time dependent on the temperature, this action allows the fiber bundles to part easily from the wood and bark as well as from each other. Having achieved this, the bacteria will go on to attack the gums cementing the ultimates together, and finally, they would attack the cellulose of the fiber itself. Thus if flax is over-retted, the ultimates separate and the flax is weakened, even damaged; if it is under-retted, the initial separation from the wood is not easy.

Jute Processing

Table 9.9 shows flow charts for medium, fine, and coarse jute yarns. Batching is an operation in which the required number of bales in various grades is selected so that their contents can be blended to produce a given quality of yarn. The medium and top qualities are passed through a bale opener, which consists of two or three pairs of fluted crushing rollers. As the material passes between the flutes, it is flexed backwards and forwards, which breaks up and loosens woody material and softens the fiber, making it more pliable.

Table 9.9. Jute-Spinning Processes

Fine and Medium Yarns		Coarser Yarns Using Lower Grade of Fiber	
Batching	M	Batching	M
Bale Opening	O, M		
Spreading	M, S	Softening	O, M
Carding (2 parts)	O, C, M, S	Carding (2 parts or 3 parts)	O, C, M, S
Drawing (3 or 4 operations)	D, M, P	Drawing (2 operations)	D, M, P
Spinning	D, T	Spinning	D, T

The material emerging is taken in bundles, known as heads or stricks, weighing 2 or 3 lb each, that are laid on the feed sheet of the spreader; they are fed lengthwise, root first, and are overlapped to preserve continuity of feed. On passing through the feed rollers they are pressed down into a bed of pins carried on an endless chain, which move slightly faster than the feed rollers; this is followed by a similar, but faster, chain of pins, and between them the fibers are drafted. The emerging sliver is a flat band about 6 in. wide; it is sprayed with a hot emulsion of mineral oil to give an oil content of 4 or 5% and rolled (under heavy pressure) into a large roll about 4 ft in diameter. The feed to this machine is regulated to ensure a uniform feed rate; thus long term irregularity of the rolled sliver is minimized. This machine therefore achieves mixing on the feed sheet and forms a sliver. The emulsion raises the moisture content to about 25%; it penetrates the fiber during subsequent storage and makes the fibers more supple, and this enables them to be manipulated more easily.

Lower grades of jute for making coarser yarns are given a different process. They are passed through a softener, a machine consisting of a series of 64 pairs of spring-loaded, fluted, cast-iron rollers that repeatedly flex the fibers, breaking up any hard material attached to the fibers and making the fibers more pliable. They are then emulsion oiled.

After the material has been softened, lubricated, and mixed to some extent, it is carded. Jute carding is carried out on roller cards, which have some distinctive features. The most important function of the first part, the breaker card (Figure 9.11) is to break up the network of fibrous filaments into entities of suitable length and diameter for drawing and spinning. Rolls of jute from the spreader are set up side by side at a feed sheet (or sticks of jute from the softener are laid on it), and the long fibers enter the card between a feed roller and a curved metal plate below it known as a "shell." The cylinder strikes downwards, and the fibers are broken off in lengths, depending on the conditions set up at the shell feed; they pass two workers and strippers placed *under* the card

ORTHODOX CARD

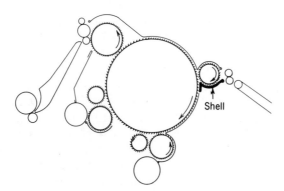

Shell

Figure 9.11. Orthodox jute breaker card (courtesy of Gilman-Fraser Ltd., Arbroath, Scotland).

JF4 LONG JUTE FINISHER CARD

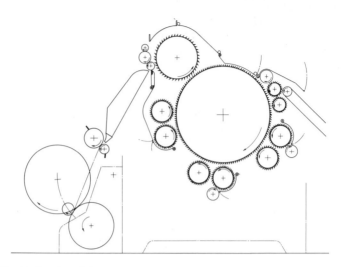

Figure 9.12. Modern jute finisher card (courtesy Gilman-Fraser Ltd., Arbroath, Scotland).

and are removed by the doffer. The card clothing is very coarse, with only three to five points per square inch. Figure 9.10 shows how a card pin may be imagined to split the fibers projecting from the shell feed.

The card sliver is again made into a roll, and a number of these are fed side by side to the finisher card, giving further mixing. This card, unlike any other card, usually employs the complete surface of the cylinder for carding, the double doffer delivery is placed just above the feed rollers, and around the cylinder are four to six pairs of workers and strippers. This card is therefore known as a full-circular card, the breaker is a half-circular card. Modern practice is to adopt half-circular cards with three pairs of workers and strippers for both breaker and finisher cards (Figure 9.12). The latter is more finely pinned than the former—8 to 15 pins per square inch.

Drawing is in gill boxes, two processes for coarse yarns, three processes for medium and fine, and four processes for the very finest yarn of which not much is made; autolevelers are used to reduce long-term irregularity. Drafts are of the order of 6 or 8, and the roller settings are from 14 in. in the first process to 10 in. at the spinning frame. The fiber length in finisher card sliver is about 3 in., and less than 1% of the fibers are longer than 10 in. A difference between the drawing of jute and the fibers discussed earlier is that doubling takes place after drawing, that is, each sliver is drafted individually, and the emerging flat slivers are combined by being laid one on top of the other. The final roving is given cohesion by being crimped in a stuffer box on leaving the delivery rollers.

Spinning on modern machinery is carried out with single-apron drafting using rollers, or a polished flat plate, as the upper control element. Spinning is by flyer, which gives a smooth yarn with a polished appearance.

Flax Processing

Flax is usually retted to loosen the fiber from the bark and woody core (the boon), but it is possible to dispense with retting and to separate the fibers by mechanical means alone; this produces what is known as green flax, which is more contaminated by particles of bark, etc., than is retted flax.

The main object in converting flax fiber into yarn is to make full use of the strength and luster of the long fibers, so the achievement of parallelization and the retention of length is of first importance. Tables 9.10 and 9.11 show the processes used.

Table 9.10. Wet Spinning of Flax

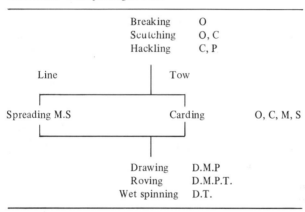

Table 9.11. Dry Spinning of Flax

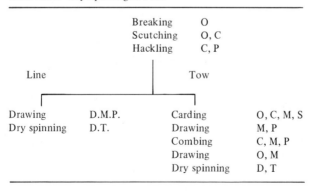

The first operation is breaking, in which bunches of material are passed between a series of fluted breaker-rollers. The straws in these bunches lie parallel, and as they pass between the flutes of the rollers (which are not in contact) they are flexed first one way then the other so that fiber is loosened from the boon without itself being damaged. These bunches of straw are next beaten in turn by revolving beaters of the scutcher; each bunch is beaten at both ends, first on one side, then on the other. In this way the boon is removed, leaving bunches of fibers, and these are put together in units of convenient size, known as stricks, for the combing treatment given by the hackling machine.

A strick is spread out, clamped, and suspended—root and downwards—in the machine. It is then combed by a hackling tool, a pair of revolving endless belts of pins into the space between which the strick is lowered. It is withdrawn and then moved, in turn, to succeeding stations on the machine, at which hackling tools with increasingly finer pinning perform combing actions of gradually increasing severity. This action, as in any combing operation, removes short fibers, entangled fibers, and vegetable matter, and lays the fibers parallel. This first treatment deals with 40% of the length of the strick—the root end; the strick is then reversed and the remainder of its length is similarly treated. The stricks of combed flax that are delivered are collectively known as "line" and the short fiber is "tow."

The long fiber from the hackling machine is processed on line systems (Table 9.10). The first machine may be either a hand-spreadboard or an autospreader that passes the line automatically to a gill box that drafts it and produces a continuous sliver. This is succeeded by a sequence of five drawframes and a roving frame, all of which have gill pins to exercize fiber control and provide drafts and doublings. The roller setting, or "reach," in these machines decreases gradually from 26 to 16 in. as the fiber structure is gradually broken down by drafting through pins. At the same time, the cross-sectional area is about halved by the same process (35).

In the line preparation process of Table 9.10, the roving is followed by wet spinning, which produces the finest, strongest, and smoothest yarns. Yarn is spun from a single roving, which is passed through a trough of hot ($150°F$) water immediately before entering the drafting zone. The penetration of the hot water into the fibers softens the gums cementing the ultimates together, enabling them to slip over each other. The process is therefore one of drafting the ultimates, and the roller setting need therefore only be of the order of 2 or 3 in. in spite of the fiber length of about 16 in., because the ultimates are no longer than 7 cm; a draft of 6 is applied. On cooling, the gums harden again.

When green flax is processed by this method, the roving must be boiled in alkali because it contains gums that have not been removed at an early stage. Some retted flaxes also give better spinning after rove boiling. Spinning from bleached roving is also now practiced, and in modern practice includes the use of a double-apron drafting system.

Table 9.11 shows the alternative line preparation system in which dry spinning is used. In this case, the spinning frame must have a "long reach," that is, a roller-setting must be comparable with the fiber length in the roving, since it is fibers and not ultimates that are now being drafted. A single-apron drafting system is used with apron and tumblers.

The tow from the hackling machine is still of a significant length, at least comparable with some wools, and therefore may itself be drawn and spun, either wet or dry.

In both cases it must be carded, which is done on a full-circular roller-type card with perhaps seven pairs of workers and strippers, and a web delivery divided to produce six slivers that are drafted in a gill box and combined to deliver one sliver. For wet spinning, Table 9.10 then shows a sequence of drawing frames (gill boxes) and a gill-roving frame as in the line process. Since drafting on the wet-spinning frame is of near ultimates, the fiber length in the roving is not of special importance. On the other hand, in dry spinning, the presence of a wide range of fiber lengths in the roving would be a disadvantage; therefore, the carded tow after drawing is combed on a rectilinear comb—similar to the French Comb used in worsted processing—and this enables the tow to be spun on a long-reach spinning frame with single-apron drafting.

In these processes, hackling occupies a central position, since it determines the quality of both line and tow. It effects the subdivision of the fiber strands, and the greater the subdivision the finer the fibers and the higher the spinning quality. Lighter hackling, to give better yields, results in a less severe action, coarser strands, and lower spinning quality. The quality, for dry or wet spinning of green or retted flax, is largely determined by three physical qualities of the hackled flax (35):

1. fiber breaking strength per unit area of cross section
2. fineness or average cross section of the hackled fiber strands (A)
3. average length of fiber strands (L).

The finer the fiber the greater its length, and the greater the fiber strength, the better is the yarn quality.

RECENT ADVANCES IN YARN MANUFACTURING

Production Limitations Associated With Ring Spinning

Of conventional spinning systems, ring spinning is the most productive and, except for yarns with special requirements, has largely supplanted other forms. However, it seems to have reached the limit of its development. For the reason given in a previous section, the output of the ring-spinning machine is determined by the rate at which the spindle (and hence the traveller) can rotate, and this is limited by the following factors:

1. The maximum speed of a traveller is bout 40 m/sec; at higher speeds the traveller burns and replacements become too frequent.
2. Yarn strength may be insufficient to withstand the tension in the balloon even at speeds within the capabilities of the traveller; higher speeds mean higher air drag on the ballooning yarn and so increased yarn breakages. (For this reason in some spinning systems balloons have been suppressed.)
3. Package size and power requirements. If (1) and (2) allow higher speed, doffing becomes more frequent; therefore, trends have been toward larger packages. Larger and heavier packages not only accentuate spindle vibration, but the rotation of heavier packages requires much more power and becomes uneconomic.

The maximum spindle speed in the ring spinning of cotton rarely exceeds 15,000 r.p.m. and is considerably less in worsted spinning. The spinning process is responsible for about half the cost of cotton spinning, and the inability to use higher spindle speeds is an impediment in the reduction of costs of production. In addition, the entire machine must be stopped for doffing, and the smallness of package size means that the yarn must be rewound on to larger packages before use.

Open-End or Break Spinning

Figure 9.13 shows at (a) the conventional system of roller drafting and how the whole yarn package must be rotated for the simple yet essential purpose of inserting the necessary yarn twist. At (b) is shown the simple principle of open-end or break spinning in idealized form. The conception is that the roving is broken at A and the fibers are transferred to a collecting point B at which they are reassembled in sequence to form a yarn. As the fibers arrive, this end of yarn, B, is withdrawn at the rate necessary to form a yarn of the required count; for example, a finer yarn would require a faster rate of withdrawal. It is then wound onto a take-up package.

Thus, unlike the conventional drafting system, there is a break in the continuity of the roving and an open end of yarn appears. This open end must be rotated to impart the necessary twist to the yarn formed.

Ideally, the fiber flow should be separated into individuals before being reassembled at B; if not, there is a possibility of the twist spreading back across the break, in which case the device would start to produce false twist instead of real twist. The fibers should presumably be kept as straight as possible and not become looped, hooked, or otherwise entangled en route, and they should not be added to the open end in groups or the yarn formed will be irregular.

This system eliminates the ring, traveller, and spindle; there is no balloon to

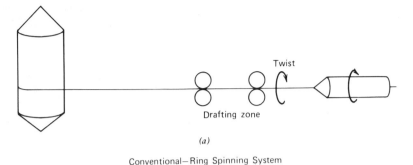

(a)

Conventional—Ring Spinning System

(b)

Figure 9.13. Break-spinning principle.

introduce problems of air drag; the need to revolve the yarn package at high speed is eliminated, and the power needed is thereby reduced. Ideally, the device twisting the open end would be small and light, and its dimensions would be independent of the size of the take-up package. The take-up package rotates relatively slowly since it does not have to twist, but merely collect, the yarn; its capacity does not affect the power consumption. It can be chosen to be as large as required, which reduces the number of knots made and eliminates the need for rewinding, unless this is regarded as necessary for breaking out weak places as a quality control measure.

The successful application of this system has resulted in the insertion of twist at much increased rates, 45,000 turns per minute being practised and 100,000 turns per minute being a possibility. It should be noted that the vital considerations in all break-spinning devices are (1) the trouble-free preparation and transport of fibers at a high, uniform, and controlled rate from A to B, (2) their continuous assembly as a uniform yarn at B, and (3) the configuration of the fibers on assembly that affects the structure of the yarn and therefore its various properties that are of importance in its future use.

Methods of break spinning have been classified in two different ways. Krause (36) classifies them according to (1) conditions of fiber transfer and (2) conditions of fiber assembly. Conditions of transfer are of two kinds: the fibers can be positively guided or move freely in transit. The conditions at assembly may be "at random" or "order maintained"; the various combinations of these conditions give four possible groups.

The Shirley Institute (37) also classifies break-spinning systems into four groups on a different basis, namely, according to the type of apparatus used for transfer and collection. The following types are identified: (1) vortex systems in which the fibers are transferred, and the open end of yarn is twisted, by a fluid vortex (usually of air), (2) axial assembly systems in which the fibers are transferred in an air current and reassembled along the axis of a rotating mechanical element, (3) discontinuous assembly systems, which take tufts of fibers from the roving, keep them under positive control during transfer, and reassemble them again by overlapping them at the revolving open end, and (4) circumferential assembly systems in which fibers are transferred in an airstream and are continuously assembled on a rapidly rotating cylindrical surface.

Break-spun yarns of sorts have been produced by all these methods, but the only one producing commercially valuable yarns is the circumferential assembly or rotor system (Figure 9.14). If assembled on the outer surface of a cylinder, centrifugal forces throw fibers off at high speeds; thus in successful devices the fibers are assembled on the inner surface of a revolving rotor, in which case the centrifugal forces help to retain the fibers in position on the collecting surface as a coherent strand. This is continuously withdrawn through an axial hole in the rotor and twisted at any rotor speed that is mechanically and economically attainable. An axial assembly in which fibers are assembled in an electrostatic field has been exhibited making excellent cotton yarns, but is not yet in general use. The first rotor-type machine in commercial operation was the Czech, BD200 using a rotor of 2-in. diameter, and a number of manufacturers now produce similar machines. At the moment their use is limited to the spinning of cotton and man-made fibers, usually of short staple (25-40 mm), although one manufacturer makes a machine with a wider diameter rotor that will accept 4-in. staple. This machine is, however, restricted to the spinning of man-made fibers; it has not yet been possible to spin wool successfully. For the short-staple machine, rotor speeds of 45,000 r.p.m. are now used that insert twist at three times the maximum rate of ring spindles and can therefore produce yarn at at least three times the previous rate if the same number of turns per inch are inserted.

In early designs rotors had a series of holes drilled around the circumference; the rotor then acted as a pump drawing air in through the entry to its hollow interior and pumping it out through the peripheral holes. This induced airstream was used to convey fibers from a suitable opening device into the rotor and

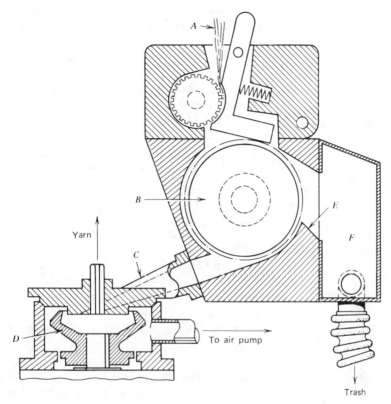

Figure 9.14. Open-end spinning arrangement.

toward its inner surface, deposition being assisted by centrifugal forces arising from the very high rotational speed. This arrangement had the disadvantage that if a hole became blocked by a particle of trash (when spinning cotton) the airstream was deflected from the blocked hole and a thin place formed in the fiber deposit at this point. The yarn then had regular, periodic, thin places at intervals of $\pi \cdot d$ inches—d being the rotor diameter. In later machines the rotors are not perforated; fibers are carried into the spinning rotor by air currents generated externally.

In most commercial machines each head is supplied with silver A, which is opened into a stream of individual fibers by a high-speed, small diameter cylinder or beater B covered with teeth (in effect a miniature taker-in as used on a cotton card). This fiber stream is picked up by the airstream, which is drawn through the unit, and is transported through a guide tube C to the rotor D, where the fibers accumulate on its inner surface, and the air spills out over the rotor edge and passes on to the pump; see Fig. 9.14.

In the machine of one manufacturer the high-speed beater is fitted with a separating edge E similar in action to separating edges set close to a card taker-in.

This removes trash from the cotton passing forward into chamber F, which is cleared pneumatically and has a considerable beneficial effect; it diminishes the amount of trash entering the rotor which otherwise in time tends to disrupt the continuity of the fiber strand, and so it dimishes yarn end breakages; it also upgrades the yarn quality.

Because of the break that is provided in this type of spinning, a completely new distribution of fibers is made possible. If V_s is the surface speed of the rotor collecting fibers, V_y the speed of withdrawal of the yarns, and λ the number of layers of fibers deposited on the inner surface before the end of the yarn comes around again to peel it off, $\lambda = V_s/V_y$. $V_s = n\pi d$ where n is r.p.m. for the rotor and d is its diameter in inches, and $V_y = n/t$ where t is turns per inch of twist.

$$\therefore \ \lambda = t\,\pi\,d$$

If t = 24 t.p.i. and d = 2 in., λ = 150, that is, 150 layers of fiber are layed down by the feed while the yarn is being formed; in other words, 150 doublings occur during yarn formation, and a new distribution of fibers arises. From what was said in Chapter 7, this should lead to a yarn approaching the ideal random fiber distribution with a very low index of irregularity. In the case of an axial fiber delivery, which must be less positive than a tangential one, the fiber arrangement might well approach a completely random one. High values of t and d increase λ and should favor the production of more uniform yarn.

It is not surprising in view of the above that break-spun yarns are more uniform than ring-spun yarns made from the same material; they also contain fewer thin places, thick places, and neps (37, 39) and are less hairy. On the other hand, they are from 10 to 20% weaker and require about 20% more twist than ring-spun yarns of the same count to provide adequate mean strength. In spite of lower mean strength, they are no more subject to breakages in spinning, winding, weaving, or knitting than ring-spun yarn.

Fibers in break-spun yarn have a more open structure and are not as well orientated as in ring-spun yarn. Most commercial units are supplied with drawn sliver that is opened into individual fibers by a high-speed, small-diameter cylinder covered with card wire, and these fibers are transported by an air stream through a guide tube to the rotor. If card sliver is fed, it will, in most cases, only spin satisfactorily if the majority of fiber hooks are trailing (40). In addition, the action of the card cylinder may well cause other hooks, or at least disturb the reasonably high degree of fiber parallelization of a drawn sliver. Because of these conditions of fiber presentation and transport, some parallelization is lost and entanglements are greater; in consequence, this yarn is about 15% more bulky than ring yarn and 20% more extensible; it is also said to have a much greater abrasion resistance. Low tensile strength is due to the fact that fiber migration within the yarn body is much less, as is the fiber extent (i.e., they tend to be buckled or hooded); they also tend to be assembled in layers (41). Tangential fiber feeds are much better in promoting fiber straightening than axial feeds, which is not surprising. It follows from these considerations that combed yarns cannot be break spun, since the advantages of fiber parallelization are lost.

Higher rotor speeds would increase production still further and might produce better fiber alignment, but the power required to drive a rotor rises steeply with speed and diameter, $P \propto n^{2.5} d^{3.8}$, and the economy of the operation depends on the relation between capital costs and power costs (37).

It might also be noted that the use of a comparatively large rotor to rotate the end of the yarn becomes a limitation of spinning speed, and with large rotors for longer fibers the correspondingly greater demands for power will reduce the economic speed still further. The size of the rotating units places an upper limit on production speed, as occurs in ring-spinning at a lower level.

Self-Twist Spinning

This method of staple-fiber spinning dispenses with continuous unidirectional twist in a yarn; instead, twist is inserted in alternating directions so that a length having S twist is followed by an equal length having Z twist (Figure 9.15a) which could be done by rotating the yarn in the directions shown at A, B, C, and D. This would insert false twist and would need little power, but such a strand has very little stability over lengths greater than one section and will untwist if free to do so. If, however, two such pieces of yarn are produced simultaneously, laid closely side by side, and released, because of the untwisting torques present in each section they will wrap round one another and will form a two-ply stable structure, as shown in Figure 9.15b.

This structure has weak places at the changeover points, X, Y, Z, since there is zero twist at these points in both single and plied yarns. This weakness is overcome in practice by allowing the two components to come together with the single twists out of phase; the two single strands and the folded structures then have zero twist at different points (Figure 9.15c).

In practice, worsted-type yarns are produced as shown in Figure 9.16a, which gives a simplified plan of the machine. Eight conventional worsted rovings are drafted by a conventional double-apron drafting system and emerge from the nip of the front rollers. The distance between each roving (0.75 in.) indicates the small scale of the machine. At a certain distance, u, from this nip they pass between a pair of rollers that oscillate or reciprocate to insert twist and also rotate at the same surface speed as the front rollers and pass the yarns forward. These oscillating rollers have a stroke d, so that as they reciprocate they roll the eight strands to and fro and insert S and Z twist cyclically. After a short distance, v, adjacent strands are brought together in pairs and they self-twist. In each pair, one strand is turned at right angles by a guide and travels further than the other to the point of union (Figure 9.16); the twist conditions in the two strands are therefore out of phase at that point as required. In principle a simple machine, engineering of high precision is required, particularly in the design and

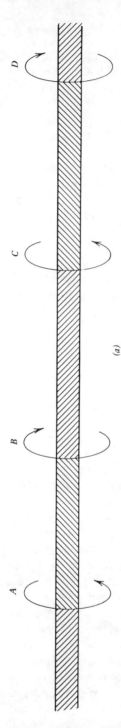

Figure 9.15(a). Single yarn-twist inserted in alternating directions.

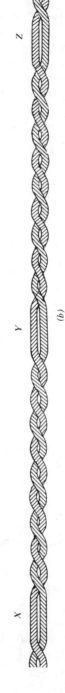

Figure 9.15(c). Self-twist yarn showing zero twist (in phase) at changeover points.

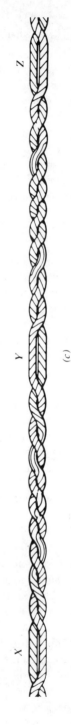

Figure 9.15(c). Self-twist yarn showing zero twist in singles out of phase.

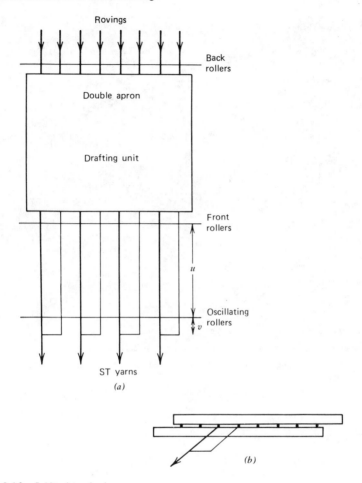

Figure 9.16. Self-twist spinning.

accuracy of manufacture of the oscillating rollers. The nip applied by these rollers is critical.* They must be loaded in such a way that the yarns between them (radius r) are rolled to and fro without slippage. Assuming that this is so and that the perimeter of the yarn is p, one turn of twist will be inserted as the upper and lower oscillating rollers each move $p/2$ in opposite directions. As the

---

*The bottom roller is in fixed bearings; the top roller is in bearings carried in a cradle that is pivoted to allow the top roller to rest on the bottom roller. Initially, the top roller is balanced so that it does not exert any pressure on the bottom roller. In use, weights are added to increase the pressure between the rollers up to about 600 g. This is the method adopted for changing the amount of twist inserted by the rollers.

Figure 9.17. Two single strands being combined, out of phase, to form an ST yarn (courtesy Platt International Ltd.).

yarn becomes finer, p will be less and more turns will be inserted for a given stroke. The following approximation is sufficiently accurate:

$$\text{Twist inserted} \propto \frac{1}{p} \propto \frac{1}{r} \propto \sqrt{\text{count}}$$

The twist multiplier therefore remains the same, and no twist adjustment need be made when the count is changed—the machine is self-compensating.

S and Z twist alternate over a cycle of length X cm and if T is the amount of self-twist appearing in each half-cycle, it can be calculated (42):

$$T = \frac{d}{p} \cdot \frac{2\pi uX}{\sqrt{X^2 + 4\pi^2 v^2} \cdot \sqrt{X^2 + 4\pi^2 u^2}}$$

In making this calculation allowance must be made for the fact that, as twist is being put in section u, it is at the same time being removed from v which is being twisted in the opposite sense.

The length of cycle X is the amount of forward rotation made by the oscillating rollers during one complete cycle of oscillation. This can be variable according to the machine design. From the above equation it can be shown that T is a maximum (as X varies) when $X = 2 \pi \sqrt{uv}$

$$\text{then} \qquad T = \frac{d}{p} \cdot \frac{1}{1 + \dfrac{v}{u}}$$

Thus the twist inserted per half-cycle, T, will be a maximum when $v = 0$ or when $u = \infty$, in practice when v/u is least. For operational convenience u should not be more than 8.5 cm nor v less than 1.3 cm. Under such conditions about 90% of inserted twist is retained, and, since T is a maximum when $X = 2 \sqrt{uv}$, a cycle of rather more than 20 cm is required—it is in fact 22 cm (43) (Figure 9.18).

Yarn production rate is 220 m/min compared with about 20 m/min for a conventionally spun 1/24s worsted-spun yarn. The machine spins a full range of counts at this speed.

It has been found that ST (self-twist) yarn has maximum strength when the phase difference in twist between the two strands is $X/10$, and this is provided in the difference in path lengths of adjacent yarns before they are united.

The advantages of this machine are listed as high rate of production, self-compensating twist characteristics, small space occupied, low maintenance costs, low power consumption, low end-breakage rates, fewer faults, less waste, much less noise, no idle spindles, low doffing time, and the production of assembly-

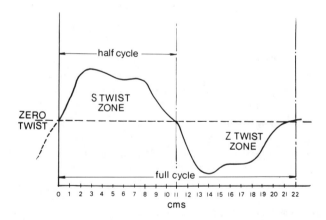

Figure 9.18.   Distribution of twist in ST yarn.

wound packages on the spinner. It seems assured of a permanent place in the spinning of worsted-type yarns of either 100% wool, man-made fibers, or blends. Wool must be finer than 50s quality and staples longer than 60 mm, and, because of the use of drafting aprons, wool must not contain more than 1.5% oil. A similar machine may well be expected in the future for the spinning of short-staple (cotton-type) yarns

ST yarns can be used for weaving and knitting: in weaving, warps must be sized. The cyclic nature of twist distribution is a disadvantage, since it can give rise to the diamond patterning already discussed. In knitted fabrics it can give rise to structural distortions. These shortcomings are overcome by an extra twisting operation in which additional unidirectional twist is added to produce what is known as STT yarn. Most ST yarns are treated in this way for high-quality woven fabrics.

As twist is added to an ST yarn it "is not preferentially added to any part of the cycle but is uniformly distributed" (44). Thus if S twist is added, twist increases in the S-twist sections and decreases in the Z-twist sections; as more twist is added the twist in the latter sections becomes zero, a conditions known as pairing. From this point, with further added twist, twist is unidirectional and patterning begins to disappear, even though the ply twist remains cylical. The reason why patterning disappears is that the surface fibers in the single strands begin to lie approximately parallel to each other at all points in the cycle. Light is therefore reflected similarly from all sections of the yarn, and there is no cyclic variation in reflecting power, which is necessary to produce optical effects needed to give rise to patterns.

The twist-factor* of a worsted yarn (TF) is defined by turns per inch = TF $\sqrt{Ne}$ where Ne is the English worsted count.

Applying similar terminology to ST yarns, average self-twist turns per inch = STF $\sqrt{Ne}$.

Defining pairing twist as "the minimum quality of uni-directional twist required to produce a yarn configuration in which ply-twist has become everywhere either zero or uni-directional" (44) and adopting a similar terminology for pairing twist factor (PRF), pairing turns per inch = PTF $\sqrt{Ne}$.

The same workers found that, for a wide range of yarns using a variety of fibers, PTF = 1.55 STF, and that an added twist factor of PTF + 0.5 eliminates patterning. To allow a margin of error, an added twist factor of PTF + 0.75 is adopted and in general there is an optimum region between PTF + 0.75 and PTF + 1.5 below which patterning is likely and above which yarn streakiness may appear. This streakinesss arises as the extra twist compresses the ply yarn and begins to accentuate any variations in its linear density.

*The term twist factor, rather than multiplier, is used here in accordance with the nomenclature used by Ellis and Walls.

Twistless (Zero Twist) Spinning

The restrictions imposed on ring-spinning production by the need to spin the yarn package can be eliminated by using an adhesive, instead of twist, to impart strength to the yarn. In the method devised by Selling and Bok (45), a cotton roving is boiled in alkali to remove any fatty substances and to bring it to a thoroughly wet state. It is then drafted in this wet state, and inactive starch grains are incorporated into the delivered strand. These are picked up from one of the delivery rollers on which a suspension of such particles is continuously poured; the surplus runs off and is recycled. The delivered strand is consolidated by a false twister and wound onto a package in a wet twistless condition.

The wound packages are steamed for 1 hr at 110°C, when the starch grains swell and spread over the fibers as an adhesive film. They are then dried in an oven at 100°C.

An interesting feature of this process is that the wet cotton roving can be satisfactorily drafted at high speeds with no mechanical means of fiber control. It appears that the increased cohesion of wet fibers increases the internal fiber control; such cohesion will anchor the floating fibers more effectively to the slow-moving fibers held by the feed rollers. It is true that they will also be attached more securely to the fast-moving fibers being withdrawn by the delivery rollers, but there are fewer of these; therefore, overall effect makes for more uniform drafting. The rate of production of cotton yarn from this process is 100 m/min. A 1/24s cotton yarn having 20 T.P.I. ring spun at 12,000 t.p.m. would be produced at 12,000/20 = 600 in., that is, 17 yd/min about 16 m/min. A break-spun yarn of the same count might require 24 t.p.i.; at a rotor speed of 45,000 r.p.m. the production rate would be about 48 m/min. Thus the twistless spinning method has more than twice the rate of production of break-spinning. The yarns produced are flat and ribbon-like, lustrous, and bulky.

This process has also been used experimentally for the wet spinning of flax with promising results. Flax contains natural adhesives such as pectin and lignin; thus no starch need be added, and the result is that there is no steaming needed after spinning and the yarn package can be dried immediately.

Other Spinning Systems

There are other, recently developed, spinning systems; two of these are briefly described because they have produced yarn with commercial applications, and both come under the definition of twistless yarns.

Fasciated Yarns

A feature of modern yarn technology has been the increasing use of false twist in yarn formation. Originally used in woolen spinning, it has found a place in the technology of textured yarns, self-twist yarns, and twistless yarns. On page 312, it was explained how, when a strand of staple fibers is passed through a rotating false-twist tube, the twist inserted in the upstream direction is removed as it passes into the downstream zone because it is canceled out by the twist imparted in the opposite sense in the downstream direction. For this action to be 100% effective, *all* fibers in the upstream zone must be completely controlled by the upstream twist; they are then *all* equally untwisted downstream.

If a few fibers escape the controlling action of the upstream twist by only having one end bound into the strand, so that the other rotates around it, they will be incompletely twisted into the strand. On passing through the false-twisting device, the bulk of the fibers will be untwisted, but the few that escaped the upstream action will now receive excess twist and will be bound around the untwisted body of the yarn. This is the principle of the formation of fasciated yarns, the name arising from the word "fasces" meaning, in Roman times, a bundle of rods wrapped around by ribbons.

To produce these yarns a roving is drawn out to a fine, but wide, ribbon that is passed through a false-twisting device (46). The edge fibers escape the control of twist in the upstream zone and form the binding fibers downstream. A particular feature of these yarns is the fine count (60s-80s cotton count) of the yarns made from 5-6-in. acrylic fiber staple. They are cleaner, more lustrous, bulkier, and more uniform than yarns spun on conventional ring-spinning systems; they are twice as strong but less extensible. These qualities can be traced to the core structure of parallel fibers with a few binding fibers to consolidate the structure (Figure 9.19). They can be spun at very high speeds—500 m/min.

Figure 9.19.   Fasciated yarn structure.

"Bob-Yarn"

This yarn is made by extruding and drawing a polymer into a pair of rollers where it is combined with an outer layer of staple fibers to form a composite strand (47). The polymer strand leaves the rollers in a semimolten condition covered with fibers, and in this condition it is consolidated by false twist inserted downstream (Figure 9.20). This embeds the fibers in the polymer matrix. It is then cooled and wound on to a large package.

The variety of polymers and fibers that can be used, the relative proportions of each, and the degree to which they are made to interpenetrate by the false twist applied provides the possibility of a wide range of yarns with widely

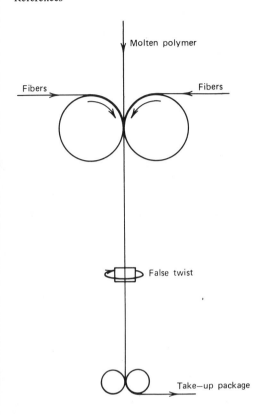

Figure 9.20. Bob-yarn process.

varying characteristics. The higher the degree of false twist the more the semi-molten polymer is squeezed through the surface coating of staple fibers.

The first commercial machine systems produced were said to spin coarse yarns (10s cotton counts) at about 350 m/min.

REFERENCES

1. *Wool Research,* Vol. 6, *Drawing and Spinning,* WIRA, Leeds, England, 1949.
2. Manual of Cotton Spinning, Vol. 3, The Textile Institute, Manchester, England, 1973, 233.
3. De Barr, A. E. and K. J. Watson, *J. Text. Inst.,* 1958, **49**, T588.
4. Charnley, F., Manual of Cotton Spinning, Vol. 3, Ch. 15, The Textile Institute, Manchester, England, 1973.
5. Cheng, R. and W. E. Morton, *J. Text. Inst.,* 1951, **42**, P442.
6. Butorovich, I. KH., *Tech. Text. Ind. USSR,* 1968, No. 1, 38.
7. Borzunov, I. G., *Tech. Text. Ind. USSR,* 1968, No. 2, 38.

8. De Swaan, A., *J. Text. Inst.,* 1951, **42**, P209.

9. Morton, W. E. and R. J. Summers, *J. Text. Inst.,* 1949, **40**, P106.

10. Morton, W. E. and K. C. Yen, *J. Text. Inst.,* 1952, **43**, T463.

11. Ghosh, G. C. and S. N. Bhadhuri, *Text. Res. J.,* 1968, **38**, 535.

12. Morton, W. E. and R. Nield, *J. Text. Inst.,* 1953, **44**, T317.

13. Simpson, J., L. B. de Lucca, and L. A. Fiori, *Text Ind.,* 1964, **128**, 78.

14. Simpson, J., L. B. de Lucca, and L. A. Fiori, *Text Ind.,* 1965, **129**, 110.

15. Simpson, J., L. B. de Lucca, and L. A. Fiori, *Text Ind.,* 1966, **130**, 90.

16. Simpson, J., L. B. de Lucca, and L. A. Fiori, *Text. World,* 1968, **118**, 58.

17. Simpson, J., L. B. de Lucca, and L. A. Fiori, *Text. World,* 1968, **118**, 817.

19. Brearley, A., *Worsted,* Ch. 6, Pitman and Sons, London, 1964.

18. Von Bergen, W. Wool Handbook, Vol. 2, Pt. 1, Ch. 1, Wiley, New York, 1969.

20. *British Wool Manual,* 1st ed. Harlequin Press, Manchester England, 1952, 126.

21. Townend, P. P. and E. Spiegel, *J. Text. Inst.,* 1944, **35**, T17.

22. Belin, R. E. and D. S. Taylor, *J. Text. Inst.,* 1967, **58**, 145.

23. Von Bergen, W. *Wool Handbook* Vol. 2, Pt. 1, Wiley, New York, 1969, 278.

24. Berlin, R. E., D. S. Taylor, and G. W. Walls, *Cirtel,* 1965, **4**, 57.

25. Belin, R. E., D. S. Taylor, and G. W. Walls, *Cirtel,* 1965, **4**, 71.

26. Martindale, J. G., *J. Text. Inst.,* 1945, **36**, T213.

27. Townend, P. P., *J. Text. Inst.,* 1948, **39**, T283.

28. Dircks, A. D. and P. P. Towend, *J. Text. Inst.,* 1964, **55**, T307.

29. Martindale, J. G., *J. Text. Inst.,* 1949, **40**, T813.

30. *Woollen Carding,* Ch. 2, 1968, WIRA, Leeds, England.

31. Ashdown, T. W. G. and P. P. Townend, *J. Text. Inst.,* 1961, **52**, T171.

32. Townend, P. P., *Text. Inst. and Ind.,* 1970, 244.

33. Towend, P. P. and J. C. Dobson, *Text. Manfr.,* 1963, 420.

34. Townend, P. P. and B. Jowett, *J. Text. Inst.,* 1951, **42**, P997.

35. Turner, A. J., *Quality in Flax.* Linen Research Institute, Lambeg, Ireland, 1954.

36. Krause, H. W., *Studies in Modern Yarn Production,* P. W. Harrison (Ed.), The Textile Institute, Manchester, England, 109, 1968.

37. *Break Spinning,* App. 6. Shirley Institute, Manchester, England.

38. Smith, P. A., *Textile Progress,* Vol. 1, No. 2, p. 18, The Textile Institute Manchester, England, 1969.

39. Locher, E. and J. Kasparek, *Textile Industries,* 1967, **131C.**

40. Krause, H. W. and H. A. Soliman, *Text. Res. J.,* 1971, **41**, 101.

41. Hearle, J. W. S., P. R. Lord, and N. Senturk, *J. Text. Inst.,* 1972, **63**, 605.

42. Walls, G. W., *J. Text. Inst.,* 1970, **61**, 245.

43. Allen, L. W. and D. E. Henshaw, *J. Text. Inst.,* 1970, **61**, 260.

44. Ellis, B. C. and G. W. Walls, *J. Text. Inst.,* 1970, **61**, 279.

45. Selling, H. J. and C. Bok, *Studies in Modern Yarn Production,* 122, The Textile Institute, Manchester, England, 1968.

46. Henberger, O, S. M. Ibrahim, and F. C. Field, *Text. Res. J.,* 1971, **41** 768.

47. Bobkowicz, E. and A. J. Bobkowicz, *Text Res. J.,* 1971, **41**, 773.

SUGGESTED READING

Wool Handbook, Vol. 2, Pt. 1., Werner von Bergen. Wiley, New York, 1969.

Manual of Cotton Spinning, Vol. 2, Pt. 2., C. Shrigley, 2nd Ed., 1969; Vol. 3, Ed. F. Charnley and P. W. Harrison, 1965; Vol. 4, Pt. 2., F. Charnley, 1964; Vol. 5, A. E. De Barr and H. Catling, 1965. Butterworth and Co. Ltd, and the Textile Institute, Manchester, England.

Wool Research, Vol. 4, *Carding,* WIRA, 1948, Leeds, England.

Wool Research, Vol. 6, *Worsted Drawing and Spinning,* WIRA, 1949, Leeds, England.

Woollen Carding, WIRA, 1968, Leeds, England.

The Łódź Textile Seminars, 2, Spinning. United Nations, New York, 1970.

Studies in Modern Yarn Production, P. W. Harrison (Ed.), The Textile Institute, Manchester, England, 1968.

Break Spinning. The Shirley Institute, Manchester, England, 1968.

Jute, R. R. Atkinson. Temple Press, London, 1964.

International Textile Bulletins, Spinning Sections. Int. Textil-Service Switzerland.

Spinning in the '70's, P. E. Lord. Merrow, Watford, England, 1970.

Twistless Yarns, H. J. Selling, Merrow, Watford, England, 1971.

Self-twist Yarn, D. E. Henshaw, Merrow, Watford, England, 1971.

10

Blending in Staple Yarn Systems

THE PURPOSES OF BLENDING

Blending has two puposes: (1) to produce a thorough intermixing of fibers and (2) to mix together fibers with different characteristics to produce yarn qualities that cannot be obtained by using one type of fiber alone.

Every yarn is designed to have certain properties—color, strength, luster, texture, etc., and it is important, even if the fibers are of the same kind, that these properties should be uniform throughout the whole of the batch. To achieve this the fibers must be as intimately and uniformly mixed as possible, and this is as true for yarns made from natural fibers, such as cotton or wool, as it is for the many blends of these fibers with man-made fibers, which have attracted a lot of attention to the subject. Perhaps one should qualify this statement by saying that it is almost as true, because in yarns of one type of fiber, for example, cotton, inequalities in mixing may not give rise to such obvious variations in quality as when the blended components have widely different characteristics, for example, nylon and wool. Nevertheless, even if only one type of fiber is incorporated, it is important that the yarn properties should be uniform throughout one batch. and in factories involved in the continuous production of a standard yarn, the constancy of its characteristics must be maintained month after month in succeeding batches.

The characteristics of natural fibers reputed to be of the same quality can vary from bale to bale, and even within one bale, for genetic reasons or environmental differences at the locations at which different parts of the contents are produced. It is therefore most important that the contents of the bales going into the blend

338

should be thoroughly mixed, preferably at an early stage of manufacture; thus subsequent opening and doubling operations can improve the homogeneity of the final product. Where long production runs of the same yarn are undertaken it is advantageous to have a number of different lots in a blend. If only one type of material were used until the supply ran out, then another lot used, and so on, there could be noticable changes in yarn properties. This danger can be minimized if a number of lots are blended together. As one lot runs out, and another similar lot is used to replace it, any small differences between these nominally similar components will have only a small effect on the blend as a whole, and this may be counteracted when another component must be replaced. For the same reason, when these long runs are involved, it is advantageous to feed material at any one time from as many bales as possible; the feed throughout is then more representative of the material supply available, and there is less possibility of variation in yarn properties.

Having emphasized the fundamental importance of mixing to obtain uniformity and reproducibility of characteristics in all yarns, we can turn next to the other purpose of blending, whose significance has greatly increased with the wide range of properties of man-made fibers now available—the design of yarns with qualities that one type of fiber alone cannot provide.

The properties of a yarn, and those of the cloth made from it, that may be affected by the choice of blend components, may be divided into aesthetic and functional. An obvious example of the first is color; a uniform blending of differently colored fibers is necessary in the woolen process to produce Lovat mixture yarns or heather mixtures; blending of fibers of different dyeing affinities can be employed to obtain differential-dyeing effects in piece dyeing. Other fibers can be blended to achieve luster or textural effects, or to affect cover and drape in the fabric (1). Properties of hand such as fullness, firmness, smoothness, and softness may be affected by blending; the addition of wool fibers with a higher crimp will increase the bulkiness of a worsted yarn, as will a mixture of acrylic fibers with different shrinkage properties in finishing; the inclusion of a small percentage (say 10%) of wool into a polyester yarn will increase the softness of fabric made from it; the addition of a small quantity of good felting wool to a blend of remanufactured wools will increase the felting when the cloth is milled, quite out of proportion to the quantity included.

One important functional property that can be modified by blending is yarn strength, a well-known example being the addition of a small percentage of nylon or polyester to wool. This improves spinning efficiency and the efficiency of winding and weaving processes because of fewer yarn breakages, and may allow finer yarns to be spun and processed into lighter-weight fabrics. Such a cloth is usually stronger and has better abrasion properties. When fibers of rather different lengths, and/or linear density, are blended together, the long or fine fibers tend to migrate to the center of the yarn and the shorter, coarser fibers to the outer layers; advantage can be taken of this to achieve favorable tensile properties—from the core fibers—and desirable surface properties—from the surface fibers. The use of man-made fibers with wool may also increase pleat and crease retention while minimizing wrinkling.

A property of a yarn for special consideration is its cost, which can sometimes be reduced by employing alternative and cheaper materials for some component of the blend. This may, or may not, have a detrimental effect on the yarn and fabric quality, but yarn cost was certainly one of the important considerations in blending long before the advent of man-made fibers. It is also worth noting that the use of a small proportion of a more costly component may so improve the processing efficiency that the increased fiber cost is more than offset by lower processing costs.

FIBER DISTRIBUTION IN BLENDED YARNS

Random Fiber Distribution and the Occurrence of Clusters

The problem of obtaining perfect blend uniformity is similar to that of obtaining perfect uniformity of linear density—it cannot be achieved. In the case of a 50/50 black and white blend of otherwise similar fibers, a uniform blend would result in a yarn with the same proportion of black and white fibers at all cross sections, with the black and white uniformly distributed throughout the section. The latter conception is difficult to define; therefore, let us consider the ribbon of fibers coming from the front rollers of the spinning frame. A uniform distribution would consist of alternate black and white fibers across the ribbon. The machinery has no inherent means of arranging for the same proportion of the two kinds of fiber to appear at every cross section, nor for the fibers to be disposed alternately across the width of the delivery of the spinning machine; it is accepted that the arguments put forward in Chapter 7 apply, and the best that it can achieve is a random distribution of fibers both longitudinally and laterally. It will be appreciated that, even if the fibers occurred alternately across the width of the delivery, the act of twisting them into a circular form in the yarn would in itself tend to mix them.

It will be remembered that no yarns are found in which the irregularity is that given by random distribution alone. The irregularity index of roving and yarn is always greater than unity. Montfort and his co-workers, in their theory of "grappes," have postulated that this is because the fibers draft in clusters that become smaller as the strand becomes finer. If this is so, the blending irregularity of yarns might also be greater than might be expected from a completely random distribution.

Coplan and Klein (2) have derived an Index of Blend Irregularity (IBI) that gives a measure of the extent to which the blend deviates from randomness.

$$\text{IBI} = \sqrt{\frac{1}{M} \sum \frac{(T_ip - W_i)^2}{T_ipq}} \qquad (10.1)$$

where

T_i = total number of fibers at a given section
W_i = number of fibers of component W at that section
p = average fraction of component W for all sections
$q = 1 - p$
M = number of sections examined.

If the blending were *perfect* along the yarn at each section, $W = T_ip$ and the above index would be 0; for complete randomness IBI = 1.0, and values greater than 1.0 mean that factors arise that indicate clustering. Coplan and Bloch (3) investigated woolen-type yarns made from blends of wool-nylon and wool-viscose of various percentage compositions as well as 100% wool, nylon, and viscose yarns. They counted the number of fibers of each kind at a number of cross sections and found values of IBI ranging from 1.2 to 2.2, indicating a degree of mixing poorer than could be expected from a random distribution.

If one assumes that the components have been drafting in clusters of C fibers, rather than as individuals, and that all components have the same cluster size, then

T_i/C = total number of clusters at a section.
W_i/C = number of clusters of component W.

If the clusters are randomly distributed, then from equation 10.1

$$\text{IBI (for clusters)} = \sqrt{\frac{1}{M} \sum \frac{(pT_i/C - W_i/C)^2}{T_ipq/C}} = 1 \qquad (10.2)$$

from which $\quad C = \frac{1}{M} \sum \frac{(pT_i - W_i)^2}{T_ipq} \qquad (10.3)$

and therefore C can be estimated.

Coplan and Bloch found in their experiments with woolen-type blends that C varied roughly between 3 and 4, and it should be noted that these results derive from woolen type yarn in which the material is effectively direct from a carding machine. This suggests that even a card does not reduce the material to a single-fiber state for mixing.

Hampson and Onions (5) described statistical tests that can be used to establish the occurrence of fiber clusters in worsted-type rovings. Their findings, as well as those of other workers concerned with cotton processing, substantiate that clustering occurs in every case, with cluster size decreasing as drawing proceeds. The data are not very extensive and therefore may not be very accurate, but a rough indication of the effect may be given by saying that cluster size in cotton-type card slivers might be about 100 fibers; 2, 3, or 4 is the unit size in yarns of both cotton or worsted types (5).

If the yarn consists of a binary blend of contrasting colors, irregularities caused by clustering can be objectionable. Lund (4) records the case of two yarns made from a 50/50 blend of black and white cotton fibers that had similar variations in the relative proportions of black and white fibers between sections, but that were so different in appearance that one was acceptable and the other quite unsatisfactory from the point of view of evenness of shade. The acceptable yarn had been made by blending the colored fibers at the picker, and fiber aggregates had been broken down in picking, carding, and drawing; in the other case, colored slivers had been blended at the third head drawframe, and such aggregates, or clusters, had not had time to be reduced.

Thus, although the uniformity of fiber proportions along the yarn is one of its important features, the distribution of fibers through the cross section of yarns is equally important, and most investigations on blending have in fact been made by cutting sections and examining the distribution of the various components within them. The technique is not without its difficulties; yarns must be embedded in a suitable matrix so that the position of the fibers is unaffected by section cutting. Examination is simplified by using black and white components. It is sufficient to study binary blends; if three or more components are present in a blend, a hypothesis regarding one of them may be tested by treating the others together as one component; each component can be treated in turn in this way if necessary.

The successive doubling and drafting of slivers in yarn making should randomize the distribution of fibers (or clusters) in a cross section, and a number of workers have suggested that this randomization can be achieved when the number of doublings used in making a yarn equals the number of fibers in its cross section. On the face of it this seems plausible, since, on the average, each doubled strand would then be contributing one fiber to the yarn, and since they are put together in a random way, this might be sufficient to produce a random radial distribution of fibers.

De Barr and Walker (7) made a series of yarns and examined their cross sections; or rather, they examined the ribbon of material emerging from the front rollers before being twisted into circular form. This enabled them to use a simple measuring technique and apply a formula derived from earlier theoretical work by Cox (8). A "perfect" mixture, when spinning a 50/50 black and white blend, would result in a ribbon in which black and white fibers alternated across its width. This of course never happens; they occur in groups, and the poorer the blend the larger in size and fewer in number the groups will be. Cox had derived a parameter, π, for measuring degree of mixing; applying it to the drafted ribbons of fibers examined by De Barr and Walker, it can be expressed as follows:

$$\pi = \frac{w - g}{w} \tag{10.4}$$

where at each place examined w = number of white fibers, and g = number of groups of white fibers (single fiber occurrences are counted as a group of one).

For the case of "perfect" blending mentioned above, each white fiber would constitute a group of one, that is, $w = g$ and $\pi = 0$. In the poorest possible mix, there would be one group of white fibers $g = 1$, $\therefore \pi = \dfrac{w-1}{w} = 1$ (very nearly). This parameter therefore ranges from 0 to 1 and increases as blending deteriorates.

In developing his theory, Cox had considered slivers laid side by side and doubled together; he derived a value of π given by

$$\pi = 1 - \frac{(1-p)(1-e^{-m})}{m} \tag{10.5}$$

where p is the average proportion of white fibers in the blend, and

$$m = \frac{n}{\Omega} \tag{10.6}$$

where n is the number of fibers in the average cross section of yarn and Ω the number of doublings used.

$$\therefore \qquad w = pn \tag{10.7}$$

From equations 10.4, 10.5, 10.6, and 10.7,

$$g = \Omega p(1 - p)(1 - e^{\frac{-n}{\Omega}}) \tag{10.8}$$

It is a reasonable assumption that the fiber distribution in yarn sections will become random as $\Omega \to \infty$ when $g \to np(1-p)$

If g in this case of a random distribution is written g_∞

$$g_\infty = np(1-p) \tag{10.9}$$

It will be noted that this is easily calculated from the blend proportion and the number of fibers in the average cross section. In practice values of g may be less than this, since we might not expect the number of doublings in use to achieve perfect randomization; thus we can take the measure of mixing in a particular test to be given by

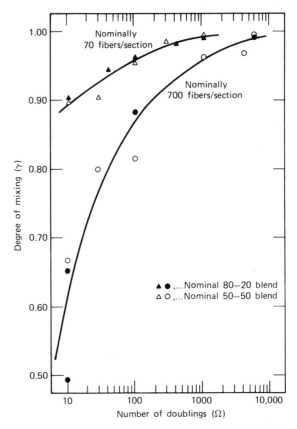

Figure 10.1. The degree of mixing, γ, as a function of the number of doublings, Ω, for two blends (DeBarr & Walker—courtesy Shirley Institute).

$$\gamma = \frac{g}{g_\infty}. \tag{10.10}$$

Very thorough random mixing will give $g = g_\infty$, that is, $\gamma = 1$. In practice, as mixing improves, $\gamma \to 1$. If "perfect" mixing were possible, $\gamma > 1$, and in practice by chance some sections will sometimes give values of $\gamma > 1$, but on the average the value is less than 1.

De Barr and Walker blended black and white viscose rayon fibers of the same denier and staple length and spun a range of yarns made by processes using different number of doublings. They did this for two blend proportions—50/50 and 80/20—each spun to two counts containing, respectively, 70 and 700 fibers in the average cross section. They examined the grouping of fibers at a number of points in the ribbon emerging from the spinning frame, determining g. Figure 10.1 shows values of γ plotted against Ω, the number of doublings used in making each yarn.

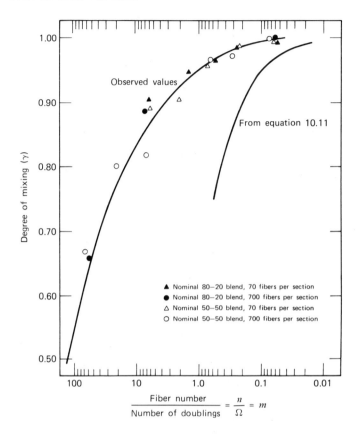

Figure 10.2. The degree of mixing, γ, as a function of the ratio of the fiber number to the number of doublings, m (DeBarr & Walker–courtesy Shirley Institute).

It will be seen that

1. γ is always less than 1 but approaches that value as the number of doublings increases.
2. γ follows the same curve irrespective of the blend proportions.
3. The finer yarns require fewer doublings to achieve the random conditions.

From equations 10.8, 10.9, and 10.10,

$$\gamma = \frac{g}{g_\infty} = \frac{\Omega}{n}\left(1 - e^{\frac{-n}{\Omega}}\right) = \frac{1 - e^{-m}}{m} \tag{10.11}$$

therefore if γ is plotted against m, all points should fall on the same curve, and it is found that they do so, as shown in Figure 10.2, although they do not lie on the curve predicted by equation 10.11. The reason for this is that Cox's theory, from which the equation derives, is based on a picture of the doubled slivers lying side by side and being drafted and attenuated without their constituent fibers mixing and intermingling with those in adjacent elements. Obviously in practice this intermingling will happen, and randomization will occur more rapidly, as shown by the graph.

Figure 10.1 provides some justification for stating that the number of doublings should equal the number of fibers in the yarn cross sections for good mixing; however, the more logical reasoning is that good mixing is reached gradually (asymptotically) as the number of doublings increases.

Many workers have commented on the fact that, in a black and white mixture, blending seems to improve to a uniform grey at a certain point in the drawing process; then, as the rovings become finer, they appear to become less uniform and more streaky. This is to be expected, because, if the fibers are randomly distributed with an average number of fibers, n, in the cross section and the fraction of white fibers in the blend is p, the standard deviation of the number of white fibers per section $= \sqrt{pn}$ (Chapter 7) and the coefficient (expressed as a fraction of the total number of fibers in the section) $= 100 \sqrt{\dfrac{p}{n}}$. As n decreases, this percentage value increases, and the variation in the amount of "whiteness" at different sections will increase, just as will the variation in the amount of "blackness," that is, the material will appear more streaky. To take this argument to the limit, if the attenuation were such that the emerging yarn was only one fiber thick, this would have to be either a black fiber or a white fiber, and the material would have been completely unblended. This would appear to be the reason why, as yarns become finer and this state is approached, the number of doublings has less effect on the degree of mixing.

Nonuniform Fiber Distributions in Yarn Sections

What has been said up to now is based on the assumption that every fiber has an equal chance of appearing at any point along the yarn or at any point in a section. This is a definition of random distribution. In all the work that has so far been described, the randomness of distribution has been verified, apart from an evident tendency of fibers to form clusters superimposed on this randomness, but it should now be noted that in all these cases fibers merely differing in color were considered. It has been found, and it is an important observation, that when fibers that differ in other characteristics are blended, fibers of one component may tend to take up their positions near the axis of the yarn and others will tend to move towards the surface. Such preferential modes of distribution are examined by cutting sections and using a microscope to study the form of fiber distribution. It is usual to divide the cross section into a central circular area and a number of concentric rings, either of equal width or equal area, outside the central core, and to count the number of each type of fiber in each area. Different experimenters have produced different types of index to express the preferential distribution or radial migration, but since they all differ, and since they depend on how the cross sections are divided for counting, no figures purporting to measure the effect will be given here.

Coplan and Bloch (3) found in the woolen type yarns they examined a predominance of wool in the outer layers of their wool/nylon blends and a less marked tendency for wool to predominate in the outer layers of the wool/viscose blends. The scale of the affect was found to be independent of the amount of twist put into the yarn. No particulars were given of fiber length or fineness that might enable these results to be compared with those of later workers, but they are mentioned because they are the only results on woolen-type yarn known to the writer.

Hamilton (9) examined yarns made from binary blends of 15 man-made fibers; Ford (10) used ten different materials; in both cases the yarns were spun on the cotton system. Townend and Dewhirst (11) examined blends of viscose fibers on the worsted system. Balasubramanian (12) processed blends of viscose fibers on cotton machinery and used a double-apron drafting system on the spinning frame. In all cases, man-made fibers were used so that fiber lengths and deniers could be selected to have suitable values. Onions, Toshniwal, and Townend (13) processed blends of nylons, wools, and nylon and wool, on worsted machinery. Blends were usually selected so that two components differed in length or denier; in some cases staples with variable lengths were used.

All found preferential fiber distribution when length and/or fineness of the two components differed, and all agreed that long or fine fibers move to the core of the yarn and predominate there, whereas short or coarse fibers migrate towards the surface. Hamilton says that comparatively small differences in length of fineness are sufficient to produce a significantly biased distribution of the two components, and the effect of differences in fiber substance is small compared with these two factors. According to Ford, differences in material can produce the effect, although this may sometimes be caused by differing amounts of fiber breakage of the two components during processing and segregation arising from resulting length differences; he also says that there is an indication that increased fiber extensibility promotes a movement toward the surface.

Earlier workers had sometimes reported findings contrary to these, for example, a short component might be reported as being found near the yarn core. It is thought that such results have been confused by differences in some other character, probably fineness. For example, one would expect a short component to migrate to the surface, but if it were also the finer component, this would tend to take it toward the core; the final result would depend on which effect predominated.

Townend and Dewhirst found that the degree of fiber migration of fibers of this kind in worsted-type yarns was unaffected by the amount of yarn twist, as Coplan and Bloch had found in woolen-type yarns. From this they concluded that the effect takes place in the drafting zone of the spinning frame, which they confirmed by finding the effect already present in the ribbon of fibers emerging from the drafting rollers. Balasubramanian, however, did not find this; he states that the distribution of fiber types across the drafted ribbon was random in his experiments, and he concluded that the result arises from the subsequent insertion of twist in the spinning operation.

Townend and Dewhirst found that, when using high drafts in spinning (40 instead of 6), either on a conventional spinning frame with carrier and tumbler control, or on a machine fitted with an A.S.D. unit, the index they used to evaluate migration was very much reduced. This gave added support to their view that the effect occurred in the drafting zone and not during twisting. They also found little migration when using a continental-type ring frame with a porcupine roller in the drafting zone; they concluded that the pins of this roller were impeding lateral fiber movements. Balasubramanian, on the other hand, reported a higher degree of migration, if any, when spinning with a high draft from a coarse roving. These discrepancies seem to be unexplained.

The types of blending deficiencies or irregularities that may occur in a yarn can therefore be summarized as follows:

1. variation along the yarn of the proportions of the different components in each cross section
2. inadequate mixing of the different components in each cross section
3. variations, along the yarn, in the degree of mixing in the cross sections

With regard to variations in blend proportions along the yarn, this can be considered as long-term or short-term. Long-term variations could be caused, for example, by changing the source of supply of one component at the beginning of the process or introducing a lot from a new dyeing if the shade were slightly different. Another possible source of long-term variations in blend composition is variation in moisture contents of lots introduced into a blend. If components are weighed out without regard for moisture regain, they may lose or gain moisture from the atmosphere during processing; thus the relative proportions of conditioned fibers change. The mixing operations throughout processing are designed to eliminate all but the longest of these variations, and doublings in drawing are designed to remove short-term variations in composition. Most authors referred to have found that blend composition along yarns is characterized by a random variation of the blend composition and that this is achieved after relatively few doublings in the drawing process. Long-term variations in blend composition may lead to visible changes in appearance that may produce barry cloth, or changes in other properties such as yarn strength, hairiness, luster, etc., which effects may manifest themselves in objectionable ways.

The work of De Barr and Walker showed that a relatively small number of doublings in processing is sufficient to ensure a random fiber distribution throughout the yarn cross sections, providing that differences in fiber properties do not cause preferential movement to core and surface. The number of doublings necessary is smaller the finer the yarn. There is no trouble in achieving randomness in either longitudinal or radial directions (given no preferential migration), and there is no evidence that this normal random distribution causes any trouble. There may be *small* fiber clusters in yarn, but as long as they are small—groups of two to four would be suggested by Coplan's work—they do not appear to be objectionable or anything other than normal.

Clearly, preferential fiber migration toward the core or the surface can be a most important consideration. On the one hand, it can be a disadvantage and give rise to faulty products; on the other hand, it might be used to advantage to design special yarn qualities. It can readily be seen that, if one fiber component moves to the surface, the surface will not have the properties expected of the average of the blend; hence it may have a color rather different from that expected, or have different dyeing properties, or a different texture. On the other hand, there may be occasions on which it might be of some advantage to design a yarn in such a way. The knowledge of the factors that cause such preferential migration is at least valuable either to use such effects or to counteract them if they arise.

De Barr and Walker mentioned the possibility of the third type of blending irregularity i.e. variations in the degree of mixing along the yarn. Townend, Harper, and Watt (14) found in the case of worsted yarns that streakiness in yarns made from fibers with contrasting colors was caused by this. On the whole, the fibers were blended in a random way, but the streaky sections contained a preponderance of one colored fiber or the other. Using De Barr and Walker's mixing factor γ, they found that it was less in the streaks; the streaks contained more large clusters than the yarn as a whole.*

These experiments by Townend and his co-workers involved a large number of worsted yarns of different colors, counts, and qualities. The streakiness of these yarns was assessed by knitting them into fabrics and making a visual assessment of streakiness using 35 observers. The concordance of the judgments of all these investigators was so high that the conclusions would, at least for these tests, appear to be very reliable. They concluded that the most important factor in determining streakiness was the color of the fiber components; the number of fibers in the cross section was the factor next in importance, and the processing

*γ was originally designed to study mixing in a ribbon of fibers. Townend and his co-workers adopted it to study the mixing in the surface of a yarn.

variables were of least account, but in practice assume an importance because they are the only ones under the control of the spinner.

When different colored fibers are blended together in binary mixtures, streakiness is most evident when there is the greatest lightness (value) contrast between the two colors, for example, black and white or purple and yellow. In these cases streakiness is not very marked if the lighter component is in very small proportion; it increases steadily and markedly as the proportion of the light component increases up to about 90%, after which it rapidly disappears. According to these workers, differences in hue are much less important in giving rise to streakiness. Streakiness was found to increase as yarn count became finer, or, for the same counts, if coarser fibers were used—this of course would also mean fewer fibers in the average cross section. Doublings had an effect on streakiness, but not as much as might have been expected; they were found to be more effective the earlier in the process doubling was initiated.

THE TENSILE PROPERTIES OF BLENDED YARNS

When a blended yarn is extended by an applied load, the fibers of the least extensible component are strained to the breaking point first. As they break, the fibers of the second component will have to bear an extra share of the load; whether they are able to do so depends on their tensile properties and the number of them present in the blend.

Consider a yarn, T tex, composed of a blend of fibers A and B whose tensile properties are shown by the stress-strain relations in Figure 10.3a.

In practice, these graphs would be either concave upward or concave downward, but only the values of stress at the two breaking strains x_1 and x_2 (the

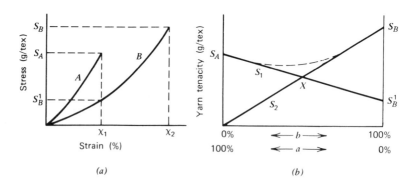

(a)   (b)

Figure 10.3(a). Tensile properties of blend components $S_B > S_A > S'_B$.
Figure 10.3(b). Change of yarn tenacity with blend composition $S_B > S_A > S'_B$.

tenacities S_A, S^1_B and S_B) are concerned in the following calculations, which are based on those of Hamburger (15).

If a is the percentage by weight of A, and b is the percentage of B, then

$$T_A = \frac{aT}{100} \quad \text{and} \quad T_B = \frac{bT}{100}$$

where T_A is the tex number for component A and T_B the tex number for B.

When the strain reaches x_1 all the fibers of A will be on the point of breaking,* and the total load P_1 supported by the composite yarn will be the sum of the loads supported by the two components P_{1A} and P_{1B}

$$\text{where} \quad P_{1A} = \frac{aTS_A}{100} \quad \text{and} \quad P_{1B} = \frac{bTS^1_B}{100}$$

If S_1 is yarn tenacity at x_1 $\quad S_1 = \frac{P_1}{T} = \frac{T}{100}(aS_A + bS^1_B)\frac{g}{\text{tex}}$ (10.12)

All fibers of component A will then break, and for the second rupture point at x_2 when all fibers of B break,

$$P_2 = P_{2B}$$

$$= \frac{bT}{100} S_B$$

If S_2 is yarn tenacity at x_2,

$$S_2 = \frac{P_2}{T} = \frac{bS_B}{100} \quad (10.13)$$

Figures 10.3a-10.5a show different combinations of tensile properties for components A and B, and Figures 10.3b-10.5b show how the tenacity of the blended yarn changes in each case as a decreases and b increases. The lines indicating these changes are derived from equations 10.11 and 10.12.

It will be seen that in the first case, where $S_B > S_A > S^1_B$ the yarn tenacity decreases to a point X and then increases as a decreases. The extent of this decline depends on the relative values of S_B, S_A, and S^1_B. The physical interpretation of this graph is that when b is small, as the A fibers break, the very few B fibers cannot sustain the large extra load applied to them, and they break too. As a increases, there are fewer A fibers to bear the load; thus the tenacity of the blended yarn decreases, and there are still too few B fibers to bear the still sub-

*Assume they all break at the same instant.

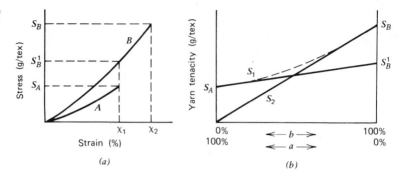

Figure 10.4(a). Tensile properties of blend components $S_B > S'_B > S_A$.

Figure 10.4(b). Change of yarn tenacity with blend composition $S_B > S'_B > S_A$.

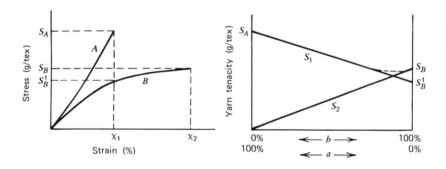

Figure 10.5(a). Tensile properties of blend components $S_A > S_B > S'_B$.

Figure 10.5(b). Change of yarn tenacity with blend composition $S_A > S_B > S'_B$.

stantial load increase when all the A fibers rupture. However, ultimately, at the point X, a has decreased and b has increased to such an extent that when the A fibers break there are sufficient B fibers to bear the extra load that falls on them, and they will survive until they reach their own breaking strain. Thereafter, as b increases to 100%, the tenacity of yarn increases.

With the fiber characteristics shown in Figure 10.4a, an increase in the proportion of component B results in a steady increase in yarn tenacity (Figure 10.4b) because, as a decreases, A fibers are being replaced by "stronger" B fibers. Figure 10.5b is somewhat similar to Figure 10.3b in that it has indications of a minimum, but the exact shape, as in all cases, depends on the relative values of S_B, S_A, and S'_B. A paper by Owen (16) gives examples of the ways in which yarn tenacity varies with blend composition for a large number of binary blends; they all fit the general picture given above.

In practice, these graphs do not show sharp changes in slope, but follow smooth curves, as shown by the broken lines. This is because all the A fibers do not break simultaneously, and because, as Kemp and Owen (17) showed, they continue to bear some of the load even after they have broken, because the two parts have frictional contacts with the other fibers and are bound to them by the yarn twist. These authors say that in some cases an A fiber may break more than once, and that if one places one's ear close to the yarn one can hear them breaking.

One should note that the above type of behavior may only be expected when the components A and B are uniformly blended. If certain cross sections have an abnormal number of fibers of one component, or if migration of one type of fiber to the surface is very marked, the transfer of load from one type of fiber to the other may not occur in the way described and may result in different behavior.

BLENDING PRACTICES

General Principles

From the experimental and theoretical investigations already outlined it may be said that, to achieve good blending, each component should be opened as completely as possible so that fibers can be mixed as intimately as possible to form a homogeneous blend; the earlier in the process this can take place the better chance there is for intimate mixing in all the processes that follow, and the greater the opportunity for breaking up fiber clusters. For this reason, blending takes place in the opening and cleaning processes prior to carding, unless there are other considerations that make this undesirable. Combing, if used, also provides a great deal of doubling that promotes mixing.

The other considerations that make blending in the early stages of opening undesirable arise if the various components require different processing conditions in opening, cleaning, and carding; if it is not possible to provide optimum conditions for the constituent components, it is better to prepare and card each separately and blend them later in sliver form. Another reason for blending in sliver form arises in the worsted process in which much fiber tow is now converted directly into top sliver without the need to employ carding or any of the other preparatory processes. This sliver is then blended with other slivers prepared by any other method, in drawframes or gill boxes. The number of drawing operations (involving drafting and doubling) following blending in this way is determined by the need to achieve the desired intimacy of blending. Sliver blending is standard practice in many cases for wool/man-made, cotton/man-made, and many wholly man-made fiber blends. The blending of dyed wool tops has always been done this way in the production of colored worsted yarns.

In the traditional cotton and wool textile industries, the old-fashioned method of blending fiber stocks was to lay down the various materials in relatively thin horizontal layers in the proportions required in the blend. The batch was then cut down in vertical sections so that each section contained the appropriate quantity of each sort. This procedure, followed by mixing in the later machines, gave an assurance of uniform blending, and, provided that the batch contained enough bales to be representative of the whole supply, succeeding batches should have produce yarn with consistent properties.

The disadvantage of the old system was that it was labor intensive and uneconomical; nevertheless, it contained the essentials of good mixing. Modern processes with automatic handling can only be equally effective if, at some point in the process, some provision is made for an accumulation of a large amount of material from which representative samples can be continuously extracted and fed into the production flow. Thus in the cotton spinning process that commences with a feed of bale cotton to a number of blending hopper bale openers, yarn consistent in count, strength, and appearance is only produced if about 100 bales are layed down as an initial supply from which material is uniformly fed into the opening line (18).

The exigencies of space prevent the presentation of a representative selection of machines for continuous or batch blending in the cotton, worsted, or woolen industries, but other books, journals, and machine makers' catalogues provide numerous examples. Figure 10.6, however, shows a recent machine produced for the cotton industry, it is typical of many that seem to aim at achieving a continuous, automatic version of the traditional stack mixing method, and their success can probably be forecast by the degree to which they can simulate it. Material from opening machines is mechanically layed down in the desired proportions in a large-capacity mixing chamber C by a feed mechanism B that traverses to and fro over the length of the chamber. When the chamber is full, the feed stops and the conveyor lattice that forms the floor of the chamber moves slowly forward; the laminar stack is then compressed by lattices D and rollers E, digested, and transferred to the next stage of the opening process.

Blending in the Cotton-Spinning System

Blending of cottons in this system has already been discussed. When cotton is blended with short-staple man-made fibers, or if blends of different man-made fibers are to be processed, the man-made fiber stock does not need cleaning; it needs carding to open the fibers and form a sliver, but the earlier preparatory processes need not be so long, or so severe, as in the processing of cotton. Having been carded, it contains little short fiber and no vegetable impurity; therefore, it does not have to be combed, and the fibers can be parallelized by drawing.

It is usual when making blended yarns such as cotton/polyester to blend after

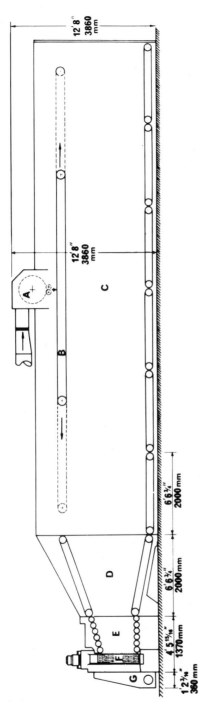

Figure 10.6. Sectional diagram of the Platt Feeder and Tuft-blender (courtesy Platt International Ltd.).

carding using two or three drawframes before the regular drawing and spinning operations; the man-made fiber is in carded sliver form, and the cotton sliver may be carded or combed. Although slivers can be blended in this way in conventional drawframes, sliver blending techniques are now being improved; for example, special blending drawframes have been designed to give greater mixing power and to facilitate the achievement of accurate blend composition. The principle adopted is illustrated by one machine in which four drafting systems are arranged one above the other; these units deliver four fleeces that are laid on top of each other in layers on a conveyor, and this sandwich is drafted, and blended, by another drafing system at the delivery head. The arrangement of slivers to each of the four drafting systems and the draft applied by each can be adjusted independently to give the required blend composition very accurately. Such a sliver-blending unit has a much higher blending power than a conventional drawframe.

Some spinners, probably a minority (20), blend in the opening line before carding. This could result in the necessity for processing under conditions that cannot be optimum for both fibers; it is practiced with the objective of obtaining maximum blend uniformity. In this system the cotton and man-made fibers are fed, in the desired ratio, from different hoppers on to a conveyor that feeds opening and cleaning machinery that in turn feeds a number of carding machines (21). Another system mentions the blending of combed cotton with polyester fiber in a similar way; the combed cotton in web, not sliver, form is transferred directly from combers to feed hoppers alongside others delivering polyester fiber; the blended stock is then lightly opened and passed to cards. Processing after carding is the normal drawing and spinning process.

It might be noted here that the high-speed rotor in modern break spinning machines is a highly efficient blending device. As explained in a previous section, something of the order of 150 thin layers of fiber are layed down consecutively on the inner surface of the rotor to form a yarn; such a high degree of doubling is a great aid to blending.

Blending in Woolen and Worsted Processess

In the woolen process, the carding machine is a long machine with a good deal of blending power. Blending takes place in the hopper, at intermediate feeds, and by the action of doffers and workers in dividing up the flow of material and causing some to be delayed and mixed with other fibers, as described in Chapter 9. There is no opportunity for blending in spinning, since yarns are spun from single ends of carded slubbing. Blending on the card probably takes care of short-term variations in blend composition, but it is essential to blend materials, whether they be different wools, different colors, or different fibers—including wools, hairs, cotton, or man-made fibers—in the processes before carding (see

Chapter 9). This should present the card with an even mix and take care of long-term variations in blend composition. If one component is present in very small quantities, say 5% or less of the total, it is advisable to premix it with 10% of another component and then blend this mixture with the other 85% in the normal way.

In the worsted process, the wools in all-wool blends are blended before scouring and all the subsequent processes (see Table 9.4) are available for thorough mixing. In this process wool is not dyed before carding; the first place at which dyeing may take place is when the wool is in the form of tops.

After dyeing, tops of fine cross-bred and merino wools are recombed. If tops of different colors must be blended to give a yarn of a particular shade, they may be blended before or after recombing. If they are blended before recombing, slivers of the different colors in the correct numerical proportion are fed to the first of a set of three intersecting gill boxes, and the doubling of slivers at these gill boxes and in the comb gives very good blending. After combing, the usual two top-finishing processes continue the mixing, and the color-blended tops then pass to the regular drawing operations. This process, although very thorough in mixing, has the disadvantage that different percentages of noils may be extracted from the different colors; this would affect the shade, and from batch to batch this behavior may vary. To overcome this preferential extraction rate, proportions before combing must be adjusted to allow for it; therefore, many workers prefer to recomb each component separately and blend the colored tops, in accurate proportions, after recombing, in a set of two or three intersecting gill boxes. This blended sliver is passed forward to the normal drawing operations. (Coarser tops are not recombed; they are blended in gill boxes before the normal drawing processes.)

Blends of wool and man-made fibers are produced in a similar way. They are not carded together because the optimum carding conditions for the two fibers are not the same, and because man-made tops can be much more satisfactorily made by tow-to-top conversion methods. There are thus again the two possibilities of blending (1) before recombing or (2) at the beginning of drawing. If recombing is done to take advantage of the considerable mixing power of the comb, it is preceded by three passages of intersectors in which mixing takes place and is followed by two top-finishing operations (19). The material is made up in batches of about 150 lb weight in which the components are accurately combined in the correct proportions—this ensures long-term consistency in blend composition. Allowance must be made for extraction of different percentages of noil from the different components—coarser and stronger fibers usually produce less—so that the desired composition of the final yarn is achieved. For this reason, some spinners prefer to blend before drawing without recombing. It should also be noted that blending before recombing also gives a mixed noil.

Wool and man-made fiber tops may be dyed, and if so, need recombing. They may be blended before or after recombing as above for the reasons given.

Similar machinery developments to those taking place in the cotton process are being introduced to improve the accuracy and efficiency of blending procedures. Blending units are made consisting of two intersecting gill boxes, each processing one of the components. The webs emerging from each are arranged one on top of another before the sliver is formed; since the draft of each head can be independently altered, blend composition can be adjusted with accuracy. The slivers formed are passed on to other gill boxes. These developments clearly lend themselves to many variations in the detailed design of sliver-blending practices appropriate to different circumstances, such as the number of blend components.

The blending of viscose rayon with other man-made fibers and with wool gives rise to other considerations, since it is usually produced as cut staple fibers. For blending with wool, it would be carded separately and blended in sliver form. In the case of blending with polyester, polyamide, and polyacrylonitrile fibers, the fibers may be blended at the card (19). Carding is followed by two stages of intersecting gills, rectilinear comb, and two finisher intersecting gill passages that produce tops for the normal worsted drawing and spinning procedures.

CORE-SPUN YARNS

A core-spun yarn is made from a central yarn surrounded by a cover or sheath of staple fibers that have been brought together at the *spinning frame* and spun into a composite structure; the two components are usually made of different materials. There is no unanimity of definition of such a yarn; it is sometimes called a core yarn. This term, however, is also used to describe a structure consisting of a number of component yarns, one of which is constrained to lie permanently at, or very close to, the central axis while the others wrap around it to act as covering yarns. This type of structure is made on a doubling or twisting machine.

In this chapter, we adopt the definition given in the first sentence, and refer to the yarn that lies at the center as the core component in order to give clear descriptions. This is sometimes called a support yarn, since it is usually there to give strength, although this is not always its purpose.

Such yarns can be made on most spinning frames, for example, cotton, worsted, or woolen. The roving is drafted in the normal way, and the core component, which can be a continuous-filament yarn, a monofilament, or even another staple fiber yarn, is fed into the system at the front roller nip (Figure 10.7). In the woolen process, it can be inserted through the false-twist tube. It emerges in close contact with the drafted fiber strand, they are twisted together, and, if conditions are correct, the resulting core-spun yarn has the core com-

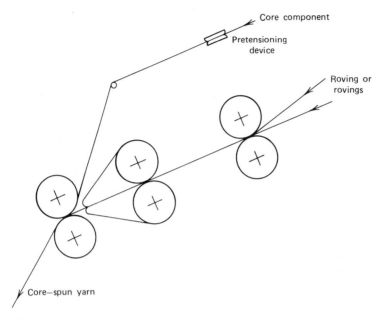

Figure 10.7. Production of Core-spun yarn.

ponent buried in a cover or sheath of staple fibers.* It is said that more uniform and more complete cover is obtained if the fibrous material is fed as a double roving and the core component is introduced between the two.

In principle, then, the technique is simple, and the choice of core component and fiber cover can be made from a wide variety of materials with a variety of purposes in view.

The core component is guided to the front roller nip through guides and a pretensioning device; it may be fed between two rovings that are kept separate until after drawing, or on top of, or to the side of two rovings, or alongside a single roving. Pretensioning is necessary to obtain suitable yarn properties. Adequate tension in the core component will cause it to adopt the axial position and to be well covered; the relative tensions and dispositions of the core component and the cover fibers also have some effect on tensile properties (22).

Core-spun yarns used in commerce are of two types—those with nonelastic cores and those with elastic cores. In the former, nylon or polyester continuous filament are the most common core components. They are used in such articles as tarpaulins, car safety belts, and net twines, which require strength and toughness provided by the core combined with such properties as softness and suitable frictional properties derived from the covering.

*By this definition, the Bob yarn is also a core-spun yarn.

Sewing threads are made from polyester cores with cotton fiber coverings. The polyester core provides high strength, resistance to abrasion, immunity to perspiration, and high resistance to chemical and bacteriological attack; the fibrous cover gives extra grip to prevent slippage of the seams, and protects the polyester core from melting at high sewing speeds. Such threads are used for sewing fabrics made from polyester fibers and blends. To prevent the cover from slipping, it may be treated with starch solutions, synthetic waxes, or the like (23). Sewing threads can also be made using a spun-glass-fiber-core component covered with suitable material such as polychlorovinyl fiber to give slow-burning threads. These are used for sewing work clothing designed for protection against high temperatures.

Core-spun yarns with a continuous elastomeric core component are used in the production of stretch fabrics, swimsuits, and other form-persuasive garments, in which they supplanted covered rubber threads. The material used for the core component is spandex, the generic name for an elastomer containing at least 85% of segmented polyurethane. This material has an extensibility of about 600-800% and has the power of rapid recovery from such extensions. The amount of spandex may be as little as 1% in the core-spun yarn; thus the resultant fabric has all the characteristics of the predominant staple fiber together with the advantages of stretch and recovery. Thus the fiber cover provides such desirable technical features as good moisture and perspiration absorbency as well as aesthetic qualities, and protects the elastomeric core from damage.

In making this type of core-spun yarn, the elastic core component is generally extended by 3.5 to 5.5 times before being combined with the drafted roving (24) just behind the delivery rollers of the spinning frame. This can be done by arranging a positive feed of the spandex from the creel package by means of feed rollers whose speed is adjusted as required in relation to the speeds of the delivery rollers. Alternatively, the spandex yarn can be incorporated unstretched at the delivery rollers of the roving frame, in which case the draft at the spinning frame must be chosen to extend the spandex rather than to draft the staple fiber (25). In either case, the core component retracts as it emerges from the spinning rollers, and the cover fibers bulk to form a rather loose sheath. The full elastic power of such yarns is not required in fabrics. The extensibility of the core can be permanently changed by heat; therefore, after weaving under tension, the finishing process contains a heat treatment during which the cloth is held to a certain width. This stabilizes or anneals the spandex core, leaving whatever amount of stretch and recovery is required in the fabric. An alternative method that has been used for controlling the amount of stretch in this type of core-spun yarn is to feed another relatively inextensible yarn in alongside an elastomeric core stretch to a chosen amount. The inextensible core component limits future stretch to this value.

Other types of core-spun yarn have been made, for example, worsted yarns with a bulked nylon core component. This can be used to give strength and stretch to a "worsted" fabric. Kruger (26) reports that the core component must be stretched at least 20% prior to spinning, that the resultant yarn is stronger than the worsted yarn without such a core, and the pilling of the fabric is much reduced. Because of the strong and elastic core, much higher spinning speeds can be attained in the spinning process.

Use is also made of the enhanced spinning qualities of core yarns for making yarns of some fibers that are otherwise difficult to spin, for example, short, smooth, stiff hairs or poor wools. A cheap cotton yarn can be spun in, using either the woolen or worsted system; there are many fewer ends down during spinning, and a stronger yarn can be produced. In this case, the cotton core would constitute about 20% of the yarn.

There is little restriction on the types of core and cover components that can be used; technology could be adapted to a wide variety of end uses that the ingenuity of the spinner might suggest.

REFERENCES

1. Bercaw, J. R., ed., W. von Bergen; Wool Handbook, vol. 1, ch. 7, Wiley, New York, 1963.
2. Coplan, M. J. and W. G. Klein, *Text. Res. J.,* 1955, **25**, 743.
3. Coplan, M. J. and M. G. Bloch, *Text. Res. J.,* 1955, **25**, 902.
4. Lund, G. V., *Text. Res. J.,* 1954, **24**, 759.
5. Hampson, A. G. and W. J. Onions, *J. Text. Inst.,* 1955, **46**, T377.
6. Coplan, M. J., C. A. Lermond, and R. A. Kenney, *J. Text. Inst.,* 1958, **49**, P379.
7. De Barr, A. E. and P. G. Walker, *J. Text. Inst.,* 1957, **48**, T405.
8. Cox, D. R., *J. Text. Inst.,* 1954, **45**, T167.
9. Hamilton, J. B., *J. Text. Inst.,* 1958, **49**, T687.
10. Ford, J. E., *J. Text. Inst.,* 1958, **49**, T608.
11. Townend, P. P. and J. Dewhirst, *J. Text. Inst.,* 1964, **55**, T485.
12. Balasubramanian, N., *Text. Res. J.,* 1970, **40**, 129.
13. Onions, W. J., R. L. Toshniwal, and P. P. Townend, *J. Text. Inst.,* 1960, **51**, T73.
14. Townend, P. P., R. Harper, and J. D. Watt, *J. Text. Inst.,* 1964, **55**, T352 and T365.
15. Hamburger, W. J., *J. Text. Inst.,* 1949, **40**, P700.
16. Owen, J. D., *J. Text. Inst.,* 1962, **53**, T144.
17. Kemp, A. and J. D. Owen, *J. Text. Inst.,* 1955, **46**, T684.
18. *Manual of Cotton Spinning,* C. Shrigley, Vol. 2, Pt. 2, p. 26, The Textile Institute, Manchester, England, 1973.
19. Józwicki, *The Lódź Textile Seminars,* 2 Spinning, p. 32, United Nations, New York. 1970.
20. Rozelle, W. N., *Text. Ind.,* 1969, March, p. 96.
21. International Textile Bulletin Spinning, 4, 1971 (English Ed.).
22. Balasubramanian, N. and V. K. Bhatnagar, *J. Text. Inst.,* 1970, **61**, 534.

23. Tozmarynowski, W., *The Lodz Textile Seminars,* 2 Spinning, p. 54, United Nations, 1970.
24. Meredith, R., *Elastomeric Fibres.* Merrow, Watford, Herts, England, 1971.
25. Cordial, I. F., *Text. Inst. Ind.,* 1966, 261.
26. Kruger, P. J., SAWTRI Technical Report, 146, 1971.

SUGGESTED READING

Von Bergen, W., *Wool Handbook,* Vol. 1, Wiley, New York, 1963.
The Lódź Textile Seminars, 2 Spinning. United Nations, New York, 1970.
British Wool Manual, 2nd ed. Columbine Press, Buston, England, 1969.

11

Continuous-Filament Yarns
and Tows

METHODS OF FIBER SPINNING

Introduction and General Considerations

Early methods of producing man-made fibers involved the extraction of molecules of cellulose from cotton waste products or from other sources, such as wood, which themselves did not provide readily spinnable staple fibers. Cellulose in combination with other chemicals was obtained in solution, and this was extruded by pumps through fine holes in a spinneret. The fine jets of liquid emerged from the spinneret into a bath containing other chemicals chosen to react with the solution so that the cellulose was regenerated and reappeared as fine continuous filaments. In the case of cuprammonium rayon, the cellulose itself was in a solution that needed only to be extruded into a bath of water for the cellulose to be precipitated in fiber form. Protein molecules were later regenerated in a similar way. Regenerated cellulose in the form of viscose and acetate rayons still maintains a large share of the fiber market, but proteins produce relatively inferior textile fibers, and, in view of the alternative uses of protein as food, they are more costly; regenerated protein fibers are therefore of little commercial importance for textile purposes, but new types of protein fiber are now being produced and are being fabricated into food products.

The production of fibers from spinning liquids in coagulating or regenerating baths gives rise to the process known as *wet spinning*. An early man-made fiber was made from cellulose acetate, which can be rendered soluble in acetone. When this solution is extruded into warm air, the volatile solvent evaporates and filaments of cellulose acetate appear. This method of *dry spinning* has been applied to the production of other fibers from solutions of fiber-forming materials in suitable volatile solvents.

Some of the more recent synthetic fiber-forming polymers are thermoplastic and have relatively high melting points. These polymers are prepared in granule (chip) form; the granules are melted and the molten polymer is forced through the holes of a spinneret; the liquid jets are cooled by a stream of air and solidify. This is *melt spinning*.

Man-made fibers of many varieties are now made; some which are classified as the same type may contain small quantities of secondary components that vary according to the manufacturer. For example, a polyester fiber is defined by the United States Federal Trade Commission as "a manufactured fiber in which the fiber-forming substance is any long-chain synthetic polymer composed of at least 85 per cent by weight of an ester of dihydric alcohol and terephthalic acid." This allows room for differences in composition, and conditions of manufacture may also result in different physical structures and properties of fibers in the same "family"; these differences are dealt with in more specialized reference books.

As in all textile yarns, uniformity of quality is essential in continuous-filament yarns. It is therefore fundamental that the filament-forming material should be consistent in quality, free from impurities, and free from air bubbles (unless these are introduced as a special feature). The problem varies in character according to the material in process, but appropriate steps are taken in each case to ensure consistency of batch conditions, and filters are employed between the pump and the spinneret as a final measure to remove impurities.

The interposition of filter between pump and spinneret draws attention to the need to protect the latter from impurities. If one of the holes is blocked, or partially blocked, the corresponding filament will be missing from the yarn, or will be too fine. This will lead to yarn irregularity. The holes in the spinneret must be of a diameter to produce filaments of the required fineness, and they should all be of the same size. Every care is taken to see that conditions at the exit side of the jet are such that accumulations of material do not gather there to clog it; the continuity of filament production is a most important consideration in the economy of the process. The spinneret must also be made of a substance that will not be affected by the liquids passing through it so that the diameter and shape of the filaments produced do not change.

The pumping of the fluid is also important. The extrusion pressure must be smoothly applied and be as free from fluctuations as possible; otherwise, the amount of material will fluctuate, and the yarn will vary in count. If pumping fluctuations occur and are regular, that is, if they are periodic, the count variations will also be periodic, with the possibility of patterning in fabrics as described in Chapter 7. For this reason, gear wheel pumps are now used in which rotating intermeshing gear wheels propel the spinning fluid in a regular flow to the spinneret. In the case of melt spinning, these pumps must be heated.

When a filament is formed, the long molecules are not usually arranged in a straight and extended form parallel to the filament axis; the normal arrangement is much more disordered, with the molecules coiled or crumpled and the filaments weak in consequence. If the filament is drawn, the molecules are straightened out and become progressively more oriented parallel to one another, and to the fiber axis, and the filament develops a higher tenacity.

In what has been written so far, the words filament, fiber, and yarn have been used to describe the product. The extruded fiber material does of course form a continuous filament, and a number of these, 20-120, may be combined side by side to form a yarn. Before coming together they may be moistened, lubricated, or given an antistatic treatment and then given a very small amount of twist that serves to keep the filaments together in yarn form rather than to give added strength. The spinnerets are made with the appropriate number of holes for the required yarn.

Continuous filaments are also combined in sliver form called a tow; as many as 500-2000 filaments are required, and the spinnerets are then larger, since they must accommodate the larger number of holes.

Wet Spinning

The equipment required for wet spinning, illustrated in Figure 11.1 consists of a supply tank for the spinning solution, pump, filter, spinneret, and coagulating bath.

Viscose rayon is the fiber that is spun in greatest quantity by this method and is produced by extruding a jet of cellulose xanthate into an aqueous bath con-

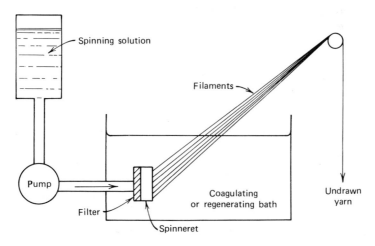

Figure 11.1. Wet spinning.

taining sulfuric acid, sodium sulfate, glucose, zinc sulfate, and sometimes magnesium sulfate. The exact function of each chemical is not known, but the cellulose is regenerated from the cellulose xanthate with the evolution of carbon bisulfide. The filaments are stretched about 100% to produce greater molecular orientation and tenacity. High-tenacity rayons and polynosic (cotton-like) rayons are produced in a similar way, with different formulae for the composition of the coagulating bath and varying stretching techniques. The detailed technology in these fields is very extensive, and the student is referred to specialized books for detailed discussion. To give an idea of some of the factors that can affect the structure of the fibers produced, one can mention the concentration of acid, the speed of reaction, and the conditions of stretch—whether during formation in the bath or afterwards; if afterwards whether hot or cold draw, and if in hot solutions, the nature of the solution.

Calcium alginate is another fiber in this category in which the material forming the fiber is produced in situ in the coagulating bath from the reaction that takes place there. In this case, the spinning solution is one of sodium alginate (obtained from seaweed), which is extruded into a calcium chloride solution; the calcium alginate is precipitated continuously and drawn away as fiber.

Many other fibers are wet spun, but in these cases the fiber-forming material is itself present in the spinning solution and is simply reprecipitated on extrusion. *Cuprammonium rayon,* for example, is made in this way by extruding a solution of cellulose in ammoniacal copper sulfate into a bath of water; this water removes the ammonia and some of the copper sulfate, and the cellulose reappears. *Regenerated proteins* such as those obtained from the casein of milk, or vegetable sources such as soya beans, ground nuts, or maize, are generally carried in solution in caustic soda and are extruded into a neutralizing bath of sulfuric acid containing formaldehyde as a hardening agent. *Dynel,* which is a modified acylic fiber, and *polyvinyl chloride* fibers are produced by dissolving the polymer in acetone and extruding into a water bath—the water dissolves the acetone and removes it from the polymer, which reappears. *Polyvinyl alcohol* fibers are produced by extruding a solution in hot water into a bath of sodium sulfate. Most acrylic fibers such as *acrilan* and *courtelle* are wet spun.

The rate of production by this method of spinning is of the order of 80 m/min.

Dry Spinning

Figure 11.2 shows the simple principle of dry spinning that is used in the spinning of cellulose acetate, cellulose triacetate, vinyon, and the polyacrylonitrile fiber orlon. The technological difficulties appear to reside more in the preparation of suitable solutions of the polymers than in the yarn production methods.

Cellulose acetate is made from cotton linters that are kier boiled with an alkali liquor, bleached, washed, and dried. This purified cotton is treated with acetic acid, acetic anhydride, and sulfuric acid, which, under suitably controlled condi-

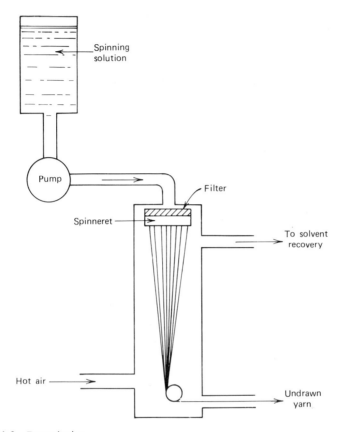

Figure 11.2. Dry spinning.

tions, convert it to cellulose triacetate. The triacetate is then converted by partial hydrolysis into cellulose acetate, which is soluble in acetone; this produces a highly viscous liquid, or dope, that is filtered and deaerated and to which pigments and the delustrant titanium dioxide can be added if required.

The method of spinning is clear from Figure 11.2; the dope is filtered between the pump and spinneret, the hot air enters the drying chamber at 100°C and removes the acetone, leaving an acetate filament. A small degree of stretch is imparted between spinneret and take-up roller that gives some molecular orientation and higher tenacity. A small amount of oil is added as a lubricant, and antistatic agents may be applied.

Cellulose triacetate is prepared as for the acetate process, but instead of being hydrolyzed, it is dissolved in methylene chloride and dry spun in a similar way. *Vinyon* is a copolymer of vinyl chloride and vinyl acetate that is soluble in acetone and dry spun; this yarn has to be stretched about 800% to give it adequate tenacity. *Orlon* is a polyacrylonitrile fiber and is also spun in this way; the solvent may be dimethyl formamide, the drying medium hot air, nitrogen, or steam, and the filaments are hot drawn.

The rate of production is about 300 m/min.

Melt Spinning

Figure 11.3 gives a diagrammatic representation of melt spinning, which is used for *nylons, polyesters, polyolefins,* and the polyvinylidene chloride fiber *saran.*

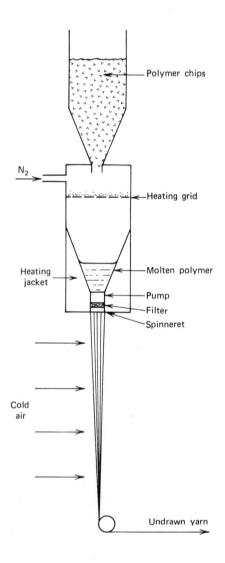

Figure 11.3. Melt spinning.

The polymer chips consist of small granules or thin flakes; the small size is necessary, since the thermal conductivity of the material is low, and large pieces would create melting difficulties. They are fed from a hopper onto a heating grid and drops of molten polymer fall through into a heated sump. At this high temperature, the possibility of decomposition and changes in the degree of polymerization arise; thus the time spent in the molten state should be short. For this reason, the amount of molten polymer in the sump is kept to a minimum. When the liquid level has risen to the required point, an electric control switches off the current to the heating grid, stopping further melting of the granules. The filaments are extruded, solidify, and pass through a cooling chamber in which they are cooled by a current of cold air (the rate of cooling can have an effect on fiber properties). In the case of nylon, this is followed by a steaming process that moistens the fibers, polyester fibers are moistened by passing them over moist rollers or discs.

In the case of nylon and polyester, the process takes place in an atmosphere of nitrogen, since the effect of oxygen on the melt is to discolor the fibers. It is also essential that water and air bubbles should be absent from the material in process. If a colored fiber is required, the polymer is dyed before the chips are made, and any delustrant is also added at this stage.

Nylon is cold drawn by about 400%; polyester, on the other hand, is hot drawn by about 500%, being passed around rollers for the purpose; the rollers are heated to 80-90°C. High-tenacity polyester yarns may be stretched more than this. Saran, polyethylene, and polypropylene are also cold drawn, and the latter is usually produced as a monofilament.

Melt spinning has the advantage of needing no chemical baths or solvent recovery systems; it also achieves the highest spinning speeds of about 900-1200 m/min.

Glass fibers are produced in a somewhat similar way; the material in the form of marbles is melted in an electric furnace, the molten glass flows through spinnerets and is wound up at high speeds, the filaments being lubricated before being combined into a yarn.

Spandex yarns are also extruded as thick monofils of the synthetic polymer, or as bundles in which the individual filaments adhere to one another.

SUITABILITY FOR PURPOSE—THE DESIGN OF FIBER PROPERTIES

Tensile Properties Developed by Drawing

As mentioned in preceding sections, man-made fibers must be drawn to orientate their constituent molecules into an orderly arrangement parallel to the fiber axis, and this leads to higher tenacity. The greater the draw ratio, the greater the degree of molecular orientation (i.e., crystallization) and the higher the fiber tenacity. Most man-made fibers must be drawn to some degree to give them adequate tenacity to fit them for commercial applications, but, that having been said, there is still room for producing a range of fibers of different tenacities from the same material to suit different purposes. For example, industrial uses require fibers of higher tenacity than normally needed for apparel fabrics or carpets.

If one uses the analogy of yarns in which fibers are parallel to one another (staple-fiber yarns) and those in which fibers are very much crumpled and disoriented (textured yarns), it is not difficult to see why drawn, and therefore strong, fibers with good molecular orientation are much less extensible than undrawn fibers with little molecular orientation. In fact, as tenacity increases due to drawing, extensibility decreases. This means that Young's Modulus for the fiber increases and the fiber becomes stiffer; this will have an effect on the aesthetic properties of fabrics made from the fiber such as handle, softness, drape, and also, probably, on wearing properties.

The length of the molecules also affects the fiber tenacity, and the means of modifying tensile properties vary a great deal from one fiber material to another. In wet spinning, the constitution of the coagulating bath may be modified to slow down the rate of polymer formation—conditions of formation affect molecular orientation; in melt spinning, the rate of cooling affects the size of the crystals that form. In wet spinning, some drawing may take place in the spinning bath; in other cases it takes place afterward; in other cases again it takes place in hot liquid baths. In the case of melt spinning, drawing may be done cold, as in the case of nylon, or hot, as in the case of polyesters. It may take place in one stage, but in some cases it is done in two stages, and these factors all affect the manner and degree of molecular orientation. Polypropylene crystallizes so rapidly that *undrawn* filaments are highly crystalline. In this it is different from other fibers that are melt spun, and the production of this fiber is therefore very sensitive to spinning conditions; the ability to control these conditions can be used to produce fibers with a wide range of crystallinity and physical properties.

The Design of Other Fiber Attributes

Many other physical properties can be modified to confer desirable properties on man-made fibers. For example, other chemicals may be added to modify their composition and properties; these can open up the structure and make the fiber more accessible to water and dyestuffs. To modify the aesthetics of fabrics made from man-made fibers, manufacturers have also altered linear density, shrinkage, crimp level and character, surface characteristics, cross-sectional shape, and many fibers can be produced with built-in color. Variable fiber shrinkage gives bulking properties; crimped fibers are springy and bulky, and straight fibers are soft and smooth to the hand. Surface friction is also important; fine fibers with smooth surfaces can provide a soft, luxurious handle.

The shape of fiber cross section can be modified by spinning from spinnerets with noncircular orifices. The degree of departure from circularity and differences of shape offer a large variety of possibilities. Flat fibers have a high luster and tend to glitter; they are said to have a harsh handle. Fibers with lobed cross sections have a high luster; this type of section affects the bending of the fibers and is said to give increased firmness and crispness of handle. Trilobal nylon fibers are similar in cross section to silk fibers, and fabric made from them is then said to closely resemble silk in handle. Such fibers, it is also claimed, do not show soiling so readily.

Thermal stability of fibers is also important for such domestic processes as ironing, and the related property of flammability is even more crucial. These matters are largely concerned with the fiber material; new types of nylon have much higher degradation temperatures and are much more heat resistant; they are used in fire-resistant clothing.

There is therefore a very wide range of technology involving many factors of material and method, and within it the properties of man-made fibers can be modified in fundamental ways that enable them to be designed for particular end uses. Also, because of the possible variations of many processing factors, what may appear to be the same sorts of fiber made by different manufacturers may have different physical properties and may behave differently in processing. If, for example, elastic properties play a part in the migration of man-made staple fibers during the spinning of staple-fiber blends, as discussed in Chapter 10, fibers that may appear to be nominally the same may respond differently in the spinning of such blends if their elastic properties are significantly different.

Bicomponent Systems

The characteristic crimp in fine wools such as merino and fine crossbreds is related to a feature unique to these fibers known as their bilateral structure. Such fibers consist of two different types of fiber-forming material that lie side by side along the length of the fiber and are called the orthocortex and the paracortex. The differences in physical properties and chemical reactivity of these two components account for many of the unique properties of wool fibers.

The bilateral structure is related to the crimp wave; the more reactive of the two components—the orthocortex—is always found on its outer (convex) side, and as the wave undulates the orthocortex moves around to keep on the outside (Figure 11.4).* The crimp wave is probably a result of the difference in properties of the two components and their adjustment to one another at the point of

*An unconstrained wool fiber has an irregular helical shape, and the orthocortex lies on the outside of the helix. The close packing of the wool fibers in the sheep's fleece does not allow the helices to form freely, and the fibers are pressed together and retain a wavy crimp.

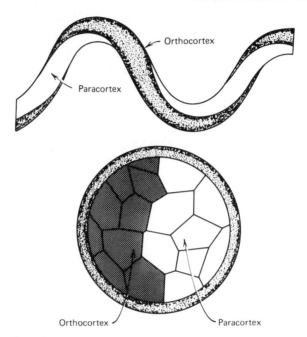

Figure 11.4. Bilateral structure of Merino wool.

formation of the fiber in the follicle. However this may be, it is certainly true that the crimp in these fibers is a result of the bilateral structure and is a valuable attribute that accounts for many of the desirable features of wool. Crimp ensures bulkiness in wool yarns; the associated air content makes cloths made from them good heat insulators and capable of holding water. When the fibers are placed in a humid atmosphere, or in water, the two components swell differently and the crimp changes, but it reappears again on drying. This is an important property because, when the fibers are distorted or flattened in use, this behavior helps the material regain its former bulk and associated qualities. For these reasons fiber manufacturers have tried to emulate the sheep by producing man-made fibers with a bilateral structure designed to have similar desirable properties.

Coarser wool fibers have less crimp. In these fibers there is the same dual structure of ortho and para material, but it is not arranged in the bilateral form; the ortho material is found in the center of the fiber as a core with the para material surrounding it as a sheath. This type of structure has also been produced in man-made fibers; both types are known as bicomponent fibers.

Figure 11.5 shows, the diagrammatic form, the types of bicomponent fiber cross sections that can be made; (a) and (b) are termed side-by-side bicomponent

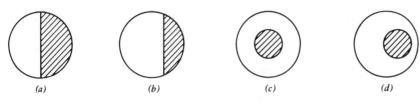

Figure 11.5. Sections of bicomponent fibers.

fibers, (c) and (d) are sheath-core fibers, (c) having concentric and (d) eccentric components. Such fibers have also been called "hetero-component," "composite," and "conjugate" fibers.

Such fibers are formed by using spinnerets that allow two types of polymer to combine at the point of extrusion to give the desired filament structure. Figure 11.6 shows the principle of formation of the sections shown in Figure 11.5.

In practice it is necessary to arrange for the simultaneous production of many of these filaments side by side, and the supply of two polymers in a convenient way to a number of orifices has led to the design of different systems of varying complexity.

One such arrangement for producing side-by-side bicomponent fibers is known as "mixed-stream spinning" (Figure 11.7) in which the two components are fed as concentric cylinders of polymer through a supply tube to rings of orifices that are positioned at the interfaces of the two components. They may be positioned

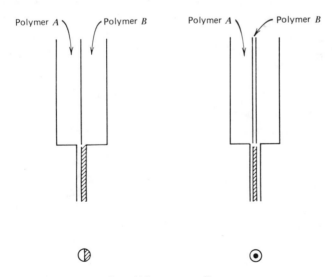

Figure 11.6. Principle of formation of bicomponent fibers.

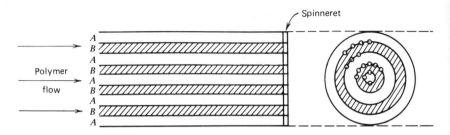

Figure 11.7. Mixed-stream spinning.

so that each spinneret receives the two polymers in the same proportion, which may be equal proportions as in Figure 11.5*a* or unequal proportions as in Figure 11.5*b*. If the orifices are not arranged in concentric circles in this way, a heterogeneous collection of bicomponent fibers will be produced with varying proportions of the two polymers. It is also clearly not impossible that fibers may be produced with varying proportions of the two components along their length.

This system, as with some others producing this type of fiber, requires that, if the proportion of the two types of polymer is to remain constant, the flow of the liquids must be nonturbulent. To maintain fiber uniformity, pumping pressures and the viscosities of the polymer solutions must also remain constant. It is also generally necessary that the two components be compatible in that they adhere strongly after the fiber has been formed.

In the case of sheath-core filaments, the eccentric type can be produced by off-setting the supply of the core component. Other methods are possible by introducing restrictions into the flow channel or by producing a concentric flow of the two polymers and then introducing a further supply of the sheath polymer before extrusion takes place (see Figure 11.8). No doubt other methods will appear as the technology develops. Fibers of the sheath-core type have also been produced by passing a filament of one type through a solution of another polymer that provides the sheath on cooling.

Up to now it has been assumed that circular orifices are in use, but this need not necessarily be so. It is possible, by using a shaped orifice, to produce side-by-side bicomponent fibers with other sections, for example, a trilobal section as in Figure 11.9*a*. In the case of sheath-core fibers, either or both of the two components can be extruded from variously shaped orifices Figure 11.9*b, c,* and *d* represent some possible fiber sections of this type, which have been described in patent specifications.

Air spaces responsible for 15-25% of the fiber volume can be introduced into bicomponent fibers; they may occur in one or both of the components, or between them, and they result in a lower specific gravity.

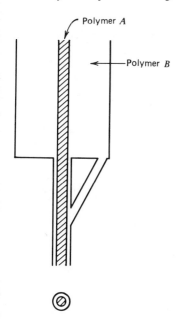

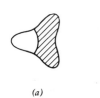

Figure 11.8. Eccentric sheath-core fiber production.

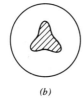

(a)   (b)   (c)   (d)

Figure 11.9. Possible fiber sections for bicomponent fibers.

Bicomponent fibers are drawn, and when relaxed the two components either shrink or have in-built latent shrinkage properties that may be released by treatment with steam or hot water, or, for those fibers unaffected by water, by a heat treatment. Since there is a shrinkage differential between the two components, side-by-side and eccentric sheath-core fibers are self-crimping. These fibers take on a helical form, the component with the smallest shrinkage being on the outside of the helix. The degree of crimp depends on the shrinkage differential between the two components, and about 20 to 30 crimps per inch would seem to provide useful characteristics. Yarns made from such fibers therefore possess bulking properties, and in many cases steam and hot water treatments can restore the bulkiness of materials that may have become flattened in use.

Some side-by-side bicomponent fibers are made with one component consisting of an elastomer. These fibers are drawn to four or five times their original length and relaxed. The elastomeric component shrinks most, and such fibers have great extensibility. The shrinkage differential between the hard and elastomeric components is so great that many of the fibers split with the elastomeric components in the center and the hard filaments bulked around them.

Concentric sheath-core fibers do not crimp but are used in other ways. The sheath is formed of a material that softens at a lower temperature than the core; under the influence of heat such fibers can be used in blends to provide bonded materials that can be used in apparel and furnishing applications. A needled web of such a fiber blended with normal fiber can be heated when fusion of the sheath occurs, and, where such fibers touch, a very strong bond results. Nonwoven nylon carpeting is made in this way; it gives a set to the needle felt and a durable product.

Biconstituent fibers (also known as matrix fibers or matrix-type bicomponent fibers) also fall under the heading of bicomponent systems. They are produced by the extrusion of a polymer containing suspended droplets of a second polymer that, on extrusion, are converted into long thin fibrils embedded in a matrix of the major constituent. Two commercial fibers of this type consist of 70% polyamide and 30% dispersed polyethylene terephthalate and are said to have a unique luster. The fibrils in such fibers are of the order of 0.1 micron diameter, and there are several thousand in the cross section of one fiber.

Since the technology of such fibers is in its infancy, considerable developments can be expected in the qualities of the yarns made from them.

YARNS FROM FILM

Thermoplastic materials, which can be made into filament yarn by melt spinning, can also be made into a sheet of film by extruding the molten polymer through a narrow slit. This film can then be cut into tapes and made into "yarns" suitable for various purposes. Paper yarns have been made in this way for many years; a sheet of paper is cut into narrow strips, and these strips are twisted to give a comparatively round section.

In the case of polymer film, the film is cut into strips, or tapes, after extrusion. Two types of yarn are made: (1) twisted tapes used as string, twine, cords, and for rope-making, and (2) flat, untwisted tapes used for the weaving of fabrics for carpet-backing, sacks, upholstery, and industrial purposes. The material used is mostly polypropylene.

In the manufacture of twines, a film about 0.006 in. thick is cut into relatively broad tapes (4-10 in 1 meter width (1)) and is immediately hot drawn in an oven to orientate the molecules and confer high strength in the longitudinal direction, which is desirable in twines, cords, etc. The amount of stretch is of the order of 1:8 or more, and this, together with the following twisting operations, causes the tapes to split up or "fibrillate," making the twine more flexible. Such twines are strong, smooth, rot-proof, and of low density.

Tapes for weaving are made from thinner film 0.003 in. thick (1), and since they need not be as strong as twines and it would be undesirable for them to split in drawing, they undergo less stretching. For this reason, and because the tapes for weaving must be quite narrow, they are cut finer, say 0.5 cm.

Thus coarse yarns and fine tapes, respectively, are made by these methods. In a development of these techniques known as "film fibrillation," the stretched tapes are subsequently acted on by a pinned roller brought into contact with, and running at a higher speed than, the tape. The sharp pins convert it into a lattice-like network of interconnected fibers, and by altering the relative speeds, or arc of contact between roller and tape, the length and character of the fiber mesh can be varied. When twisted, the yarns produced have the appearance of being made from continuous filaments, but the expensive procedures of extruding individual filaments through a spinneret have been avoided.

Badrian and Choufoer, in a paper describing fibrillated polypropylene yarn (2), outline three methods of tape production. In the first, a flat sheet of extruded film is cooled in a water bath; in the second, a flat sheet is passed around a pair of chill rolls; in the third, the polymer is extruded through a circular ring-shaped orifice producing a tubular film—this is air cooled. In each case the film is slit into tapes by knives and is hot drawn in a stretching oven. Since this leaves the tapes with some internal strains, they are annealed by a second heat treatment and then fibrillated. Figure 11.10 shows the process diagrammatically using the film tube method.

Yarns made by this method can be used for soft twines, sewing yarn, carpet yarn, knitting, or weaving, and the process is so cheap that ring spinning cannot compete with it.

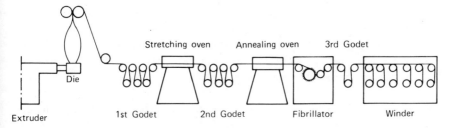

Figure 11.10. Production of fibrillatedpolypropylene (courtesy of Shell Chemical Co. Ltd.).

It may be noted that bicomponent fibers can be produced by coating a film with a second film and either splitting the film to produce bicomponent tapes or fibrillating it.

THE PROCESSING OF TOW

A tow is a collection of thousands of parallel, continuous filaments of man-made fibers in rope form processed so that individual filaments are cut, or broken, into staple fibers of suitable length, which are processed into yarn on conventional spinning machinery.

The earliest, and simplest, method of doing this was to cut the tow into uniform, predetermined lengths; these chopped-off lengths were baled and had to be opened before passing through the conventional spinning processes for staple fibers described in earlier chapters. Rather different procedures are adopted for the cutting of different types of fiber; for example, viscose rayon may be cut when wet immediately after coagulating and drawing; it is then washed, given suitable purifying chemical treatments, and dried. On the other hand, it can be drawn, purified, and dried before cutting. Other man-made tows have similar treatments appropriate to their properties; thermoplastic fibers such as polyesters are drawn, crimped, heat set, and cut.

Although no outstanding difficulties are now experienced in processing cut staple fiber in conventional machinery, either on its own or in blends with other fibers, the processing necessarily entails sliver formation and fiber parallelization processes, which seems to be an unnecessarily wasteful effort when one considers that the material was originally a sliver of parallel filaments. Machinery has therefore been developed that cuts, or breaks, the continuous filaments in the tow while preserving the latter's continuity and evenness and the parallel arrangement of its fibers; that is, the tow is converted into a sliver with the characteristic fiber arrangement of tops used in worsted processing. Tow-to-top methods therefore eliminate the need for opening, carding, and combing operations; the tops are introduced into drawing, as described in Chapter 10.

Tow-to-top converters operate (1) by rupturing the fibers by stretch breaking or (2) by cutting them. If a tow is passed between drafting rollers and a draft is applied that is greater than the stretch capacity of the fibers, providing that the rollers can grip the tow sufficiently securely to prevent slippage, the filaments will be broken in a random way to produce fibers of variable length. This length variability is a disadvantage in further processing, and the excessive stretch, which must be applied to cause breakage, produces structural changes in the fibers and reduces their extensibility. The large forces involved also make machine design more difficult, and thick tows cannot be processed.

Other stretch-breaking systems, typified by the Turbo stapler, control the breaking point of the filaments by using a pair of breaker rollers in the stretching zone. These rollers have sharp teeth like gear wheels that intermesh but do not

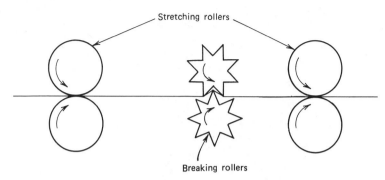

Figure 11.11. Stretch breaking of a tow fiber.

touch, and the stretched filaments pass between them. As they pass between, they are sharply flexed between the rotating gear-like teeth; this produces a high local stretching and breaks any filament already tautly stretched between the two sets of stretching rollers (see Figure 11.11). Thus all filaments are broken in turn as they pass through, the fiber length is more uniform than in random breakage, and, since all fibers are not broken at the same time, the continuity of the silver is maintained.

Fiber-cutting systems take a tensioned tow and cut it on the bias using a rotating upper roller, around which spiral knives are wound. These spiral cutting edges are pressed with great force against a lower, smooth anvil roller. By this means fibers are cut to uniform length, the fiber length being equal to the pitch of the cutting spirals. This process has certain possible disadvantages, which have been mentioned in the chapter on blending, that is, the cutting is really a crushing action, and groups of fiber ends can be welded together at the crushing point; some fibers may not be completely cut.

It is probably true that cutting converters produce much more man-made fiber top than stretch-breaking systems. The Pacific converter is typical of the type and is shown in Figure 11.12. This shows tow that is first passed through a section where it is hot drawn; following this another undrawn tow may be introduced. The tow then passes between the cutting roller and the anvil roller and through fluted debonding rollers which flex it and are designed to break up any interfiber adhesions produced by the high pressures in the cutting device. The band of slivers is then passed through two drafting zones in tandem with drafting rollers, aprons and shuffling rollers, all of which shuffle the fibers and destroy the uniformity of the bias cuts. Another type of top can be blended with the man-made top at this point, being brought in by the unit and introduced to the material under process between the two apron-drafting zones. On emerging, the web is rolled up on the diagonal roller and crimped in a stuffer box before being packaged as a top.

Tow-to-top systems have the advantage over cut-staple processing in that

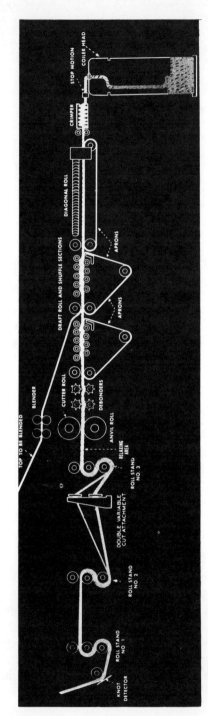

Figure 11.12. Pacific Converter.

380

further opening, carding, drawing, and combing operations are avoided, the yarns made from them are more uniform and stronger, and they can be spun to finer counts.

REFERENCES

Since many of the sources of information for this chapter are patent specifications, a list of which would be unduly long, the reader is referred to the publications in the suggested reading list where several hundred such references will be found if required. The writer acknowledges the value of these publications in the preparation of this chapter.

SUGGESTED READING

S. R. Cockett, *An Introduction to Man-made Fibres.* Pitman, London, 1966.

R. W. Moncrieff, *Man-made Fibres,* 4th ed., Wiley, New York, Heywood Books, London, 1966.

H. Ludewig, *Polyester Fibres–Chemistry and Technology.* Wiley, New York, 1964.

R. Jeffries, *Bicomponent Fibres.* Merrow, Watford, England, 1971.

Textile Progress, P. W. Harrison (Ed.), Vol. 3, No. 1. The Textile Institute, 1971.

Little has yet appeared in textbooks regarding fibrillated yarns; the following two references are therefore given.

J. E. Ford, *The Shirley Link* p. 4 Winter 1969/70.

Badrian, W. H. and J. H. L. Choufoer, *Text. Manuf.,* 1971, 97, 263.

12

Textured Yarns

INTRODUCTION

The texturing of thermoplastic yarns has been one of the most exciting developments in the field of textile processing. This process has completely revolutionized the use of nylon and polyester yarns in the men's and women's apparel, rug, and carpet, and hoisery and knitting industries. This can be realized from the fact that the total volume of textured yarns produced in the United States increased from 63 million pounds in 1968 to 744 million pounds in 1972. It is expected (1) that by 1978, stimulated by such developments as improved machinery, introduction of draw texturizing techniques, new and improved yarn types, and the expected population growth, the total demand for textured yarn will increase to almost 2 billion pounds. This chapter deals with the various texturing processes currently used in industry as well as the behavioral characteristics of the textured yarns.

TEXTURING AND ITS PURPOSE

Most natural fibers have a certain amount of crimp and waviness that helps to impart some degree of bulkiness, a property considered very desirable in textile fabrics. However, because of the way they are manufactured, most synthetic polymer fibers have a smooth surface and circular cross section, and when these continuous-filament yarns are woven into fabrics, they feel slippery to the touch and clammy when worn next to the skin. Nevertheless, synthetic filament fabircs have an attractive appearance, offer excellent wear, and can be washed and dried easily. But they lack warmth, comfort, and dimensional stability properties characteristic of fabrics made from natural fibers. Some of the aesthetic and comfort characteristics could be developed in fabrics by using synthetic fibers in crimped and staple form. Spun yarns (from staple fibers) are inherently bulky and have fuller appearance. Staple yarns also have good heat insulation and comfort characteristics. These properties result mainly from the large volume of air entrapped between the interstices of the yarn.

As is evident, achieving such characteristics involved an additional process of crimping and then cutting or breaking the continuous filaments into staple fibers to make yarns with improved bulk and warmth characteristics. To utilize the synthetic fibers more advantageously in their continuous-filament form, new techniques have been developed that modify their structural characteristics to make them bulkier (imparting texture) and fuller in appearance without converting them into staple fibers. Consequently, continuous-filament yarns are subjected to a modifying process, the purpose of which is to introduce permanent waviness (crimp), loops, coils, and wrinkles and thereby to modify the geometry of the constituent filaments. In addition, this process also imposes certain structural changes that become apparent in the modified tensile and other related properties of the filaments.

This process has the effect of modifying the geometry and structural characteristics of the constituent filaments of a yarn. In other words, the geometric shape of the filaments is deformed by either bending them or inserting crimps, loops, and curls into them by a number of means to provide a higher specific volume to the yarns (to make them bulkier). Thus they acquire many of the characteristics associated with spun staple yarns. The process by which this kind of modification in the character of continuous-filament yarns is achieved is known as "texturing" or "texturizing."

Concomitant with the changes in the bulk characteristics of yarns are other advantageous effects:

1. Texturizing helps improve the pill and crease resistance and confers better dimensional stability, fuller and better appearance, and greater durability.
2. Textured yarn fabrics have better shape retention ability than those made from straight filament yarns.
3. Flexibility is a very important property desired in textile yarns and fabrics. It depends on the ability of the filament (in the outer layer) to deform (extend) easily. A straight filament is comparatively less flexible than a crimped, looped, and curly filament. This ease of deformation of filaments is a function of their bending and torsional rigidities determined by the linear density of the filament and the geometry of the filament in the yarn. In a texturized yarn, filaments lie in all sorts of geometrical configurations and are loosely packed so as not to inhibit their relative movement. This technique confers better flexibility and extensibility to yarns and ulti-

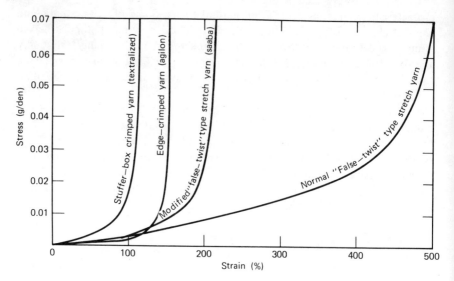

Figure 12.1(a). Stress-strain curves of stretch and bulk yarns (steam-relaxed 2-ply 70/34 nylon yarn) in the crimped region only.

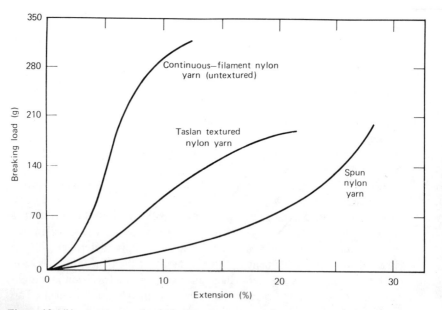

Figure 12.1(b). Load-extension behavior of untextured and textured continuous filament nylon yarn.

mately to fabrics made from them. The extensibility of various textured yarns is illustrated in Figure 12.1.

4. Texturizing results in softer yarns. Other factors being equal, the softer the yarn, the better its covering power.
5. Textured yarn fabrics are easy to wash, and they dry readily.
6. They reflect the high abrasion resistance, strength, and toughness of the thermoplastic polymer fibers from which they are made.
7. Most of the thermoplastic synthetic fibers have very low moisture content as compared to cellulosics and other natural fibers. However, the air entrapped between the interstices of the textured yarns retains moisture and thus makes the texturized yarns have fairly good comfort characteristics.
8. They provide higher covering power because of increased bulk.

CLASSIFICATION OF TEXTURED YARNS

Textured yarns may be classified into three major groups:

1. stretch yarns
2. modified stretch yarns or set yarns
3. bulk yarns

Stretch yarns are characterized by their high extensibility and good recovery, but possess moderate bulk in comparison with the other two classes of textured yarns. Stretch yarns, used extensively in stretch-to-fit garments in which extensibility and recovery from stretch are of primary importance, are produced mainly by the false-twist and by the edge-crimping (Agilon®) processes.

Modified stretch yarns may be defined as those with characteristics intermediate between stretch and bulk yarns. They are produced by modifying the stretch characteristics, usually by an additional heat treatment, after which they retain some stretch but have increased bulk. This modification treatment stabilizes the yarn structure and its bulk and stretch characteristics. The modified stretch yarns may be produced from stretch yarns by either of the following two methods:

1. In one method, stretch yarns may be modified by overfeeding them by as much as 30% or more into a heated zone in which they are stabilized in dry heat (process used on the yarn texturing machine to produce modified "stretch" or "stabilized" yarns).
2. The stretch yarns may be first soft wound in packages and then heat set or stabilized either in an autoclave in steam or during the dyeing process (the resulting yarns are usually called set yarns).

The modified stretch yarns are used in applications in which a lower degree of stretch and better hand and appearance are demanded. These yarns are generally used in knitted fabrics because they impart excellent stitch clarity, greater smoothness, and softness of handle, as well as better bulk and low extension under low loads. The low level of shrinkage of modified stretch yarns has favorable consequences when their dyeing behavior is considered.

Bulk yarns are characterized by their high bulk with moderate stretch and are used where bulk and fullness of hand are of greater importance than extensibility. However, bulked yarns generally possess adequate recovery characteristics. They are mostly used in carpets, upholstery, and garments requiring warmth and comfort characteristics. Bulked yarns are produced by air texturing, stuffer box, knit-de-knit, gear crimping, twist texturing, and various other types of crimp texturing processes.

Following is a brief discussion of the various types of filament yarns used for manufacturing textured yarns and the implication of the processing factors involved in modifying the resultant yarn structure. For detailed discussions of the various processes and the structural and morphological changes brought about by texturing, the reader is referred to the source material listed at the end of the chapter.

FIBERS USED IN TEXTURING

In the early years of development of textured yarns, nylon was the only fiber used in quantities that might be termed significant in terms of the total production of all fibers. In the early 1950s, processing techniques were developed to crimp nylon continuous-filament yarns by using the conventional throwster's machinery to insert twist, subsequently setting the twist by heat treatment techniques, and finally untwisting the yarn. This method of twist-heat set-untwist in separate operations is very expensive and has low productivity. In subsequent years, new continuous methods of texturing continuous-filament yarns, such as false twist, stuffer box, edge crimping, etc., were developed, which increased productivity as well as improving the quality of the yarn produced. With the addition of other thermoplastic fibers in the field of textile applications during the following years, the scope for the use of texturized yarns was enhanced. Today, in addition to nylon, such fibers as polyester, acetate, and polypropylene are extensively used in the texturized form.*

One major problem in the use of false-twisted textured (continuous process) thermoplastic yarns is their dye nonuniformity, which produces an effect known

---

*More recently, a patent was granted to the Japan Exlan Company and Toyo Boseki K.K., both of Japan, for the production of textured acrylic yarns.

as "barré" in the fabric. Acetate, on the other hand, suffers from the two major disadvantages of "voids" and "shading" effects in the yarn. Voids are caused by twist slippage through the spindlette or by broken filaments trapped by the twist. Shading may result from variations in yarn quality or from uneven processing conditions.

Polyester offers good aesthetic properties as well as excellent wear characteristics. Although polyester yarns also suffer from the problem of "barré" in dyed fabric, those invovled in the texturing industry are making a concerted effort to ensure the production of quality yarns. The major factors that affect this nonuniformity in the material are end-to-end temperature variations, tension in the thread line, heater zone parameters, and take-up package density. Some improvements in the quality of the textured yarns have resulted from post-treatments and from various other stabilization techniques now being employed in the industry.

In addition to the conventional types of thermoplastic fibers presently used in the production of textured yarns, there have been some efforts made recently to introduce new varieties of yarns in this area. One such development is the introduction of reinforced textured acetate yarns. This yarn is produced by reinforcing the freshly extruded acetate fiber with either a nylon or polyester or polypropylene filament at the cup of the spinning chamber. This process results in a stabilized acetate yarn that has excellent resistance to laundering and possesses good bulk retention as well as hand characteristics.

The filament denier and the total denier of the yarn used will depend on the type of machine used and the desired end-use applications. The lower denier yarns (ranging between 15 and 30 denier with 3 to 5 denier per filament) are extensively used in hosiery applications, whereas medium (40-200 denier) denier yarns are used in men's and women's outerwear. The most commonly used deniers are 70, 100, and 150. Heavy denier yarns (500 denier and above) are used in industrial applications.

METHODS OF PRODUCTION

Basically, three major processing techniques are used in the production of stretch yarns:

1. The Helenca® process—This is the conventional batch-type method of producing stretch yarns. Helenca is the registered trademark of Heberlein and Cie A-G, Switzerland.
2. The false-twist method—This is the most versatile and most widely used method of producing stretch-type textured yarns. Some of the false-twist machines presently being used in the industry are those made by Scragg,

ARCT, Barmag, Leesona, Heberlein & Co., Mitsubishi, Berliner, and Platts, to name a few (2).

3. The agilon process—The agilon process is highly suitable for processing low-denier hosiery yarns.

The Conventional (Helenca®) Twist-Heat Set-Untwist Process for Producing Stretch Yarns

The earliest method to produce permanent torque-crimp thermoplastic yarns was developed and patented by Heberlein Patent Corporation of Switzerland. This is a discontinuous process and involves the following three basic steps:

1. uptwisting to insert a high twist
2. heat-set or yarn-set (by the application of heat) to set the yarn in its twist-ed configuration while still on the bobbins
3. detwisting to remove the twist

The complete process for discontinuously texturing thermoplastic yarns is achieved by five primary steps followed by two or three secondary steps, as illustrated in Figure 12.2.

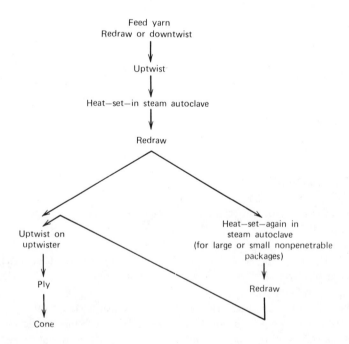

Figure 12.2. Helenca Process.

unbalance torque created through untwisting will impart stretch characteristics to the yarn.

DOUBLING

In most cases, the stretch yarns produced by this process are plied in such a way that an end previously thrown in an S direction is generally plied with another end twisted in a Z direction with the same number of turns per unit length. This plying process results in a fully balanced, no-torque yarn (one having no tendency to rotate when hung freely) that does not lose its bulk and stretch characteristics. The plying or doubling twist generally varies within the range of 3 to 7 tpi (1 to 3 turns per centimeter) depending on the denier of the yarn. The direction of the doubling twist (whether S or Z) in yarns depends on such considerations as their intended use in weaving twill constructions. To achieve the best results in yarn bulk and stretch, it is preferable that the final twist in single yarn after doubling be either zero or in the direction opposite to the processing twist.

CONING

To facilitate further processing, the yarn is treated with 2-5% of lubricating oil during coning.

Stretch yarns produced by this method appear as in Figure 12.3. The modifications imposed by this process generally result in the yarns having the following characteristics:

1. Heavy-denier yarns acquire stretch up to 300% and more, whereas the fine denier yarns may extend from 400-500%.
2. Their bulk increases up to 300%.
3. They have excellent recovery from stretch.
4. They generally have crisper hand and possess higher crimp, and greater elasticity and elongation.

Figure 12.3. View of a typical stretch nylon yarn textured by the conventional twist-heat set-untwist technique.

These properties are generally affected by such factors as filament denier, total yarn denier, filament cross-sectional shape, and whether the yarn is used in

Methods of Production

REDRAW

The feed yarn is wound or downtwisted on a redraw machine t
packages suitable for use on an uptwister.

UPTWIST

The packages from the redraw machines are creeled on the uptwist
yarn is uptwisted at high speed. The twist inserted in the yarn is very
ing from 47 to 114 turns per inch (1850 to 4448 tpm) depending o
denier. The twist in a 70-denier nylon yarn is in the neighborhood o
3000 tpm. The approximate twist can be calculated by the followin
ship:

$$TPI = \frac{7000}{Denier + 60} + 20$$

It is essential that the thread line tension during twisting and windi
as uniform as possible. However, the nature of the process is such tl
variations in tensions between bobbins and within the various locati
bobbin are bound to occur. Tension in the yarn is maximum near the
when it approaches the top of the bobbin. These variations in tension
changes in the quality of texture, which in most cases show up as fa
dyed and finished fabric.

HEAT SETTING

The twisted yarn packages are placed in an autoclave in which they are
steam under pressure to high temperatures for a considerable length of
temperature generally varies from 115 to 130°C at pressures between
pounds above the atmosphere for a period of approximately 1 hr. The
of this particular step is to impart to the yarn the highest possible "me
the highly twisted state.

REDRAW

After the yarn has been heat-set in its twisted condition (under h
developed as a result of thermal shrinkage), the bobbins are allowed t
the yarn can stabilize. It is then redrawn onto large packages. The ya
stage may be heat set again and then redrawn and transferred onto uptw

UNTWISTING

The redrawn heat-set yarn is untwisted on an uptwister to remove the t
pletely (zero twist) or to some other desired level. In the latter case, the
torque* caused by unremoved twist will cause the yarn to bulk,

*"Torque" yarns are characterized, by their tendency to rotate when hung freely.

singles or plied form. The effect of various factors is discussed in more detail in another section. However, it must be pointed out that the lower total denier yarns generally have higher stretch potential; the fewer the number of filaments in a yarn of a given total denier (i.e., changing filament denier), the greater its stretch and resistence to extension, and the better its recovery. Filament number (as well as filament denier) also affects the hand and strength-retention behavior; higher-filament denier yarns have crisper hand and retain a higher proportion of the original strength.

In addition to the detrimental effect of variation in tension during uptwisting on the quality of textured yarn, the heat-setting process itself, for many reasons, can cause variations in crimp, crimp retention, and dyeing behavior of the finished yarn (resulting in barré effect). The application of moisture and heat tends to cause shrinkage in thermoplastic yarns. If the yarn is not allowed to shrink freely, the shrinkage tension (force) would cause the yarn to elongate, thereby affecting the stress-strain characteristics of the yarn. These variations are distinctly visible between that portion of the yarn near the core (barrel of the bobbin) and the layers on the outside. The layers near the barrel of the bobbin are unable to shrink, whereas the outside layers can shrink by moving inwards.

The other cause of variation in this process is uneven heating of the outside and the inside layers of the yarn. The yarns wound next to the flanges of the bobbin would be subjected to a different heat treatment.

False-Twist Method

The false-twist method is the most important and the most widely used technique for producing textured yarns. The yarns produced by the conventional twist-heat set-untwist (three-stage) process generally have excellent characteristics; however, there are certain inherent drawbacks in the process itself. These disadvantages include;

1. Slowness of the process. Table 12.1 shows the production rates of various texturing processes used in the industry (11). It can be seen that the batch process has the lowest production rate when compared to all the other processes. This is because of the limitations imposed by the spindle speed of the uptwister and the batch orientation of the process itself (an uptwister spindle running at 10,000 rpm will produce approximately 1 lb per 168-hr week).
2. The process requires extra care in handling the packages to avoid mixing of yarns.
3. Uniformity is difficult to control between packages and lots.

To overcome the above-mentioned objections, the continuous false-twist method is employed extensively in the production of stretch and modified

Table 12.1. Typical Rates of Production for Various Types of Textured Nylon Yarns (11)

Type Textured Yarn	Yarn Output Speed (ft per min)	(m per min)	Production 70 den./ Spindle or Position/ 168 hr. Week (lbs)	(kg)
Agilon (stabilized type) yarn produced on	180	54.9	2 ply-19	2 ply-8.6
the *Hobourn* No. 2110 edge-crimp machine			5 ply-24	5 ply-10.9
C-B-Tex[a] (knit-de-knit) yarn	1983	604.4	134	60.8
Taslan yarn	300	91.4	19	8.6
Textralized yarn	1200	365.8	63	28.6
Turbo Duotwist yarn	1000	304.8	52.8	24.0
Conventional *Helanca* stretch yarn	—	—	Less	0.45
produced by the twist-heat-set-			than 1	
untwist batch method				
False-twist type stretch yarn[b]				
(assuming a twist level of 78 tpi or				
3071 tpm):				
45,000 rpm	48	14.6	2.5	1.1
100,000 rpm	107	32.6	5.6	2.6
150,000 rpm	160	48.8	8.4	3.8
180,000 rpm	192	58.5	10.1	4.6
210,000 rpm	224	68.3	11.8	5.4
240,000 rpm	256	78.0	13.5	6.1
270,000 rpm	288	87.8	15.2	6.9
300,000 rpm	321	97.8	16.9	7.7
330,000 rpm	353	107.6	18.6	8.4
360,000 rpm	385	117.3	20.3	9.2
400,000 rpm	427	130.1	22.5	10.2
500,000 rpm	534	162.8	28.1	12.8
750,000 rpm	791	241.1	42.2	19.1
1,000,000 rpm	1068	325.7	56.2	25.5

[a] Trademark of Textile Machine Works, Reading, Pa.

[b] Output of a false-twist spindle is directly proportional to the spindle's speed and to the denier being textured. Thus for any given spindle speed, a yarn of lower denier (which requires more turns per inch) must be run at a slower yarn speed and therefore at a corresponding lower rate of production.

stretch yarns (set). The false-twist method combines all three stages, namely, twisting, heat setting, and untwisting in one continuous operation. Furthermore, the twist in the yarn is inserted with the spindles revolving at very high speeds, in certain cases up to 800,000 rpm. In friction twisting, the yarn rotates at a speed equivalent to a spindle type machine at up to 4 million rpm. Such high-speed operations permit very high rates of production.

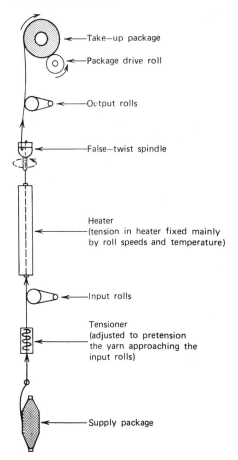

Take—up package

Package drive roll

Output rolls

False—twist spindle

Heater
(tension in heater fixed mainly
by roll speeds and temperature)

Input rolls

Tensioner
(adjusted to pretension
the yarn approaching the
input rolls)

Supply package

Figure 12.4. Line diagram showing the path of a yarn through a false-twist texturing machine.

It is out of the scope of this book to describe in detail all the principal false-twist-type yarn machines currently used for manufacturing stretch and modified stretch yarns. Although there are many variations in the design and the overall layout of different types of machines, they perform essentially the same function. The general principle of the false-twist texturing process is illustrated in Figure 12.4. The yarn is drawn from the supply package, fed at controlled tension over the heater and through the false-twist spindle, and finally wound on a package. The twist in the yarn is set when it is between the input feed roll and the false-twist spindle, by heating and cooling before it leaves the false-twist spindle.

To explain the principle of the false-twist process, reference is made to line diagrams in Figure 12.5a. Here the yarn is held at both ends and twisted in the center by a false-twist spindle. This action would impart S twist on one side of

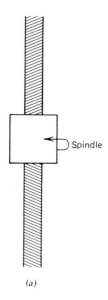

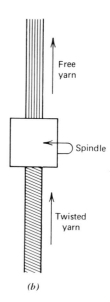

(a)

(b)

Figure 12.5(a). Insertion of false twist in a stationary yarn.

Figure 12.5(b). False twist in a moving yarn.

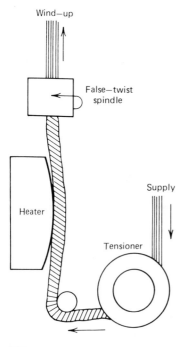

Figure 12.6. Illustration showing the principle of false-twist texturing.

the spindle and Z twist on the other side. However, if the yarn is running continuously while the false twist spindle is inserting twist continuously, the twist on the input side of the spindle (after passing through the spindle) will cancel the twist on the output side (because of the equal and opposite nature of the twist). This is shown in Figure 12.5*b*.

The next step in this process is to apply heat to soften the thermoplastic material to make it easily pliable; it is then deformed by twisting it in its softened state (to allow the stresses in the filaments to decay) and then by letting the yarn cool before untwisting it under relatively high tension. This is illustrated in Figure 12.6. The torsional forces are high enough to cause some significant changes in the geometry of the filament cross-sectional shape (Figure 12-7) in addition to the modifications in the morphology (orientation and crystallization) of the polymer material. Textured yarn thus produced has some residual torque in either the S or Z direction, depenting on the direction of rotation of the spindle. The extent of crimp and the latent crimping power of individual filaments in a false-twist yarn vary along the length of the yarn, and the crimp amplitude (size) is a function of radial position of the filament within the yarn. The radial position of the filament in turn is decided by the migration pattern imposed by the degree of twist and by the tension at which the twist is imparted. An idea of the crimp dependence, that is, of frequency and the ampli-

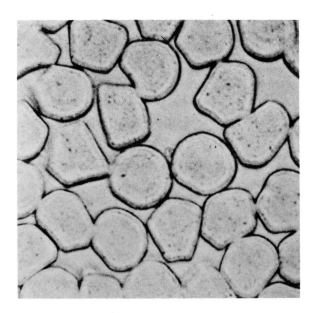

Figure 12.7. Deformed cross section of nylon filaments caused by the application of twist during their passage over the heater.

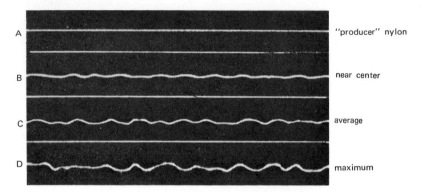

Figure 12.8. Longitudinal views of filaments from a 70 den/32 fils, nylon yarn, showing the differences in crimp due to the radial position in the yarn. A filament from untextured yarn; B, near center; C, middle layer; D, near the surface.

tude in a filament on its radial position in the yarn, can be obtained from the illustration in Figure 12.8; it shows variation in crimp in filaments caused by their location in the hot zone. The overall appearance of a 70-denier, 17-filament nylon stretch yarn produced by the false-twist yarn method is shown in Figure 12.9.

The productivity of a false-twisting machine depends on the spindle speed, provided the twist is kept constant. It is for this reason that much effort over the past twenty years has gone into the development of a spindle that can rotate yarn at ultrahigh speeds and still meet high performance standards. There are basically two types of twisting techniques used on false-twist texturing machines:

1. twist tube (spindle) or magnetic pin
2. friction twisting
   a. bush type
   b. stacked disc type

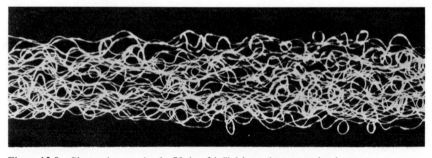

Figure 12.9. Photomicrograph of a 70 den-34 fil false-twist textured nylon yarn.

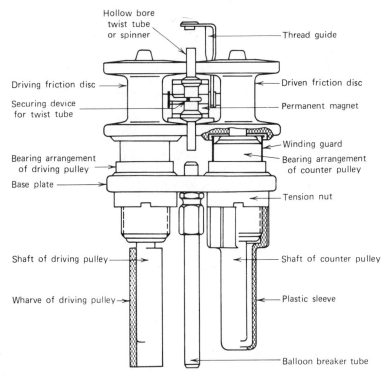

Figure 12.10. Schematic showing Type MFD FAG single-unit magnetic spindle. (Courtesy: Schäfer Industriegesellschaft M.B.H.)

Spindle or Pin Twisting

The earliest spindle used on the Fluflon[®] (trademark of the Leesona Corporation) machine was of the ball bearing type, which rotated at a speed of 45,000 rpm. Today, spindles of the magnetic pin type run around 800,000 rpm (MFD-800 Kugel-Fischer type). One such magnetic pin type spindle, Type MFD FAG (Schafer Industriegesellschaft GMBH, Schweinfust), is shown in Figure 12.10. A twisting spindle consists of a tube with a pin or a pulley fixed at one end. The yarn passes through the tube along its axis and is threaded around the pin, as shown in Figure 12.11. This arrangement ensures positive insertion of twist when the spindle rotates. The theoretical twist in the yarn can be calculated by the simple relationship between spindle (twist-head) and yarn speeds (wind-up).

$$\text{Twist per unit length} = \frac{\text{spindle speed (revolutions per unit time)}}{\text{yarn speed (length per unit time)}}$$

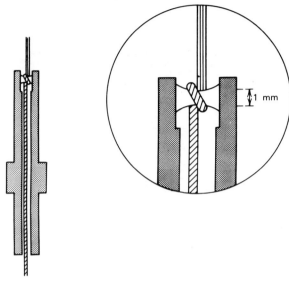

Figure 12.11. Yarn threading through a pin-type spindle.

The twist thus obtained is called "flat" twist. However, the actual twist in a yarn when it is still in the heater is somewhat different (generally more) than the calculated one. The fact that the amount of twist inserted in the yarn controls the production can be seen in Table 12.2, in which some typical production figures for three twist-head speeds are given (3).

Table 12.2

Denier	Twist (flat)	30,000 rpm		60,000 rpm		140,000 rpm	
		ft/min	lb/168-hr	ft/min	lb/168-hr	ft/min	lb/168-hr
30	100	25	056	50	1.28	117	2.61
60	80	31.3	1.28	62.6	2.56	147	6.00
100	70	35.7	2.14	71.4	4.28	167	10.00

FRICTION (BUSH-TYPE) TWISTING

Friction twisting, as the name implies, inserts twist into the yarn by frictional contact with a rotating head or a bush, as shown in Figure 12.12a and b. The yarn attains twist by rolling inside the moving inner surface of a hollow tube of plastic or rubber. The inner diameter of the cylinder is considerably larger than that of the yarn, which means that one revolution of the tube produces a very large number of rotations in the yarn, neglecting any slippage. Consequently, the yarn can be fed at relatively high linear speeds at comparatively low rotational speed of the twisting unit. Twisting rates of up to 4 million rpm are claimed commercially, which would correspond to the bush speeds of 20,000-25,000 rpm. However, one great disadvantage of this system is the large amount of slippage, up to 50%, that occurs during twisting. This results in uneven crimping of the yarn as compared to the pin-type spindle (positive twisting) false-twist process. But, because of the higher production rates possible, this system is of increasing commercial importance in producing fine-denier textured yarns.

A major drawback of the bush is its gradual wear, which ultimately affects the yarn quality. The wear of the bush material is affected by higher running speeds and the processing of heavy denier yarns.

STACKED DISC-TYPE FRICTION TWISTING

A second type of friction twisting device commonly used in the industry is the stacked disc-type friction unit illustrated in Figure 12.12b. Other disc-type units commonly used in the industry are also shown in Figure 12.12. In the stacked disc principle, the yarn is twisted by frictional contact with the outer surface of the disc sets located on three shafts. The axis of the shafts are positioned on the apex points of an equilateral triangle.

The diameters of the stacked discs are such that they overcap in a fashion that imparts to the yarn desired angle of contract with discs for the most efficient twist insertion. There is a critical relationship between the diameter, the thickness, and the position of discs that must be satisfied for the most efficient twisting operation. The stacked disc-type generally has nine discs. With this type of arrangement, the angles of contact of 1000-1300 degrees can be realized effectively. In other words, in principle it should be possible to twist yarns by stacked discs using wear-resistant friction surfaces with a low coefficient of

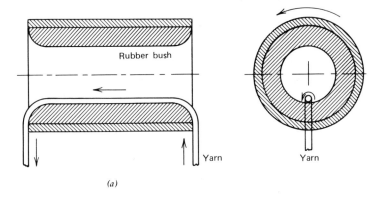

(a)

Figure 12.12(a). Principle of friction twisting.

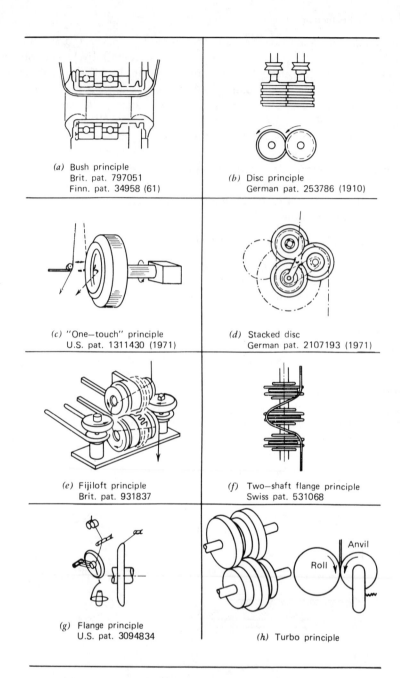

(a) Bush principle
Brit. pat. 797051
Finn. pat. 34958 (61)

(b) Disc principle
German pat. 253786 (1910)

(c) "One—touch" principle
U.S. pat. 1311430 (1971)

(d) Stacked disc
German pat. 2107193 (1971)

(e) Fijiloft principle
Brit. pat. 931837

(f) Two—shaft flange principle
Swiss pat. 531068

(g) Flange principle
U.S. pat. 3094834

(h) Turbo principle

Figure 12.12(b). Construction principles of friction twisting devices.

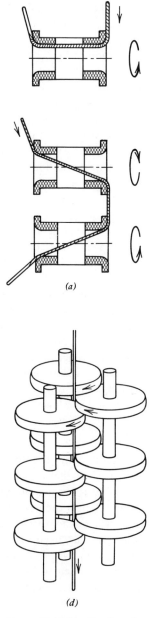

(a)

(d)

Figure 12.12(b). Continued.

friction such as metals and ceramic (the torsional moment α to the production of the coefficient of friction of the surface and the contact angle). However, in practice this does not work, and a friction surface with a high coefficient of friction (rubber or a plastic) gives desirable results. Some reasons that account for the unworkability of low-friction surfaces are:

1. If the last friction surface does not have a high initial moment, the emerging false twist textured yarn will have a disturbed cross-sectional shape— ribbon-like instead of the more desirable circular shape.
2. The initial contact that the yarn makes with the twisting device is critical. It is believed that a large portion of the twist is inserted in the yarn at the instance of first contact. This is true for both the bush- and the disc-type friction units.
3. Generally, all metal, metal oxides, or ceramic surfaces cause extensive damage to the filaments.
4. The pores in the metal and ceramic surfaces are clogged by monomers and spin finishes stripped off the yarn, thereby altering the frictional character- istics of the surface and thus the twist insertion.

Friction surface materials such as rubber and polyurethane are commonly used for both bush- and disc-type twisting. Discs have a tendency to wear slower than bushes. Heavy-denier yarns can be processed easily on disc-type units, since they offer less wear and more flexibility (in denier range).

Recently, Turbo Machine Co. announced the development of a simple device that has proved highly successful in controlling the yarn slippage and thus the twist uniformity during texturing. They claim that the yarn produced with this unit installed on the texturing head has a lower tendency to have tight spots.

Principle of Friction Twisting

The principle of friction false twisting has been described by Arthur and Weller (3). They have discussed two devices that may be used in friction twisting, as illustrated in Figure 12.12. The first is of the bush type as mentioned earlier, and the second uses a moving band. Using a rubber-lined bush of approximately 5/8 in. internal diameter, they carried out experimental studies to investigate the effect of the principal variables that might influence twist during friction twisting. These variables included (1) twist head speed, (2) yarn speed, (3) yarn tension (4) wrap angle, and (5) yarn denier. The yarn twist values reported are based on the untwisted yarn length. (The twist was determined by removing a length of yarn and then untwisting in a twist tester under a tension equal to that used in the operation.) Their results showing the relationship between the twist in the yarn (t) and the ratio of yarn speed (μ) and twist head speed (the speed of the inside twist-tube surface, s) for various yarn speeds are plotted in Figure 12.13.

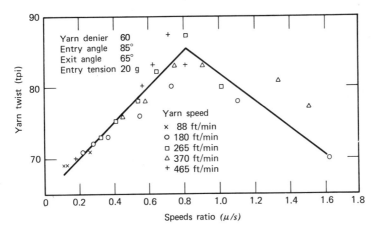

Figure 12.13. Yarn twist versus speeds ration. From Arthur and Weller.

It can be seen that there is an optimum value of (μ/s) at which twist is at a maximum level. The optimum value of (μ/s) would depend, among other factors, on the surface characteristics of the lining material used in the twist tube. Arthur and Weller have reported the optimum twisting efficiency for nylon 6 and nylon 66 yarns at $\mu/s = 0.81$ for natural rubber with carbon black filler as the lining material. The significance of the maximum is that the torque required to twist the yarn is just sufficient to overcome the frictional force on the surface of the yarn. Any deviation from this point would cause instability in the torque-twist characteristics of the yarn: increasing torque-twist (movement toward lower μ/s value) behavior caused by a decreasing value of the coefficient of friction would result in lower twist. Another advantage of the bush-type (friction) twisting head is its self-adjusting property over a wide range of yarn deniers, which is evident from the straight-line relationship observed between twist and the reciprocal of the square root of the linear density (twist factor = twist $\times \sqrt{\text{tex}}$) over the range of 30-840 denier. These authors also report a linear relationship between twist and the sine of the entry angle (defined as the angle made by the ingoing yarn path with the twist head axis) and suggest that, for optimum crimp, an entry angle close to $90°$ is suitable.

Thwaites (4) has recently reported a theoretical analysis of the mechanics of friction twisting and has derived expressions that relate variations of twist factor, yarn tension, and yarn path (entry and exit angle) through a false-twist spindle. His experimental results and theoretical predictions show a reasonable agreement and confirm the findings of Arthur and Weller. However, the effect of the change of the frictional properties of the bush material on the processing behavior and the torque-twist characteristics of yarn during texturing is still not fully understood and needs further study. This is essential if progress is to be made in improving the regularity of the friction-twisted textured yarns.

Factors Affecting the Stretch Characteristics of False-Twist Textured Yarns

The false-twist texturing process is used to produce crimped yarns covering a very wide range of properties. These include such yarn characteristics as rough appearance to softer and smoother silk-like qualities, high skein shrinkage, and varying amounts of stretch before and after relaxation. However, specific yarn properties, such as texture, softness, cover, and stretch, among others, could depend on a number of materials and processing variables. These variables can be grouped into the following three categories:

1. Material variables
    a. filament denier
    b. filament count and yarn denier
    c. cross-sectional shape of filament
    d. spin finish
    e. luster
    f. mechanical properties of parent yarn
2. Machine variables: this includes the machine design factors
    a. heater type
    b. heater length
    c. heating mode: conduction, convection, radiation
    d. feed system
    e. type of spindle—pin or friction (bush) type, fluid twisting, duo-twisting
    f. winding or take up mechanism
3. Processing (texturing) variables
    a. heating and cooling rate
    b. amount and direction of twist
    c. spindle speed
    d. yarn take-up speed
    e. heat zone thread line tension
    f. post-treatments such as second heater, autoclave (steam) setting

MATERIAL VARIABLES

Filament Denier and Number

The individual filament denier and the number of filaments for a given total denier of yarn have a profound effect on the textured yarn crimp rigidity, as well as on the stretch and bulk characteristics. The crimp rigidity, that is, the latent retained energy or torque, increases directly with the diameter of individual filaments. This is because the coarser the filament, the higher the torsional energy required to distort (or twist) it and the greater the retained energy after twisting and setting. The torsional stress in a filament is proportional to the fourth power of its radius (square of denier). It follows, therefore, that

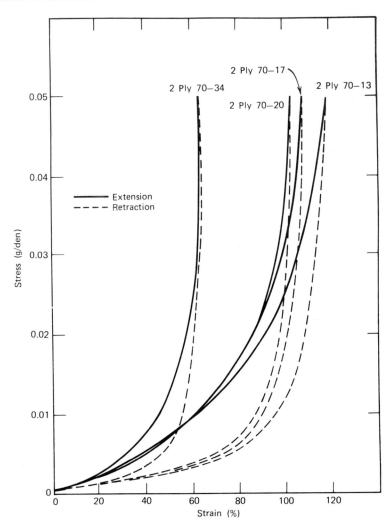

Figure 12.14. Extension-retraction curves (third cycle) for 2-ply, 70-den, false-twist textured nylon yarns.

the fewer the number of filaments in a stretch yarn for a given denier, the higher its stretch potential and resistance to extension, and the faster its recovery from extension. On the other hand, larger-diameter filaments produce textured yarns with crisper or harsher hand but retaining a higher percentage of their original strength, as can be seen from Figure 12.14 (5). This shows the effect of filament denier on the stress-strain characteristics of a typical 2-ply 70-denier false-twist stretch nylon yarn. It is clear from the illustration that the extension at the point of crimp removal and the energy of extension and retraction increases with decreasing number of filaments (increasing filament denier).

Total (Yarn) Denier

Yarns are generally twisted with a constant twist factor during false twisting. It follows that the finer-denier yarn will be twisted to a higher twist angle and consequently will have a greater stretch potential. For example, for a given filament denier, a 70-denier yarn will develop higher stretch potential than a 100-denier yarn.

Cross-Sectional Shape

In addition to the regular circular type, synthetic filaments come in various cross-sectional shapes, such as trilobal, triangular, and dog-bone. The torsional rigidity of a filament varies directly with the shape factor, which has a value of unity for a circular cross section and decreases as the polygonal shape approaches a triangle. Textured yarns made from multilobal cross-sectional shapes tend to have silk-like hand and appearance. They also have greater covering power than the corresponding circular filament yarns with the same individual filament and total denier.

Spin Finish

Thermoplastic yarns are generally given a spin finish after spinning to make them amenable to further processing. The requirements from the point of view of processing, such as texturing, include:

1. correct fiber frictional properties at all stages of processing
2. control of static electrification at all stages of processing

It is believed that a spin finish, generally created by a lubricating additive, affects the deformability of filaments during twisting because of decreased friction between the filaments. An absence of spin finish would cause high frictional resistance between filaments and thus would prevent them from being twisted as desired and would result in very high thread line tension. This could then result in filaments with kinks and snarls, which are undesirable for a smooth appearance. On the other hand, spin finish can affect the free movement of the yarn around the pin and can hamper the performance of the twist trapper.

Properties of the Parent Yarn

The physical properties and the chemical nature of filaments (polymer type) would affect such factors as the thread line tension, temperature, the twist that must be adjusted during false twisting, the setting or the dwell time on the heater, and the degree of permanency of crimp. The role of the physical properties of the parent yarn in influencing the behavior of textured yarn can be best understood from the relationship between the yarn retractive force (5) and the yarn parameters, as indicated by the following expression:

$$\frac{F}{D} = \frac{CEd}{A^2} \tag{12.1}$$

where

F = the retractive force of the yarn
D = the total yarn denier
C = a constant
E = filament elastic modulus
d = filament denier
and A = diameter of the filament helix.

It can readily be seen that the retractive force is directly proportional to the product of modulus and denier. Thus a material with lower tensile modulus will have lower retractive force, provided all other factors remain constant. For example, since the modulus of polyester (PET) is generally higher than nylon 66, which in turn is higher than nylon 6, the retractive force for a given total denier and filament would vary in the order $F_{PET} > F_{N66} > F_{N6}$. It is clear from equation 12.1 that, by manipulating the filament denier, it is possible to produce yarns from nylon 6, nylon 66, and polyester with similar retractive force for a given total yarn denier. It is known that a 70/24 nylon 66 yarn has retractive force similar to a 70/24 nylon 6 yarn.

It is recognized that torque-crimped nylon 66 as well as nylon 6 yarns have very good stretch and recovery, texture, covering power, etc. On the other hand, polyester yarns present some difficulty in retaining their torque-crimp memory if the parent yarn is not properly drawn and the textured yarn not adequately relaxed. These aspects are discussed in more detail in the following sections. However, it must be pointed out that the fabrics made from properly textured and relaxed polyester yarns are believed to have better texture or hand, better cover, and better crease-resistance characteristics than similar fabrics woven or knitted from nylon yarns.

Polypropylene yarns generally require longer dwell time in the heat zone, necessitating the use of longer heating, which to some extent slows down the production rate. However, polypropylene yarns have very good torque-crimp memory, provide excellent covering power, because of low specific gravity, yield a very pleasing "apparent" effect of lightness of hand.

MACHINE VARIABLES

There are more than a dozen manufacturers of yarn-texturing machines. Each machine has its unique design features, which include the heater type (contact, radiation), tensioning system, yarn-feed system (apron or roll), and yarn take-up mechanism to name a few of the most important ones. All these factors taken singly or collectively create a complex situation in the production of textured yarns.

In addition, the following factors further complicate the situation when considering the effect of machine variables on the textured yarn quality:

1. simultaneous and sequential draw texturing
2. use of partially drawn polyester yarns
3. type of twisting technique, such as, duo-twist, friction (bush) twisting, fluid-drawn texturing
4. crimp type

There is no one machine that is ideal. To get a better picture of the innovativeness of the various machinery manufacturers, the reader is referred to Table 12.3, in which the most important features of the latest equipment (shown by the manufacturers at ATME-I-1973 at Greenville, S.C., U.S.A.) currently available for yarn texturing are tabulated. A detailed discussion on the subject of machine engineering design is out of the scope of this chapter. However, some novel ideas in texturing shown for the first time at this international machinery exhibit are discussed later.

Nevertheless, it is deemed appropriate to discuss briefly the importance of heater design, since heat setting is perhaps the most critical aspect of the texturing process. In modern texturing machines, the setting time is relatively short; the whole process of twisting, heat setting, and untwisting takes less than 1 sec.

The basic requirements of a heater system include:

1. reliability
2. reproducibility
3. ease of control
4. flexibility
5. economy of operation

The attainment of these requirements would depend to a certain extent on the type of heater employed. According to the mode of heat transfer, the heaters can be grouped in three basic types, that is, radiation, conduction, and convection. A further subdivision can be related to the heating medium, for example, electricity, hot air, oil, steam, Dow-Therm, etc. There is not much information available regarding the influence of the heater type on the quality characteristics of textured yarn. However, there are some studies reported in the literature on the effect of heater temperature and heater length on the bulking and stretch behavior of yarns, and these are discussed in the appropriate sections.

Table 12.3. Details of Machines Exhibited at the ATME-I, 1973

| Manufacturer | Spindles | | | Magnetic Spindle Drive | Heaters | | | |
| | No. | Type | RPM | | Heater 1 | | Heater 2 | |
					Lgth. (mm)	Type	Lgth. (mm)	Type
ARCT FTF 483	192	Friction or magnetic pin	800,000 600,000	Bi-roll or monoroll	1500	Contact	1000	Radiation and convection
Barmag FK5C	216	Friction or magnetic pin	4 million 400,000 600,000 800,000	Bi-roll	1200-1500	Contact	1000	Convection
Cognetex SFT	192 216	Magnetic pin	800,000	Na	1000	Na	1200	Na
Giudici TG4.V/72	120	Friction	1 million 800,000	Na	990	Contact	–	–
Giudici TG4.F/2	120	Magnetic pin	800,000	Bi-roll	1500	Contact	1000	Radiation
Heberlein FZ 27S	160	Magnetic pin	600,000	Bi-roll	Na	Contact Convect.	Na	Contact convection
Leesona 556	156	Pin	500,000	Direct Drive	1000	Contact	520	Contact
Platt Kosmotex Type 984	132	Friction	40,000	Na	1200	Na	300	Na

Table 12.3. (continued)

Manufacturer	Spindles			Magnetic Spindle Drive	Heaters			
					Heater 1		Heater 2	
	No.	Type	RPM		Lgth. (mm)	Type	Lgth. (mm)	Type
Scragg Super Draw Set 2	216	Friction or magnetic pin	800,000 600,000	Bi-roll	1000 or 1500	Contact	995	Radiation contact
Sotexa SW30	216	Friction or Magnetic pin	800,000 600,000	Bi-roll	1500	Contact	1200	Convection
Spinner Oy VK-VTS/A	160	Friction	40,000	Na	2000	Contact	2000	Contact
Toshiba J-202	108	Air jet	—	—	420	Convect.	420	Convection

Take-Up

Manufacturer	Temp. Tolerance	Cooling Zone (mm)	Speed (m/min)	Type Pkg.	Pkg. Size (mm)	Type Doff.	Yarn Size Range (d tex)	Pkg. Wt. (kg)
ARCT FTF 483	±1°C	1050	400	Straight or bi-con.	250X250	Semi-auto	20-200	NA
Barmag FK5C	NA	460-910-1700	400 (frict.) (mag. pin) 250	Straight or bi-con	NA	Auto.	15-200	4.5
Cognetex SFT	NA	750	NA	Straight or bi-con.	NA	NA	NA	NA
Giudici TG4.V/72	± 1%		150-200	Straight or bi-con.	185X220	Manual	15-50	NA
Giudici TG4.F/2	±1°C	1500	300	Straight or bi-con.	290X300	Manual	17-300	NA
Heberlein FZ 27S	NA	NA	150	Straight or bi-con.	200X200	Manual	7-300	3.0
Leesona 556	±1°C	430	NA	Straight	139X270	Manual	8-1388	NA
Platt Kosmotex Type 984	NA	1600	1000	Straight	235X250	Manual	15-200	NA
Scragg Super Draw Set 2	NA	750 or 900	166-205	Straight or bi-con.	250X245	Manual	15-250	3.7-4.6

Table 12.3. (continued)

Manufacturer	Temp. Tolerance	Cooling Zone (mm)	Speed (m/min)	Type Pkg.	Pkg. Size (mm)	Type Doff.	Yarn Size Range (d tex)	Pkg. Wt. (kg)
						Take-Up		
Sotexa SW30	±0.5%	1800	400	Straight or **bi-con**	280X290	Semi-auto.	15-300	NA
Spinner Oy VK-VTS/A	NA	1500 700 mm cooling plate	800	Straight or bi-con.	250X250	Auto.	15-200	6.0
Toshiba J-202	NA	NA	300	Straight or bi-con.	200X240	Auto.	15-750	NA

Manufacturer	Fiber Type	Feeder Yarn Charact.	Type Creel	Type of Texturizing	Method of Texturizing	Thread-up Time (min)	Twist Capabilities	Noise and Fume Controls	Process Monitoring Devices
		Supply							
ARCT FTF 483	Polyester Nylon	Drawn Partially Drawn Undrawn	Over-head	Convent. Simul. Sequen.	FT or FTF	1.5-2.0	S, Z or SZ Combin.	Fume-blower aspirator system	Central location for overall conditions. Indiv. heater monitor
Barmag FK5C	Polyester Nylon	Drawn Partially Drawn Undrawn	Side creel	Convent. Simul. Sequen.	FT or FTF	2.0	S, Z or SZ Combin.	Noise-transparent shield	Indiv. Temp., spdl. sp., Yarn tension at central location

Machine	Material	Feed yarn	Creel	Threading	FT or FTF	(T/M)	S, Z or SZ Combin.	Fume/noise	Monitor/control
Cognetex SFT	Polyester Nylon	P.O.Y. Undrawn	NA	Simul.	NA	NA	S, Z or SZ Combin.	NA	NA
Giudici TG4.V/72	Nylon	Drawn	Side creel	Convent.	FT	0.5-1.0	S, Z or SZ Combin.	Fume exhaust	Portable heater monitor. system
Giudici TG4.F/2	Polyester Nylon Acetate	Undrawn P.O.Y	Side creel	Convent. Simul.	FT or FTF	1.0	S, Z or SZ Combin.	Fume exhaust	Indiv. heater monitor.
Heberlein FX 27S	Polyester Nylon	Drawn Undrawn P.O.Y.	Side creel	Convent. Simul. Sequen.	FT of FTF	1.5	NA	Fume removal	Part. monitor for spdl. sp.-cent. loc. monitor available
Leesona 556	Thermo-plastic	Drawn	Side creel	Convent.	FT or FTF	NA	NA	Fume duct. No roller bearings	Central location for overall conditions
Platt Kosmotex Type 984	Polyester Nylon	Undrawn P.O.Y.	Over-head	Sequen.	FTF	NA	NA	Fume exh. top and bottom	Individual heater control monitor
Scragg Super Draw Set 2	Polyester Nylon	Drawn Undrawn P.O.Y.	Side creel	Convent. Simul. Sequen.	FT or FTF	2.0-3.0	NA	Noise-shield Fume—fan	Central individual control module with master temp. setting control
Sotexa SW30	Thermo-plastic	Drawn Undrawn P.O.Y.	Side creel	Convent. Simul. Sequen.	FT or FTF	1.0-1.5	NA	Noise prot. Fume vac.	Central location for temp.
Spinner Oy VK-VTS/A	Thermo-plastic	Drawn Undrawn P.O.Y.	Side creel	Convent. Simul. Sequen.	FT or FTF	NA	NA	Fume-air suction	—
Toshiba J-202	Polyester Nylon	Drawn Undrawn P.O.Y.	Over-head	Convent. Simul. Sequen.	FT or FTF	1.0	NA	Fume-exhaust fan. less noise	Central location for temp.

NA not available.

PROCESSING VARIABLES

It has been pointed out that the heating and cooling of the yarn is the most critical operation encountered during the process of false-twist texturing. This aspect of the process attains further dimensions when viewed in the light of technological advancements made in the field of increased spindle speeds. However, for effective heat setting, the yarn must be maintained at the setting temperature for a sufficient length of time—an event controlled by yarn speed and heater length—to yield the desired stretch characteristics. To a certain extent, the problem of dwell time has been overcome by incorporating longer heaters (in some modern machines the heater length may go up to 2.0 m).

Heating and Cooling of Yarns

It has been reported by Arthur and Jones (6) that the setting process, in addition to being dependent on the temperature attained by the yarn, is also governed by a rate process that is probably associated with the modifications in the molecular structure at the setting temperature. These authors have reported a detailed study on the heat transfer from hot air to a twisted yarn and the temperature attained by the yarn in a heat zone. They conducted experiments on a model nylon 66 yarn of 15,000 denier (1666 tex) composed of 6 denier per filament. The core temperature of the yarn was measured by a fine thermocouple. They used an analytical expression to calculate the temperature profile in a yarn cross section while it was being transported through a heating medium (air, in this case). The thermal diffusivity of the model nylon 66 yarn was found to vary from 0.95×10^{-3} c.g.s units at 20°C to 0.57×10^{-3} c.g.s. units at 220°C. In the case of highly twisted, low-denier nylon yarns (up to 70 denier) heated in hot air at temperatures between 240 and 500°C, the difference between the surface and yarn axis temperatures in the vicinity of the setting temperature (230-240°C) was essentially found to be negligible. The heat transfer coefficient of highly twisted yarns was approximately constant for a given yarn denier over the temperature range 20-500°C and was independent of yarn speed from 6 to 300 m/min. The values of heat transfer coefficient varied from 3.7×10^{-3} cal/sec cm² °C for 30-denier (3.3 tex) yarn to 3.15×10^{-3} cal/sec cm² °C for 70 denier (7.7 tex) yarn. The heat-transfer coefficient values were used to calculate the yarn temperature at any time during heating at a particular temperature. These experiments showed that the surface temperature of a 70-denier (7.7 tex) yarn twisted to 75 tpi (2952 tpm) could be raised to 230-240°C (setting temperature) in 0.35 sec, but the yarn produced did not have adequate set. However, a heating time of 0.7 sec, obtained by reducing the processing speed and heating rate, was found to produce a well-set yarn. The investigators conjectured that the poor set obtained at heating times less than 0.35 sec was caused by either (1) incomplete conduction or (2) incomplete molecular rearrangement.

The effect of the process of heating and cooling of false-twist textured yarns has been reported by Morris and Roberts (7). They studied the relation between contact time and yarn temperature and between contact time and the degree of set produced in false-twist 70-denier, 34-filament nylon 66 yarns, by using a Scragg-Shirley Minibulk machine. The degree of set was characterized by the crimp rigidity test devised by the Hosiery and Allied Trades Research Association, England. The procedure consists of suspending of hank of a yarn in water at 20°C tensioned under a load of 0.1 g/den. The relaxed length L_1 is measured after 2 min; the load is then decreased to 0.002 g/den, and the relaxed length L_2 is measured after further lapsed time of 2 min. The crimp rigidity is defined as $100 (L_1-L_2)/L_1$. The modified C.M.G. heater was used on the bulking machine, and the spindle speed could be varied from about 60,000 to 375,000 rpm. The heater length used was 24 in., assembled in six sections, each of 4 in. length. By using two different types of spindle assemblies in the section heater, the contact time could be varied from 0.15 to 3.5 sec. The relationship between crimp rigidity and spindle speed for nylon 66 yarn is shown in Figure 12.15. The yarns were heat set at 230°C and 80 tpi. On the other hand, Figure 12.16 illustrates the effect of contact time on crimp rigidity. These results indicate that the crimp rigidity drops rapidly when the machine speed is increased, whereas the increased contact time increases crimp rigidity exponentially. It must be pointed out that the actual contact time, because of contraction caused by twist (as well as filament linear denisty), is generally higher than would appear from the ratio of heater length and yarn speed. The contact time for 70/34 nylon yarn textured at a nominal twist of 80 tpi and a heater temperature of 240°C has been reported to be 1.45 sec within the spindle speed range of 62,000-250,000 rpm.

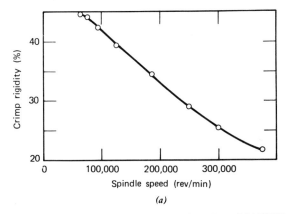

Figure 12.15. Effect of spindle speed on crimp rigidity of twofold 70/34 nylon 66 yarn processed at 230°C (after Morris and Roberts).

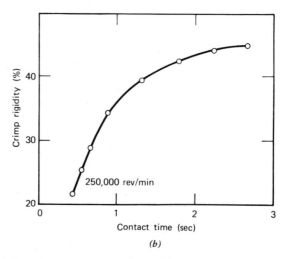

Figure 12.16. Effect of contact time on crimp rigidity (after Morris and Roberts).

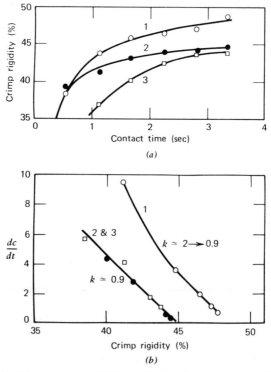

Figure 12.17. Variation of crimp rigidity with time for yarn processed at: (1) constant heater temperature (230°C); (2) constant yarn temperature (218°C) after two sections (the results refer to twofold 70/34 nylon 6.6 yarn processed on the multistage heater at a spindle speed of 48,000 rev/min). After Morris and Roberts.

416

Studies on the effect of yarn speed and heater temperature have indicated that crimp rigidity is determined not only by the yarn temperature and contact time, but is also significantly dependent on the rate of change of temperature with time. These conclusions are based on the observations that, for a constant heater temperature and yarn speed, the crimp rigidity increases exponentially with time, and there are two distinct regions with different "time constants" in this process. During the first phase, when the yarn temperature is increasing rapidly, the crimp rigidity changes likewise; in the second period, after the yarn has reached the heater temperature, the crimp rigidity continues to increase, but at a much slower rate until a maximum is reached. This is shown in Figure 12.17 (7). This illustration demonstrates that the yarn temperature reaches the heater temperature after about 0.6 sec; however, the crimp rigidity still shows an increasing trend, at a slower rate, and shows no sign of leveling off when the yarn temperature reaches a constant value. This relationship (for temperature rise) has been shown to have the analytical form (8):

$$T_y = T_h \left(1 - e^{-kt}\right) \tag{12.2}$$

where

T_y = yarn temperature
T_h = heater temperature
k = heat transfer function
and t = heating time

T_y and T_h are temperatures measured above ambient level. (When $kt = 3$, T_y is 95% of T_h) and

$$\frac{dT_Y}{dt} = k \left(T_h - T_y\right) \tag{12.3}$$

(which is identical to the relationship derived by Morris and Roberts (7) for the change in crimp rigidity: $-\dfrac{dc}{dt} = k \left(Cm - C\right)$, where C is the crimp rigidity, and Cm is the value of crimp rigidity when the yarn reaches the heater temperature).

The value of factor k is the slope of the plot of dT_y/dt vs. T_y, which is found to be approximately equal to 5 in practice. In other words, the yarn temperature reaches 95% of its final value at $t = 3/5 = 0.6$ sec. From the plot of crimp rigidity against contact time (Figure 12.17), a k value of 2 is obtained. When the yarn temperature is rising and at constant yarn temperature, $k = 1$. This type of analysis would suggest that a time of 1.5 sec is required when $k = 2$, or 3 sec for a value of $k = 1$, for a yarn to attain 95% of its final crimp rigidity value.

The next important step in the processing of textured yarn is the cooling of the yarn before untwisting, because it is during this stage that the setting effect

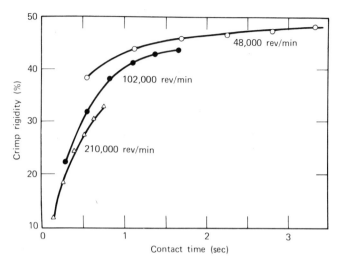

Figure 12.18. Relationship between contact time and crimp rigidity at three spindle speeds. After Morris and Roberts.

takes place. Denton and Morris (8) have considered this aspect of the process. Some idea of what happens to the crimp rigidity when the cooling time is changed can be obtained from the illustration shown in Figure 12.18 (7). These experiments were conducted with a multistage heater, at three different spindle speeds, while the cooling distance was kept constant at 10 in. An examination of the crimp rigidity at each heating time included in all three curves (e.g., 0.7 sec) shows that the crimp rigidity tends to decrease with increasing yarn speed; in other words, crimp rigidity increases with cooling times, which, according to the authors, are as follows:

Spindle Speed	Cooling Time
48,000 rpm	1.6 sec
102,000 rpm	0.8 sec
210,000 rpm	0.4 sec

Further studies on the effects of cooling times covering a much wider range have been reported (8); here the heater length was varied while the heating time was kept constant. The cooling distance above the top of the heater was fixed at 11 in., and the cooling time varied from 0.2 to 1.3 sec. The results of this experiment, shown in the lower curve of Figure 12.19, demonstrate that the crimp rigidity increases as the cooling time increases. This effect is highly pronounced in the initial part of the curve (lower cooling times-representative of times used in some present-day equipment) and tends to level off at higher cooling times.

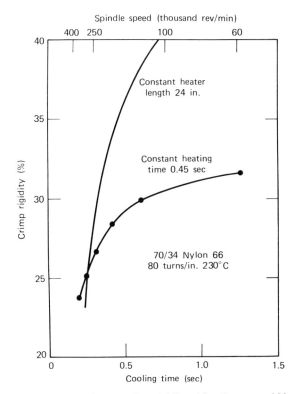

Figure 12.19. Effect of cooling time on crimp rigidity. After Denton and Morris.

The upper curve shows the joint effect of heating and cooling time. It is obvious that the effect of heating time alone is greater; however, the influence of reduced cooling time becomes progressively more important as the spindle speed increases.

The importance of the influence of cooling time has been further discussed by the authors when the process is considered in the light of thread line temperature profile above the heater. This approach is extremely helpful in determining the cooling time required for processing different linear density (denier) yarns. This can be best understood by looking at the value of k in the cooling exponential $T = T_o\, e^{-kt}$ derived from results shown in Figure 12.20. The value of k obtained by substituting the temperature of the yarn as it leaves the heater (230°C) and the temperature after cooling for 0.3 sec (130°C), both taken from Figure 12.20, is found to be 2.3, which is substantially independent of yarn speed. Calculations by Denton and Morris (8) suggest that, for a yarn to cool down to a temperature of 60°C before untwisting, a cooling time of 1 sec would be required, as compared to a time of about 0.25 sec allowed on modern

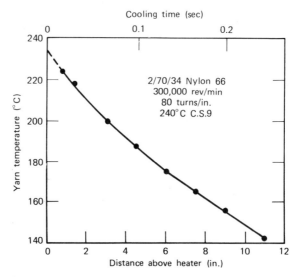

Figure 12.20. Cooling of yarn after leaving the heater. After Denton and Morris.

machines. Figure 12.20 shows that at 0.25 sec the yarn is at a temperature of 140°C. These results relate to a 70-denier yarn. On the other hand, the value of k' would be different for heavy-denier yarns as obtained from equation :

$$k' = \frac{2h}{\rho c r} \tag{12.4}$$

where h is the heat transfer coefficient in air

ρ = yarn density

c = specific heat

and r = radius of the yarn

The specific heat of conditioned nylon (65% r.h., 20°C) is constant at 1.01 ± 0.02 cal/g/°C over the range 60-200°C, at heating rates between 100°C/sec and 1000°C/sec, as reported by Jones and Porter (9).

As the yarn denier is increased from 70 denier to 150 denier, r increases by a factor of 1.4, and the value of k' decreases by 1.4; thus the cooling time for the heavy denier yarn must be increased by the same factor to cool it down to the same temperature. Yarn with lower specific heat will come down more rapidly.

In addition to the crimp rigidity, the tensile characteristics of textured yarns are also influenced by the heater temperature (10). The tensile modulus and breaking tenacity generally show a linear decrease, whereas the extension-at-

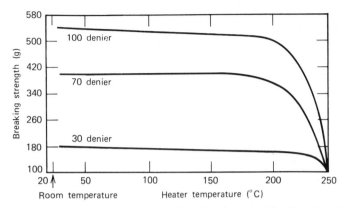

Figure 12.21. Effect of heater temperature on breaking strength. After Burnip and Hearle.

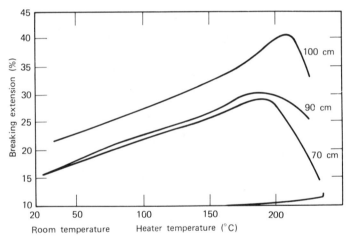

Figure 12.22. Effect of heater temperature on breaking extension. After Burnip and Hearle.

break and the effective stretch (measured at the point of intersection of tangents drawn to the low- and high-modulus regions of the load-extension curve) increase with increasing heater temperature. Figures 12.21 and 12.22 show the kind of relationship observed between heater temperature and breaking load and breaking extension. These changes in the tensile behavior of textured yarns are manifestations of the changes occurring in the morphological structure because of the heat treatment. There are changes in the size and distribution of crystallites brought about by the deformation processes (tension and bending) in the heated state, and these are temperature dependent.

There is not much information available in the literature on the relative suitability of various heat transfer modes such as conduction, convection, or radiation in the heating of yarn during texturing. But most of the heating systems have inherent capabilities that make them greatly flexible for whatever purpose they are employed.

Yarn Twist

The amount of twist inserted during texturing influences such characteristics as the bulk, crimp, strength, elongation-at-break, and the retractive power of the end product. Twist affects the yarn structure and is manifested in the deformation of individual filaments because of the helical path they follow and the radial disposition (migration) in the yarn cross section. Excessively high twist has a tendency to cause the yarn to have characteristics similar to crepe-spun yarn. On the other hand, low levels of twist cause the yarn to have a "stringy" appearance, probably because of the inability of the insufficiently distorted

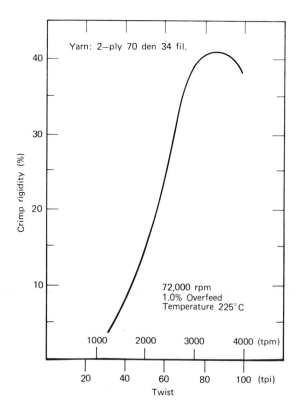

Figure 12.23. Effect of twist on crimp rigidity.

filaments to coil up in loops. The crimp rigidity of false-twist textured yarns in relation to yarn twist is affected in the manner shown in Figure 12.23 (11). The crimp rigidity increases sharply with increasing twist, but at very high levels it tends to either level off or show a downward trend. At excessively high twists, the yarn tends to have tight spots that are detrimental to its bulking and dyeing behavior. Twist also has a profound influence on the tensile properties of textured yarns. The breaking strength and elongation both show a decreasing trend. The basis for the selection of the processing twist is dependent on the consideration that the tenacity of the textured yarn does not fall below a certain desired level. It has been pointed out by Krause (12) that high processing twist levels cause the yarn to develop high torque, which is enough for it to overcome the frictional resistance on the guide pins of the false twist spindle, thus enabling the yarn to roll over the pin, causing slippage and resulting in a reduced amount of absolute twist in the yarn. This type of behavior can also cause reduction in yarn tension. To achieve softness and extra-fine pebble appearance in fabrics, higher twists are employed, whereas low twist levels would produce course and relatively harsh crepe-like handle.

Threadline Tension in the Heat Zone

Low thread line tension conditions during processing can result in the appearance of coiled filaments because of yarn ballooning, loss of twist, or formation of tight spots. On the other hand, high tension may result in reduced bulk, and it may also cause excessive yarn breakage, thus adversely affecting the production efficiency. During the false twisting process, the yarn tension may be controlled in two different ways:

1. constant tension device—used on the earlier Fluflon-type machines, in which the yarn is passed through a gate tensioner before entering the heater so as to impose a predetermined constant tension.
2. constant extension—in which the positive yarn feed system is employed, and the tension is varied by manipulating the relative speeds of delivery and take-up.

The processing tension in the constant-tension type systems is mainly dependent on the initial imposed tension, and the heater temperature has no significant effect (13). However, variation in twist may cause some fluctuations in tension below the heater, but the tension remains essentially unchanged above the heater.

In the constant-extension type machines, the effect of thermal shrinkage, that is, the development of the contractile force, is the dominant factor in determining the threadline tension. The contractile force increases as the temperature is increased up to about the second-order transition of the polymer, and then tends to decrease. The other complicating factors include the tangential force on the yarn because of twisting, softening because of heat, and contraction because of twist. This situation can be modified, to a certain extent, by overfeeding the yarn, however, excessive overfeed can nullify the process of torque-crimp. Contractile stresses (including the tension caused by yarn friction) of 1.28 g/tex at 100°C and 1.33 g/tex at 180°C have been reported (13) for 70 denier, 34-filament nylon 66 yarn with 0% overfeed and 0.3 g pretension (without twisting). However, when the same yarn was processed on a false-twist texturing apparatus (45,000 rpm spindle speed), an overfeed of 7% caused instability in the threadline, and 4% underfeed tended to produce frequent yarn breakages above the spindle. There must be, therefore, a compromise in the adjustment of heater temperature and yarn feed rate (under- or overfeed) with some concern for twist and initial tension. Nylon yarns processed at low tensions (because of overfeed) are generally known to have increased dye uptake, presumably because of changes in the morphology of the polymer.

With increasing thread line tension, yarn stretch increases initially, passes through a maximum, and subsequently decreases at very high tensions. Yarn denier and yarn bulk decreases with increasing tension, but the residual shrinkage increases.

Post-Treatment

Textured yarns produced by the false-twisting technique generally have higher stretch and retain some twist liveliness. Before use in knitted or woven structures, the textured yarns are generally stabilized and post-treated by heat treatment to reduce their stretch behavior, to control crimp, to attain increased softness of hand and bulk, and to suppress their torque liveliness. This is accomplished by relaxing or stretching the yarn in a second heated zone. The relaxation may vary up to about 30% of the fully stretched length; thus the yarn develops crimp, and the heating allows the decay of internal torsional and bending stresses.

In recent years, the use of polyester yarns in the United States alone has jumped to approximately 1.3 billion pounds, which accounts for nearly 80% of all the fibers used in texturing, with an expected growth rate of nearly 7% over the next 5 years. Polyester yarns present some special problems during setting; consequently, the post-treatment (including efficient cooling) attains even greater importance in the processing of textured polyester yarns. The post-treatment of yarns can be accomplished by any of the following two methods used in practice. In the early stages of the processing of textured yarns, the most widely used method was that of overfeeding the yarn by a desired amount; the yarn was then soft-wound on the take-up packages. The setting operation was then carried out either in (1) steam under pressure in autoclave or (2) during the dyeing operation. This process is used in making Helenca® SS (smoother sweater) and

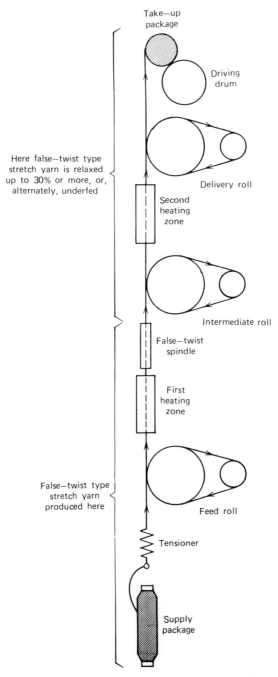

Take—up
package

Driving
drum

Here false—twist type
stretch yarn is relaxed
up to 30% or more, or,
alternately, underfed

Delivery roll

Second
heating
zone

Intermediate roll

False—twist
spindle

First
heating
zone

False—twist type
stretch yarn
produced here

Feed roll

Tensioner

Supply
package

Figure 12.24. Arrangement used for producing post-treated (modified stretch) textured yarns.

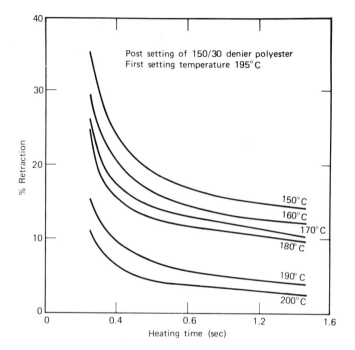

Figure 12.25. Post treatment; yarn retraction versus heating time at various second heater temperature. After Denton and Morris.

"Crimplene"[®] type yarns. The second method, which is continuous in nature is also known as the "dry method," has gained great importance in the industry. In this process the yarn is generally overfed (up to 30% or more) through a second heater situated immediatley after the spindle on the false-twist machine, as shown in Figure 12.24. The relaxation shrinkage of yarns produced by the dry and the autoclave methods generally varies between 8 and 15% and 3 and 5%, respectively, compared to 45 and 70% for regular stretch yarns. The yarns set by the autoclave method are exposed for a very long time, and consequently they are completely set and are essentially free from twist liveliness. On the other hand, heating times in the dry method are relatively short and are of the order of 1 sec; therefore, the efficiency of heating rate, in addition to the time and temperature, are of extreme importance of producing yarns of desired stretch and bulk by this method.

Denton and Morris (8) have reported some preliminary results of their investigation on the effects of time and temperature of dry post-treatment on the properties of a 150-denier postset textured polyester yarn. The temperature of first setting was 195°C. The property measured to characterize the bulking

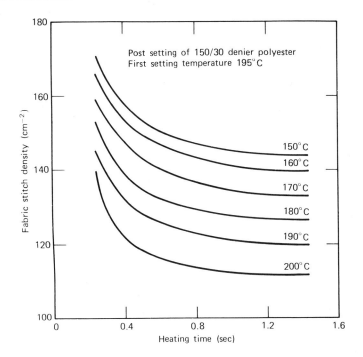

Figure 12.26. Post treatment; fabric stitch density for yarns post set at various times and temperatures. Fabric stitch length 2.9 mm (Denton and Morris).

power of these post-treated yarns was the percent retraction. The bulking power is measured on a hank of yarn after it has been heated for 5 min at 120°C with a load of 0.005 g/d. The results for the influence of heating time on percent retraction are shown in Figure 12.25. It is clear from this illustration that a heating time of approximately 1 sec is required to complete the setting process. Similar times were obtained when the effect of post-treatment parameters on the stitch density of plain knitted structures from these yarns was studied, as shown in Figure 12.26. This higher time of 1 sec suggests that longer heater lengths would be required to produce desirable stretch characteristics. This development is evident from the fact that the length of the second heater used on the most modern false-twisting units approaches 2 m.

Setting and Yarn Properties

Arthur (26) has suggested that the false-twist textured yarns develop the stretch and bulking characteristics because the individual filaments are set in a twisted form. In other words, the yarn behaves as a twist-free bundle of twist-lively fila-

ments. However, the situation in real false-twist textured yarns is more complicated because the filaments are also set in the twisted yarns in the form of
helices. These helices have varying radii because of the migration imposed by the
twisting operation. Consequently, the simple model suggested above fails to
explain fully the behavior of these yarns. On the basis of the observation of the
geometric structure of these yarns, Denton (27) suggests a half-reversed migrating helix model, in which individual filaments are held around the yarn axis in a
helix, varying in diameter as they move from the yarn center to the surface and
back again. After untwisting, the helical coils show occasional reversal of direction, but the basic geometric shape essentially remains the same. Denton and
Morris (8) have carried out a theoretical analysis of such a model and found that
this model explains the structure and properties of real yarns. A veiw of the
false-twist textured yarn is shown in the diagram (Figure 12.26(A)).

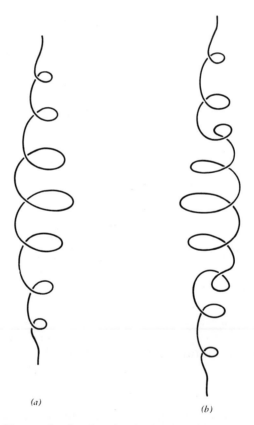

(a)

(b)

Figure 12.26(A). Diagram showing the migrating helix model of a filament in a false-twist
textured yarn. (a) Setting and before untwisting, (b) after twisting, showing reversal. After
Denton (27)

Draw Texturing

Until recently, the yarn texturing operation was performed solely by the throwsters. The feed or parent yarn supplied by the fiber manufacturers was fully drawn and oriented. More recently, the fiber manufacturers realized the opportunity of making (by increased take-up speeds) the production of filament yarns more economical and claiming a share of the fast-growing textured yarn market. The production figures reported in the published literature (14) indicate that, by the year 1978, the United States production of producer-textured (fiber manufacturers) yarn would account for nearly 36% of the total domestic shipments, compared to 15% reported in 1973. This growth pattern has been mainly due to the development of draw texturing technology. In addition to the economic considerations mentioned earlier, there are some technological aspects that were also responsible for prompting developments in the draw-texturizing field. These include:

1. Yarn quality, for example, uniformity of dyeing. This is an important aspect, particularly in warp knitting, in which a very large number of yarn ends are used to construct a warp beam. The uneven dyeing can result because of tension variations (a) within a parent yarn (from inside to outside diameter of bobbin), and (b) during the texturing operation. On the other hand, these variations in tensions are practically eliminated because of the nature of the draw texturing process.
2. Yarn processability. The production efficiency of the conventional texturing process is influenced by the spindle speed and the flat yarn quality. The parent yarn generally contains defects that are introduced during the winding operation performed on the draw-twister (ring travelers) and during the packing and unpacking of the bobbins. These defects, when viewed in terms of the machine parameters, such as the high spindle speeds (in excess of 800,000 rpm), pin diameter (1 mm), and the feed rates of 100 m/min, can compound the situation.

Draw texturing is a process in which an undrawn (fully or partially) flat yarn is drawn to the desired size and then textured in a continuous manner. The texturing process pertaining to the false-twisting method has already been discussed in detail. Some important aspects of the feeder yarn qualities are mentioned earlier in the text. However, it is deemed advisable to throw some light on the quality requirements of the feeder yarn meant for processing on draw-texturing equipment. In the conventional method of texturing, the feeder yarn is generally fully drawn, and the method of production of continuous-filament yarns is discussed in Chapter 11. However, it must be pointed out that the fully drawn yarn is produced in two steps, which makes the process expensive. This inspired the use of completely undrawn yarns (during texturing) in order to make the whole process more economical. However, the use of undrawn yarn in the draw texturing process caused some technical difficulties, and the processed yarn had some undesirable characteristics. Some of these are:

1. The undrawn yarn has poor shelf life. Particularly, the polyester undrawn yarn has a shelf life of approximately 100 hr.
2. The undrawn yarn packages are generally unstable, which causes variations in tension as well as filament cross-sectional shape.
3. It has low zero tenacity temperature, that is, the temperature at which the filament completely loses its strength.
4. The undrawn yarn is very sensitive to moisture and temperature changes in the environment, and this can result in the production of different effects from the inside to the outside of the bobbin.
5. It presents some difficulties as to the quality of the transfer knots that are essential for continuous uninterrupted operation and reduced down time.

To overcome the aforementioned drawbacks and to satisfy the demand for better-quality feeder yarn, the attention of technologists shifted to the use of partially oriented yarn (POY is the abbreviated term used in the trade). This development was the result of a compromise between increased production and yarn quality. The partially oriented yarn is produced in a single-stage process with take-up speeds of 3500 m/min (increasing productivity by 50-60%), compared to the conventional two-step process in which take-up speeds of 1300

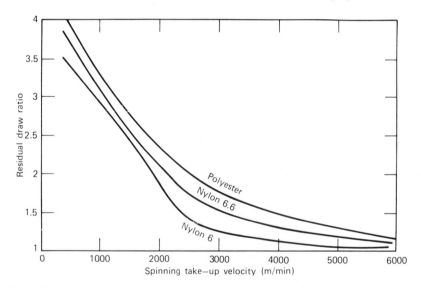

Figure 12.27. Relationship between spinning speed and residual draw ratio.

m/min are common. The partially oriented yarn spun at these speeds is quite satisfactory during storage for several weeks, and, in general, there is little difference between nylon and polyester. The partially drawn yarns meant for textuing generally have a draw ratio varying between 1.5 and 2.2. Figure 12.27 shows the relationship between the spinning take-up speed and the residual draw ratio for polyester and nylon yarns (15). It is clear from the illustration that, at spinning speeds up to 1500 m/min, the draw ratios are higher than 2.5:1. At medium speeds up to 3500 m/min, the draw ratios are less than 2:1. At spinning speeds higher than 4000 m/min and up to 6000 m/min, the curve for production flattens off (draw ratios approach 1:1), and the increase is only about 10-15%. The resultant advantages of partial drawing, which modifies the molecular orientation, are the increase in temperature tolerance and higher zero strength temperature, as shown in Table 12.4 (15). The increased orientation increases the birefringence. The yarn number 9 is a fully drawn yarn. The partially drawn yarns can be processed on almost all types of commercial draw-texturing equipment. A partially drawn yarn denier of 270-300 (residual draw ratio 1.8 to 2.0) is preferred for producing 150-denier textured yarns.

Table 12.4. Zero Strength Temperature of Various Polyester Yarns

Yarn	Birefringence $(n_\parallel - n_\perp)$	Crystallization Exotherm Centered at	Temp. of Zero Strength
1	0.004	138°C	145°C
2	0.003	135	145
3	0.010	131	143
4	0.008	131	145
5	0.038	107	250
6	0.050	106	240
7	0.056	97	234
8	0.055	98	232
9	0.218	–	251

Draw-Texturing Process

The drawing operation can be combined with any of the existing texturing processes. However, there are two possible ways in which this combination can be achieved.

1. Single-zone or simultaneous or in-draw process, in which the drawing and texturing are carried out simultaneously in one zone between the feed roll and the drawing godet, as shown in Figure 12.28a.
2. Two-zone or sequential or out-draw process, in which the drawing is per-

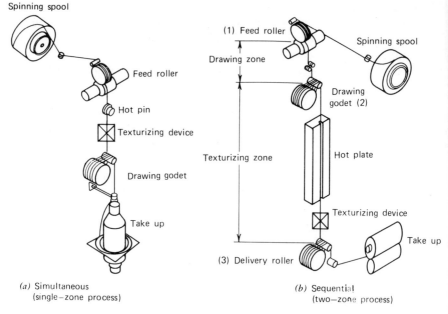

Spinning spool

(1) Feed roller

Spinning spool

Drawing zone

Feed roller

Drawing godet (2)

Hot pin

Texturizing device

Texturizing zone

Hot plate

Drawing godet

Take up

Texturizing device

Take up

(3) Delivery roller

(*a*) Simultaneous
(single–zone process)

(*b*) Sequential
(two–zone process)

Figure 12.28. Arrangements for draw texturing. (*a*) Simultaneous (single-zone process), (*b*) sequential (two-zone process). (Courtesy: Enka Glanzstoff AG)

formed between feed assembly one and two and the texturing between two and three, as shown in Figure 12.28*b*.

The most important system of texturing combined with drawing is the false-twist process (either spindle or friction twisting). The two-zone method produces yarn with characteristics similar to conventionally textured yarns. However, the sequential system that requires the incorporation of an extra set of feed rolls, and in the case of polyester, temperature-controlled hot pins or plates, suffers from the high cost of the development of drawing-false-twisting machines. Nevertheless, systems of the sequential type have been developed; one such machine has been described by Mattingley (16) and is known as the texDror system. Yarn output speeds of 175 m/min have been claimed for this system. On the other hand, the single-zone or the simultaneous process seems to be more attractive from the point of view of economy. This process is identical to the conventional system, the only difference being in the speed of the input feed roll, which is considerably slower in the draw-texturing process (required to accomplish drawing). Nevertheless, there are some subtle differences in yarn quality between the conventionally produced and the draw-textured (sequential as well as simultaneous) yarns that deserve attention.

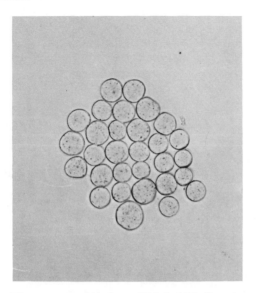

Figure 12.29(*a*). Filaments of a 150 d tex (135 den)/30 fils polyester yarn in different drawing state in the draw-textured zone.

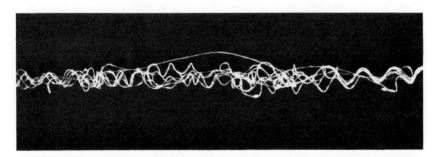

Figure 12.29(*b*). Untextured filaments formed into loops (nylon 66 hosiery yarn, 22 d tex (20 den)/7 fils false-twist textured in the first zone). (Courtesy: Enka Glanzstoff AG)

1. It has been observed (17) that, in draw texturing, broken filaments occur in the yarn. The explanation for this defect lies in the unevenly drawn state of the filaments immediately above the hot plate, as shown in Figure 12.29*a*. These filaments are generally fully drawn before or after the twist spindle. In other words, some of the drawing would occur in the cooling zone, causing some filaments to break. To avoid this, some adjustments in the false-twisting processing parameters are needed. Uneven drawing can also cause some fully drawn filaments to lose their crimp; as a result, some filaments may end up in looped form and protrude out of the filament

bundle, as shown in Figure 12.29*b*.

2. Filaments in draw-texturing processes suffer greater cross-sectional shape deformations than are observed for conventionally textured yarns (Figure 12.30). This modification in the cross-sectional shape imparts certain undesirable glitter to the resultant yarn and fabric.

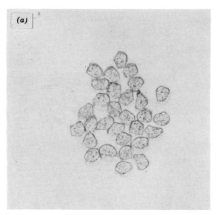 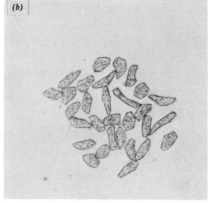

Figure 12.30. Deformed cross section of textured polyester filament yarn—150 d tex (135 den)/30 fils. (*a*) conventional; (*b*) draw-textured (Courtesy: Enka Glanzstoff AG)

3. Surprisingly, the draw-textured yarns, dye deeper than the conventional yarns.

4. The fabrics made from draw-textured yarns generally have a crisper hand than the conventional yarns. This may be attributed to the greater filament cross-sectional deformation, as mentioned earlier.

5. Because of the use of larger supply yarn packages (20-25 vs. 7-10 lb), draw texturing offers reduced processing cost because of less frequent creeling and fewer yarn breakages associated with fewer knots. Higher spindle speeds (600,000 vs. 400,000 rpm for conventional) are another plus for the draw-texturing process.

6. The efficiency of subsequent processes is improved because of the longer yarn length (larger packages) that are knot free.

7. There are essentially no differences in the tensile properties of conventionally and sequentially draw-textured yarns. However, the simultaneous process produces yarns that show lower tenacity but slightly higher elongation-to-break when compared to conventionally textured yarns. This is shown in Table 12.5 (15) for a 150-denier textured polyester yarn. Mey (18) has reported similar effects of draw ratio on the properties of conventionally false-twisted textured nylon yarn.

Table 12.5. Comparison Between Properties of Conventional and Draw-Textured Polyester Yarns

Property	Conventional/Sequential	Simultaneous
Tenacity (g/den)	3.2-4.5	2.7-3.5
Breaking elongation (%)	26-35	27-40
Leesona skein Shrinkage set method (%)	15-25	15-25
Thickness index (FAK Fabrik)	60-70	60-66
Torque (FAK Fabrik)	6-18	4-10
Fiber shrinkage	–	Variable
Fiber cross section (compared to feed yarn)	Similar	Deformed
Luster	Similar	Greater

Recent Advances in False-Twist Texturing

This section is designed to familiarize the reader with the latest developments in the field of conventional and/or draw-false-twist texturing processes. Most of the equipment shown at the last international textile exhibition (ATME-I-1973, Greenville, S.C., USA) indicated that machinery manufacturers placed the accent on lower maintenance cost and better yarn quality (improved heater design), in addition to the most obvious factors, such as productivity and labor-saving devices. The other important observation made by a large section of the texturing trade was that simultaneous draw-finishing coupled with friction spindles would constitute the major thrust in yarn texturing. There was one completely new and revolutionary concept called "fluid texturing" unveiled at the show, and it promised some advantages over the existing texturing devices. Some features of these new developments are discussed. The main features of most texturing equipment shown at the exhibition are summarized in Table 12.3.

The machine marketed by BARMAG (American Barmag Corporation, Charlotte, N.C., USA), called Autocrimper FK 5C, is designed to adapt both to conventional texturing and to sequential and simultaneous draw texturing (Figure 12.31). An important feature of the machine is the new Dow-Therm Vapor Condensation primary heating system, in which the temperature of the heater plate is controlled accurately throughout the entire surface within close tolerances. The cooling zone can be adjusted between 18 and 30 in. The machine can be supplied with either friction (the equivalent of 4 million rpm) or

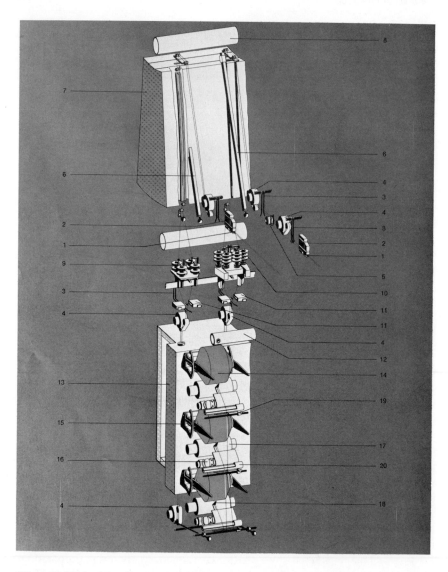

Figure 12.31(a). Line diagram showing set up of Barmag's Autocrimper FK 5C. (Courtesy Barmag Corporation) (1) Creel. (2) Yarn tensioners. (3) Yarn cutter. (4) Threadguide fork. (5) Apron feed system. (6) Threading device. (7) Primary heating. (8) Fume extraction. (9) Magnetic unit. (10) Friction unit. (11) Yarn break sensor (12) Yarn aspirator. (13) Setting equipment. (14) Take-up unit. (15) Swivel mounted tube holder. (16) Traverse motion. (17) Drive rolls. (18) Traversing thread guide. (19) Guide. (20) Yarn oiler.

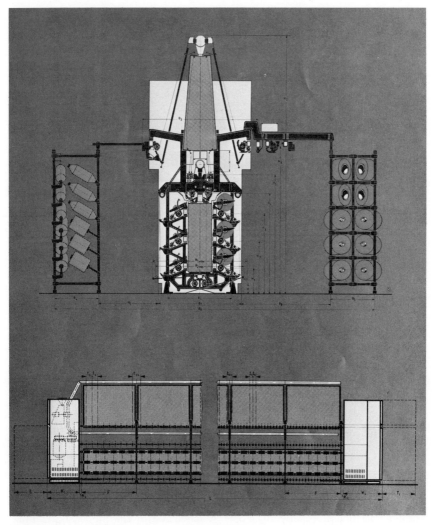

Figure 12.31(b). Side view of Autocrimper FK 5C.

magnetic pin spindle (800,000 rpm). It incorporates a relatively new tape-feed system that practically assures slip-free yarn transport even at high yarn tensions.

Spinner OY of Finland, a pioneer in the use of the friction-twist principle and the first company to combine drawing and texturing into one process, showed their latest frictional-spindle false-twisting machine, called 32VK-VTS/A. This machine has two heaters and is capable of processing both high-elasticity and set

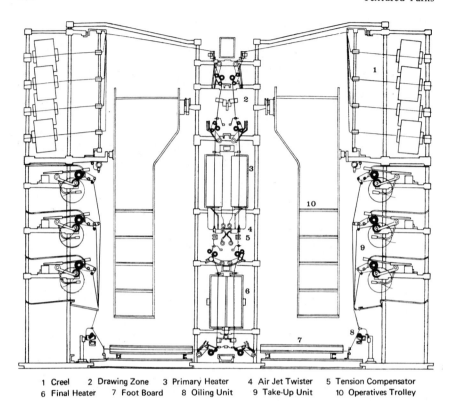

1 Creel	2 Drawing Zone	3 Primary Heater	4 Air Jet Twister	5 Tension Compensator
6 Final Heater	7 Foot Board	8 Oiling Unit	9 Take-Up Unit	10 Operatives Trolley

Figure 12.32. Fluid draw-texturing set up model J-202 (Courtesy: Toshiba Machine Co. of Japan).

yarns over a large denier range (15-200 denier). A 245-denier, 34-filament polyester feed yarn can be processed into a nominal 150-denier yarn at 500 m/min.

The concept of fluid texturing, originally developed and patented by Uniroyal Corporation in the United States, has been exclusively licensed and perfected by Toshiba Machine Company of Japan. A schematic of the model J-202 Fluid Draw-Texturer is shown in Figure 12.32. This concept can be applied to single- or double-heater false twisters, with or without continuous predrawing, to process textured yarns at practically double the present production rates. This is achieved mainly by the complete elimination of the conventional false-twist spindle or friction devices and using a compressed-air jet, as shown in Figure 12.33. The use of an air jet ensures rapid, smooth, and quiet (very low noise level) operation at low yarn tensions. The machine can process 150 denier textured polyester yarn from a 260-denier, partially drawn yarn at an output speed of 240 m/min. The air comsumption at standard 100 psi pressure is

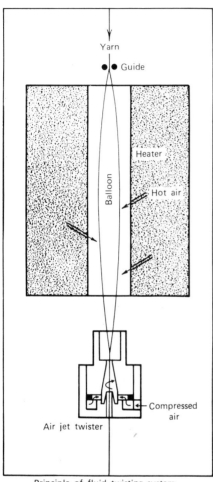

Principle of fluid twisting system

Figure 12.33. Principle of fluid-twisting system.

extremely low (101/min/position). The special type of heater is of the non-contact type, and the use of hot blown air into the path of the yarn makes the thermal setting more effective.

The Kosmotex yarn-texturing machine manufactured by Platt International Company of England is of the sequential type (separate draw, texturing, setting, and take-up zones). This machine is capable of processing 150-denier (167-decitex) set polyester yarn at 600 m/min. A friction spindle of the bush type is used for twisting. The successful operation of friction twisting as applied to the process of draw texturizing depends on the following factors:

1. Frictional characteristics of the bush surface must provide stable twist conditions
2. the application of spin finish is critical to optimum fiber-to-fiber friction
3. fiber-to-fiber surface friction must be optimized by applying a desired amount of spin finish to feeder yarn
4. stability to the threadline demands careful adjustment of processing conditions, for example, there should be no tension variations with twist variations (twist bleed or surging).

The developments in false-twist texturing described above and those apparent from the machine details given in Table 12.3 should not be accepted as a limit to the technological development in the sophisticated area of texturing. The area that needs most urgent attention is the conservation of power. In texturing, a large part of the heat input to yarn heaters is used to make up losses rather than to heat the yarn. Higher output rates may be considered to make more efficient use of power. Noise level is another important factor for which most countries are setting mandatory limits to protect workers. The use of friction aggregates with individual motor drive at each position, and air support bearings, or the use of the air-jet twisting principle, offer inherently low noise levels.

Edge-Crimped Yarns

Stretch and modified stretch yarns can also be produced by a process known as "edge crimping." Edge-crimping processes have been developed and licensed by Deering Milliken Research Corporation (USA) and ICI Fibers Ltd. (U.K.). The yarns produced by the edge-crimping process are marked Agilon®. The commercial use of the edge-crimping process has grown slowly and steadily, mostly using 15-denier monofilament and 15-denier/3-filament yarns for use in the ladies stretch hosiery area. Since its introduction, the use of edge-crimped yarns has been extended to heavy-denier yarns for use in such diversified areas as carpets (1000 denier and heavier), knitted outerwear, men's hosiery, etc.

Thermoplastic yarns are edge crimped by a continuous process in which a yarn is tensioned, stretched, heated, bent, and drawn around an edge, followed by shrinking and cooling steps. The basic principle of edge-crimping is illustrated in Figure 12.34. The crimp development in a filament is caused by the bilateral compressional and extensional strain deformations induced in the heated filament as it is being pulled over a blunt knife edge. The part of the filament touching the knife edge is under compression, whereas that on the outside experiences extension. Consequently, the region pressed against the knife edge is deformed into a flat, and the filament cross section appears as shown in Figure 12.35a. The final crimp development is thus a function of temperature, bending curvature, and the dwell time on the edge. The bending curvature is determined by the filament or yarn diameter, threadline tension, the angle of inclination of

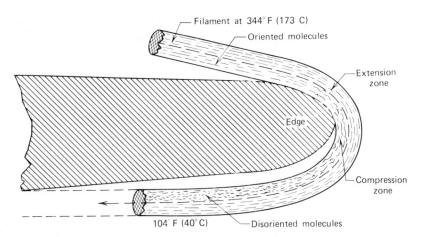

Figure 12.34. Principle of edge crimping.

the yarn as it bends around the knife edge, and to a certain extent on the geometry of the edge itself. The dwell time on the bend is influenced by the speed at which the yarn or filament is withdrawn. When a filament or a yarn is heated above its second-order transition (which varies with the polymer type), the polymer material in the compression zone undergoes morphological changes such as rupture of crystallites and disruption of molecular orientation. On the other hand, the region farthest from the edge is stretched, with the intermediate segments being compressed or stretched in direct proportion to their distance from the edge. To develop crimp in the filament, the strains thus induced are relaxed by subsequently relaxing and heating the yarn under controlled conditions.

Relaxation is carried out either in steam or in dry heat. The release of the strains causes the unstable compressed region to shrink, whereas the stretched zone is relatively unaffected. In this process the filaments tend to form coils in almost perfect helical configuration reversing at more or less regular intervals, as illustrated in Figure 12.35b. This inherent rotational process of helical reversal produces a yarn that is balanced or torque-free and does not require any further plying operation to modify its torque-liveliness characteristics.

Mechanism of Edge Crimping

Consider a filament that is being deformed by a tensile strain on one side and a compressive strain on the other. Analysis of such a model (19) predicts that the filament will take up a helical form, and it is possible to calculate the number of helical turns per unit extended length of the filament by the following expression:

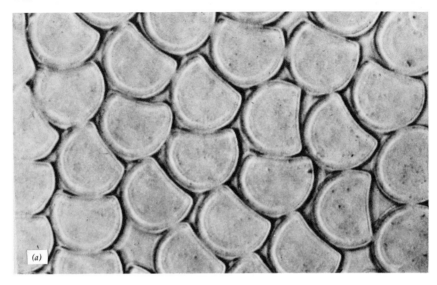

Figure 12.35(a). Nylon 66 edge-crimped yarn showing flattened filament cross sections.

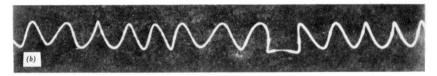

Figure 12.35(b). Edge-crimped monofilament. (Courtesy: Deering Milliken Research Corporation)

$$C_f = k \ C_p \ \Delta \sigma / h \qquad (12.5)$$

where C_f = frequency or number of crimps per unit extended length of filament
    k = constant
    $\Delta \sigma$ = strain differential
    h = fiber thickness
and C_p = a measure of strain asymmetry.

However, in the above expression, C_p also denotes crimp potential function and is a measure of the original texturizing stress, which is a function of the relative depth of penetration of the compressive forces into the fiber.

In edge crimping, the compressive forces vary along the yarn cross section, being maximum at the contact point and gradually decreasing as the distance increases from this reference point. The application of the above analysis can be applied to the edge-crimped yarn in which the bilateral symmetry is set by the sharp bend of the filament around the edge. The crimp potential function C_p for

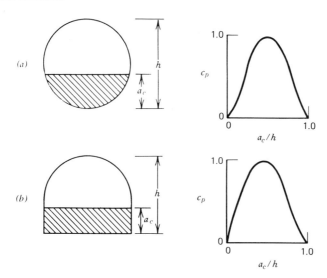

Figure 12.36. Edge-crimping model showing the relationship between crimp potential function and the compressive forces.

such a yarn is a function of the depth of penetration A_c of the compressive forces, and for a circular cross-sectional model, it is shown in Figure 12.36. For this model the crimp potential is at a maximum when the penetration depth is equal to the radius of the filament. The other model (Figure 12.36b) is a representation of the edge-crimped yarn. Here the edge deformation, which is flattened, shifts the optimum penetration of the compressive forces toward the deformed edge. Thus the depth of penetration of compression is determined by the temperature of the filament and to a lesser extent by the threadline tension. The maximum strain differential $\Delta\sigma$ between the compressive and tensile components is determined by the geometry of the edge and the angle that the filament makes with the edge. Obviously $\Delta\sigma$ will be highest for the smallest edge angle. Consequently, the highest crimp will be obtained using the smallest angle; however, the angle of $10°$ is considered to be the practical lower limit. The crimp frequency is inversely proportional to the filament diameter (thickness) and for a circular cross section to the one-half power of denier. In other words, for a given crimp potential, the crimp frequency will be higher for finer denier filaments.

It was determined earlier that the development of crimp is determined by shrinkage and recrystallization during relaxation in the compressive zone. If the longitudinal restraint during relaxation is extremely high, it prevents the formation of coils during crimp development. There is generally some restraint present in practice, and this causes the formation of an open coil helix.

Modified or stabilized Agilon® yarn with lower stretch and high bulk can be produced on the same machine by adding a bulk-development section. When a stretch yarn is overfed to the development heater, the crimp in the yarn is developed, and it stabilizes in an open bulked configuration. Yarns varying from 0 to 50% shrinkage may be produced by this process. Fabrics made from stabilized Agilon® yarns have a soft, full hand, good surface texture, very good dye uniformity, moderate stretch, and excellent recovery from stretch.

Stuffer-Box Crimping

The process of texturing by the stuffer-box method is based on the principle of heat setting filaments held in a confined space in compressed state and then withdrawing them in their crimped form. The chamber in which the filaments are stuffed is known as the "stuffer box." There are essentially three commercial stuffer-box texturing processes:

1. Pinlon®
2. Spunize®
3. Textralized® yarn

The Pinlon® process can be applied to a single or multi-ends, and the texturing is carried out on a Pinlon machine (KMG Machinery, Ltd., England) that is capable of processing 800-3500 denier and above at an ultrahigh speed of 600 m/min. The types of yarns produced by this process can have a wide variety of crimp configurations.

The Spunize® (Spunize Company of America, Inc.) process imparts three-dimensional crimp at very high speed and heat sets the yarns at the same time. The feed generally consists of multiple ends, and denier varies from 30 to 5000 and heavier. These yarns are generally used for carpet and upholstery materials.

The third commercial process, known by the Ban-Lon® and Textralized® trademarks, is a bulking process exclusively owned and controlled by Joseph Bancroft and Sons Company, Wilmington, Delaware, USA. Ban-Lon® is the trademark identifying garments and fabrics produced from stuffer-box bulked yarns called Textralized® yarns, which are required to meet the standards set by the licensing company. The crimping unit used in producing Textralized® yarns is shown in Figure 12.37. The process itself is carried out on special models of Foster cone winders. The thermoplastic feed yarn is positively fed by two feed rollers into the heated tube-stuffer box. On the output side, the yarn is passed through a weighted hollow tube or slug that impedes the progress of the crimped yarn traveling up the tube, thereby causing the yarn to back up inside the stuffer-box tube. The feed rolls at the same time keep delivering fresh yarn against the backed-up aggregate in the tube. This operation causes the filaments of the yarn to stack up in bent or crimped form. The crimp resembles the

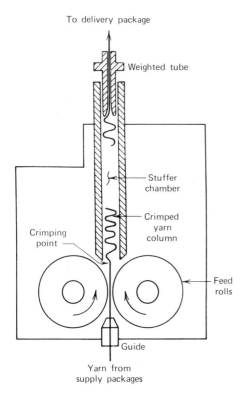

To delivery package

Weighted tube

Stuffer chamber

Crimped yarn column

Crimping point

Feed rolls

Guide

Yarn from supply packages

Figure 12.37. Stuffer-box principle used for producing Textured yarns. (Courtesy: Joseph Bancrost & Sons Co.)

saw-tooth configuration. While the feed yarn is being delivered to the stuffer-box, the crimped yarn is removed at a predetermined rate and wound onto the take-up package, and the aggregate in the tube moves up and is heat set at a required temperature in the stuffer box, which is jacketed by heaters. The yarn is oiled as it emerges from the weighted tube, it is then coned.

The ratio of feed to take-up speed and the weight of the tube determine the crimp amplitude and frequency. The temperature of setting and the feed yarn characteristics (filament denier and total yarn denier) are additional factors that determine the crimp quality of the bulked yarn. The tension applied during the heating and cooling cycle undoubtedly affect the degree of stretch and the recovery characteristics that the resultant yarn possesses.

The hot stuffer-box yarns generally have a wiry appearance, a soft feel, and high cover and bulk characteristics. Of course, the amount of relaxation would modify the bulk, as shown in Figure 12.38. These yarns also have good moisture absorption properties because of the minute spaces (interstices) created between the filaments (because of crimp) that can hold moisture. In general, the Textralized® yarns that possess wavy, random, and zigzag crimpness impart the following characteristics to fabrics made from them:

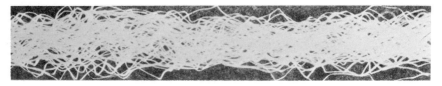

Figure 12.38. Textralized 2-ply 70 den-34 filament yarn.

1. soft and bulky hand
2. increased volume in the range of 200-300%
3. sufficient elasticity to provide a snug comfortable fit in garment form
4. high moisture sorption, which is good for comfort

Yarn deniers ranging from 15 to 4000 and above are processed by this method. The retractive (uncrimping) power of stuffer-box crimped yarn is governed by the geometry of the crimp (planar in this case), the sharpness of crimp, and the crimp frequency and can be roughly expressed by the following relationship:

$$P = KC_f D^2 / N \tag{12.6}$$

where

P = retractive (uncrimping) power
K = constant
C_f = crimp frequency
D = total yarn denier
N = number of filaments in a yarn.

Note that the uncrimping power is proportional to the first power of C_f, whereas in the case of yarns having helical crimp (false twist and edge crimp), it varies as the square of crimp frequency. Thus much of the stretch in stuffer-box crimped yarns can be removed by a relatively small pretension. However, their bulking power is generally excellent.

Air-Textured Yarns

The air-jet texturing process for producing bulk yarns was developed by E. I. du Pont de Nemours & Co., Inc., USA. Bulk in continuous-filament yarns can be produced by blowing a stream of air into a twisted yarn while it is being delivered at a higher rate than it is being taken up by the winding process. The air stream creates a turbulance that causes the formation of random loops in overfed individual filaments. In this process, the yarn contracts in length, and as it emerges, the loops are locked in place to impart bulk to the yarn. The yarn thus produced has an appearance like a staple yarn but possesses higher bulk, greater covering power, reduced opacity, and a warmer hand compared to flat

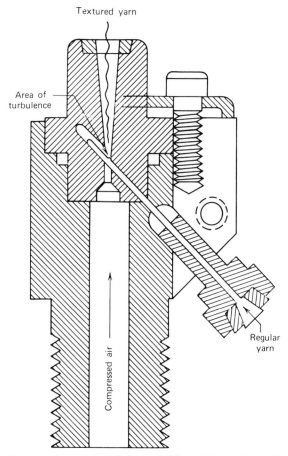

Figure 12.39. Cross section showing air-jet assembly used for producing Taslan-type yarns. (Courtesy, E. I. duPont de Nemours & Co.)

continuous-filament yarn. The textured yarn produced by this process (commercially owned by E. I. du Pont) is designated by the trademark Taslan®.

The air-jet assembly used in the production of Taslan® yarns is shown in Figure 12.39. This assembly can be fitted onto conventional throwing machines; however, there are specially designed machines (Figure 12.40) available on the market that can process yarns at higher speeds and that employ larger take-up packages (Garfield Machine Works, Inc., USA). A schematic illustration of the structure of Taslan® yarn is shown in Figure 12.41; Figure 12.42 shows the actual loop formation and the appearance of a nylon textured yarn. The feed yarn used in air texturing is generally twisted and steam set for the reason that will become obvious from the following discussion.

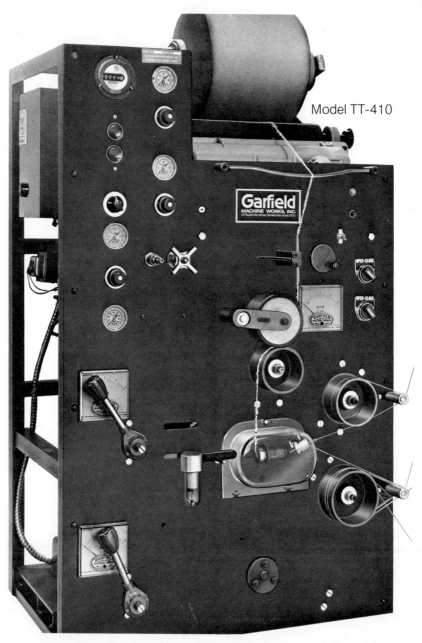

Model TT-410

Figure 12.40. Garfield Model TT410 Taslan texturing machine (Courtesy, Garfield Machines Works, Inc.)

Figure 12.41. Structure of an air-textured yarn (courtesy E. I. Du Pont de Nemours & Company, Inc.).

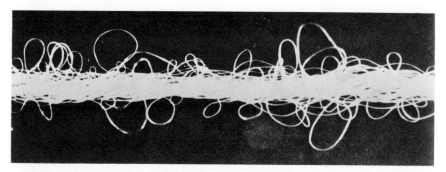

Figure 12.42(*a*). Photomicrograph of an air-textured yarn.

Figure 12.42(*b*). Photomicrograph showing fine-denier, multifilament Taslan textured nylon yarns. (courtesy E. I. Du Pont de Nemours & Company, Inc.).

An analysis of these questions requires that we must first define the processes involved during air texturing. There are two essential features of this process:

1. the yarn overfeed, which makes available an excess yarn length so that the individual filaments can form loops
2. the turbulant air flow, essential to disrupt and rearrange the looped filaments so that they are interlocked with sufficient interfiber friction to render the yarn stable.

The yarns used in air texturing generally have some initial twist. It has been reported that, if the pretwist in the feed yarn is low, to produce a stable textured yarn, the process needs either higher air pressures to generate greater filament entanglement or, alternatively, additional twist to secure stability. Wray (20) hypothesizes that, if the feed yarns are twisted before entering the jet, the effect of twist will be a contribution to the entanglement after the opening of the filament structure by the air turbulence. He further conjectures that this phenomenon may be explained as follows: First, the turbulence in the air stream helps to open up the twisted bundle, thereby allowing loops to be formed, partially because of the deviation of individual filaments by the air forces, and partially because of the snarling of the twist-lively filaments slackened by the overfeed. Second, interlocking is achieved because of the redistribution of twist in the yarn bundle as it is withdrawn abruptly at right angles to the air stream after leaving the air jet.

Wray further elaborates on the displacement of individual filaments in the "turbulent wake" caused by the obstruction of the feed needle in the path of the air stream and suggests that the tendency to form loops would be increased by

(i) higher twist, and thus greater spirality to effect looping of the tousled coil owing to the snarling of the filaments;

(ii) higher rate of overfeed, since this presents more slack yarn to promote loop formation;

(iii) higher air pressure, since this causes the turbulent action to be more intense and thus expands the yarn structure to allow the combined effect of (i) and (ii) to take place, and may actually assist in forming convolutions in the filaments by deflection and entanglement.

The filament displacement, the looping mechanism, and the interfiber friction further influence rearrangement of filaments caused by tension variations. The filaments that are in the turbulent wake in the Venturi throat become slack and subsequently looped; this action results in the preceding and succeeding sections of the filaments in the immediate vicinity of the loop becoming taut. These tensioned portions of the filaments tend to move toward the core, replacing the neighboring slackened filaments. This sequence of events continues as fresh over-fed yarn is delivered to the needle point, resulting in the formation of loops from slackened filaments.

The next major event, interlocking of the loops, occurs as the yarn is being pulled out of the tube at right angles under a slight threadline tension. Wray suggests that this action may have two effects in affecting interlocking:

1. The rapid removal of the filaments from the vertical exit stream may help the slack filaments that are in the compression zone inside the bend to be blown through the gaps between the highly tensioned filaments (by virture of being on the outside of the bend).

2. The withdrawal tension may cause straightening of the structure. However, this is very unlikely because the entanglement caused by coiled loops, interfiber friction, and air turbulence would hinder it. The tension would probably affect the compacting of the structure, however, because of the interlocking of those loops and convolutions that are directed inward (toward the yarn axis) at the time they happen to be in the jet. The reason for the occurrence of the inward-directed loops could be attributed to the twist liveliness of the filaments and the "uncontrolled rotation of the filaments," since they are free of axial constraints.

Wray has reported the results of an experimental investigation on a 70-den/34-filament nylon 66 yarn bulked by air texturing to confirm some aspects of the postulated hypothesis. He used a specially constructed transparent air-bulking unit with overfeeds

$$(\text{overfeed } \% = \frac{\text{input speed} - \text{output speed}}{\text{output speed}} \times 100)$$

varying from 0 to 60%, yarn pretwist from 0 to 20 turns per inch, and air pressure from 10 to 80 lb/in^2 (gauge). While investigating the effect of a parti-cular processing condition, the other parameters were maintained at constant levels of 15% overfeed and 15 tpi pretwist at 50 lb/in.^{2} (gauge) air pressure. He observed that the angular position of the needle (step on the feed needle) and the relative position of the yarn at the exit end relative to the air jet are quite critical in producing optimum bulk. The angular position of the needle is measured in degrees from the vertical position that occurs when the open end of the needle is facing the exit end of the Venturi. The interactions of angular posi-tion of the needle for essentially two identical needles with various processing parameters are shown in Figure 12.43a and b. It is clear from Figure 12.43a that for up to 25% overfeed a more oblique positioning of the needle produces a better

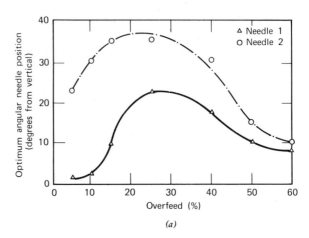

(a)

Figure 12.43(a). The difference in optimum angular needle settings caused by interchanging "identical" needles (for varying overfeed).

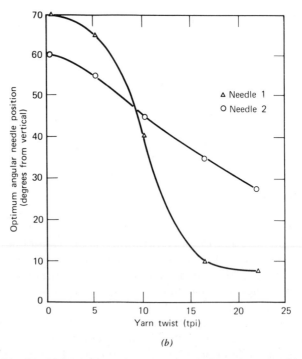

(b)

Figure 12.43(b). The difference in optimum angular needle settings caused by interchanging two "identical" needles (for varying input-yarn twist). (after Wray)

yarn. On the other hand, at overfeeds beyond 25%, satisfactory results are obtained if the needle position is moved toward the vertical. The explanation of this behavior lies in the argument that the total yarn volume (or surface area) exposed to the air stream is lower in the former case, whereas it is much higher in the latter condition, which can cause irregular movement of the yarn out of the needle.

When initial twist is increased (Figure 12.43b), the angle of the needle point must be reduced to obtain satisfactory bulking. For twistless yarns (in which the filaments are parallel), a steeper angle of 60-70° is required, whereas at higher pretwist (20 turns/in), when the filaments follow a helical path, the reduced angle of inclination ensures maximum exposure to the air stream. Air pressure in the region of 50 lb/in.^{2} requires the needle to be in the horizontal position; any deviations, either toward lower or higher air pressures, would require steeper inclinations.

Furthermore, by using both high speed cine photography and flash-still photography, Wray observed that there was some sliding of the filaments during bulking. The tracer fiber technique used in the study of loop formation confirmed that part of the hypothesis that states that the formation of a loop in a filament is followed by some tensioning of the part immediately following it. This effect, in cooperation with interfiber friction, generates U-shaped waves in the adjoining filaments, and, with the assistance of the bending of the yarn and the overfeed, the residual torsional energy in the filaments form stable loops. The stability of the yarn is strongly affected by the amount of pretwist. Lower pretwist yields yarn with poor bulk, and higher twist helps in the formation of a locked structure (because of the temporary removal of twist by the false-twisting action caused by the turbulent wake producing rotating eddies, and the reassertion of this twist before the yarn leaves the jet). An additional factor accounting for the stability of the textured yarn structure is the "filament slippage" effect mentioned earlier.

Factors Affecting Air-Textured Yarn Properties

When filament yarns are textured by the air-jet process, the most apparent change that occurs in the structure is the formation of a large number of small fiber loops distributed along the length of the yarn. This yarn has a well-defined core, as shown in Figure 12.44. There is a marked increase in overall yarn diameter, and the yarn has increased bulk. The bulked yarn structure is affected both by the feed yarn parameters and by processing variables. The yarn parameters include:

1. filament denier
2. number of filaments and total yarn denier
3. yarn twist

$$\text{Loop size} = \frac{\text{overall diameter} - \text{core diameter}}{2}$$

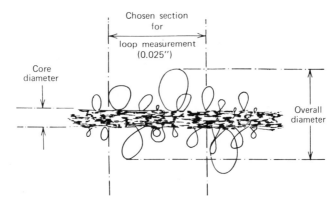

Figure 12.44. Definitions of overall diameter, core diameter, and loop size. (after Wray)

The processing factors include:

1. the amount of overfeed
2. texturing speed
3. the air pressure
4. the angular position of the needle

The effects of the various factors on the properties of air-bulked yarns have been reported in two major studies (21, 22). Some observations made in these studies are discussed below.

Textured Yarn Linear Denisty

When the filament yarns are air textured, the resultant yarn changes in linear density because of bulking. Generally, variation in pretwist and air pressure do not have a significant effect on the linear density of air-textured yarns. However, an increase in overfeed produces a proportional change in linear density. Wray (22) has reported a linear relationship between the overfeed and percent change in bulk yarn linear density.

Yarn Bulk

The bulk of an air-textured yarn is a function of the frequency and the size of the loops protruding from the yarn core. The du Pont method (23) used for assessing the bulkiness involves comparison of the package densities of the yarn before and after texturing. A package is made from parent yarn weighing 3 oz (85 g), and a volumetrically similar package of textured yarn is then wound at the same tension. The ratio between their weights then gives a measure of bulk:

$$\text{percentage bulk} = \frac{\text{net weight of feed or parent-yarn package}}{\text{net weight of textured-yarn package}} \times 100$$

The overall diameter, core diameter, loop size, and loop frequency characteristics are measured by microscopical and graphical techniques (Figure 12.44). The loop size is defined by:

$$\text{loop size} = \frac{\text{overall diameter} - \text{core diameter}}{2}.$$

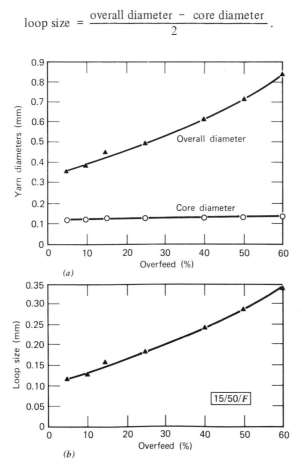

(a)

(b)

Figure 12.45. The effect of overfeed on overall and core diameters and on loop size. (after Wray)

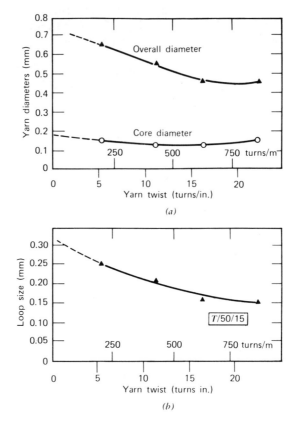

Figure 12.46. The effect of twist on overall and core diameters and on loop size. (after Wray)

The overall diameter and loop size increase with increasing levels of overfeed, as shown in Figure 12.45a and b. Yarn twist, on the other hand, retards the process of loop formation. As the yarn twist increases, the overall diameter as well as the loop size decrease, as shown in Figure 12.46a and b. Wray has derived analytical expressions for making theoretical predictions of the frequency and the size of loops based on the considerations of the amount of yarn twist and overfeed. He has observed that there is generally a good agreement between the theoretically predicted trends in loop dimensions for the various processing conditions and the experimentally measured values. A comparison of the theoretically calculated average loop size and number of loops with those obtained experimentally as reported by Wray for nylon 66 (70 den/34 fil twist set at 88°C for 45 min before air bulking) are shown in Figure 12.47a, b, and c. The theoretical loop size has been calculated from the analytical expression $h =$

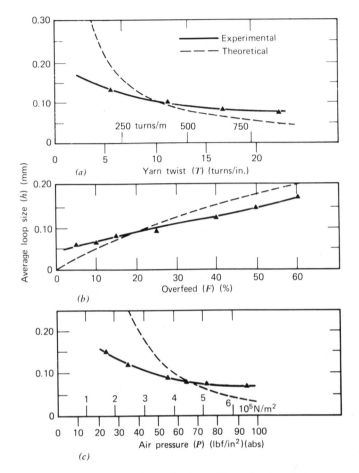

Figure 12.47. Theoretical and experimental values of loop size for varying processing conditions. (after Wray)

$$\frac{0.540\,F}{100+F}\,mm,$$ where h = loop diameter and F = % overfeed. The number of loops per centimeter is given by:

$$\text{loops/cm of yarn} = nT\,\frac{(100+F)}{100}\cdot\frac{1}{2.54}$$

where n = effective number of filaments, assumed to be 15/2 in a 34-fil yarn, T = turns per inch, and F = percent overfeed. It is clear from these relationships that there is a good agreement between the theoretically predicted trend for loop size and loop frequency with varying yarn twist and percent overfeed with those obtained experimentally by Wray.

Yarn Stability

The usefulness of an air-textured yarn depends a great deal on the stability of the loops formed during the bulking process. In some air-textured yarns, such as those made at high overfeed, even a small amount of hand tension can remove some loops. Once these loops are removed, there is no built-in mechanism for their reformation, as occurs in stretch bulked yarns. The loss of loops during subsequent processing, such as weaving, can be detrimental to fabric quality. The yarn stability is highly affected by the amount of pretwist. At low pretwists, the bulked yarn is generally stringy, loose, and easily extensible because of the low degree of entanglement of long floats of looped filaments. Yarns produced from parent yarns with 10-15 turns/in. (which is within the commercial limit) are most stable. Higher pretwist also tends to yield yarns that are less stable because of the formation of a large number of small loops and lower number of tensioned filaments in a given cross section. (The tensioned filaments are the ones that take up the strains during yarn extension.) Increasing overfeed and air pressure both tend to produce less-stable yarns.

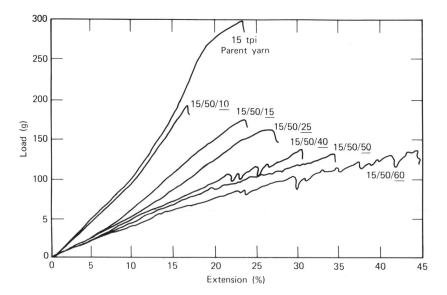

Figure 12.48. Typical load-extension curves (on Instron Tensile Tester) at various overfeeds (Yarn designation–first figure refers to twist per inch; second to % overfeed; and the third to the air pressure lb/in.²). (after Wray)

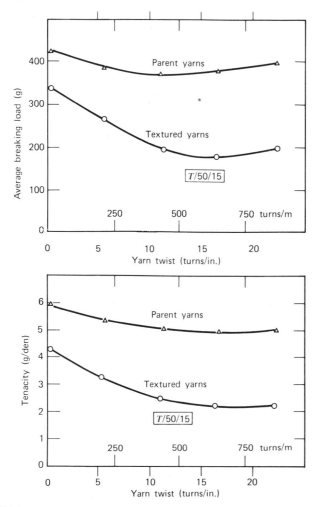

Figure 12.49(a). The effect of twist on breaking load and tenacity (determined on Uster Automatic Single-thread Tester). (after Wray)

Tensile Properties

Air-textured yarns generally exhibit lower tenacity and elongation-to-break than that of the parent continuous-filament yarn before texturing. Processing conditions such as twist, overfeed, and air pressure significantly modify the tensile behavior of air-textured yarns, as shown in Figures 12.48 and 12.49.

Tenacity tends to decrease with increasing twist, overfeed, and air pressure. On the other hand, elongation-at-break decreases with increasing yarn twist and

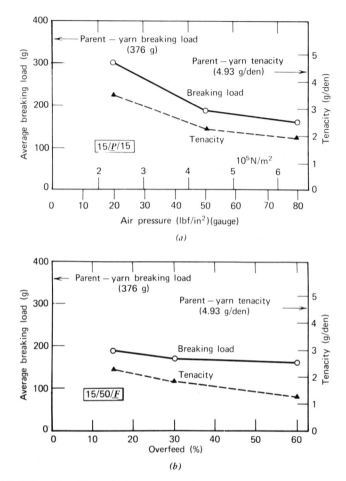

Figure 12.49(b). The effect of air pressure and overfeed on breaking load and tenacity. (after Wray)

increases with increased overfeed and air pressure. It must be remembered that the elongation of air-textured yarns is the result of the extension contribution due to the opening of the overfed loops in addition to the normal elastic extension of the straight filaments.

Air-textured yarns are used in a number of end-use applications, such as apparel, furnishings, and some industrial fabrics.

Mechanically Produced Air-Textured Yarns

A very significant development, which resulted from the observation made by

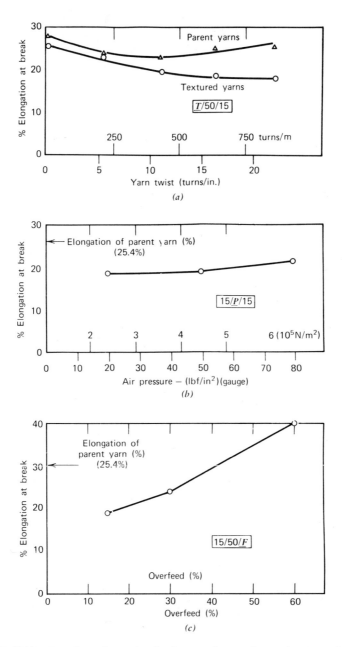

Figure 12.49(c). The elongation at break of textured yarns for varying processing conditions. (after Wray)

Wray (22) concerning the nature of the turbulent wake in the Taslan[®] jet, was the modification of the design of the jet so that it could operate at lower air pressures. The effect of the design modifications of a Taslan[®] jet on its overall performance regarding air consumption and bulking efficiency have been discussed by Wray and Entwistle (24). The major change in the design of the jet is the blocking of the otherwise ineffective secondary air stream (depending on the tilt of the needle) by a plug lodged in the needle on the upstream side. Such a change resulted in savings of as much as 65% in air consumption, with little or no effect on the bulk characteristics of air-textured yarns. It was also observed that, with the use of the modified jet, the feed-needle angle was far less critical for satisfactory texturing performance when compared to the original jet design.

The basis of the improved design of the jet and the observations made of the mechanism of air bulking prompted the development of a new method called the mechanical-bulking method, reported by Sen and Wray (25). This method can be used to produce textured yarns with characteristics similar to the air-jet bulked yarns. A line diagram of the mechanical bulking system is shown in Figure 12.50. The system utilizes purely mechanical means for the untwisting process in

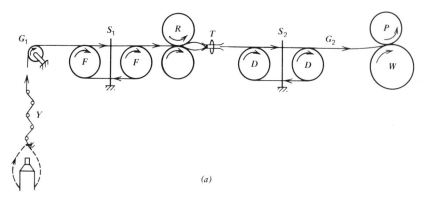

(a)

Figure 12.50(a). Mechanical builking apparatus.

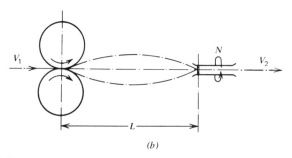

(b)

Figure 12.50(b). Bulking zone.

place of the air jet. The concept of manufacturing air-jet-type textured yarns by mechanical means stems from the observation of the mechanism of bulking during the passage of the yarn through the air jet. The mechanism of the air bulking process as proposed by Wray and later confirmed by Wray and Entwistle is discussed in an earlier section of this chapter. However, the most important aspect of the mechanism of bulking is the temporary untwisting of the pretwisted yarn caused by the rotational nature of the turbulent air stream (resulting in a false-twisting type action) when it is passing through the air jet. When this is combined with the yarn-slackening effect caused by overfeed, it results in the snarling of twist-lively filaments into the loops and coils that are characteristic of air-textured yarns. In the mechanical bulking process, the temporary untwisting of the pretwisted yarn is achieved by mechanical means (false-twist spindle) under conditions of overfeed; consequently, the filaments snarl into loops, and after they have passed through the false-twisting unit, the initial twist reasserts itself, thereby locking the already snarled loops.

In addition to such factors as (1) initial feed-yarn twist, (2) total yarn linear density and number of filaments, and (3) overfeed, the bulking characteristics of the mechanically bulked yarns are also affected by the following parameters: (1) feed-in speed, (2) false-twist spindle speed, and (3) bulking zone length.

The authors suggest that, for the yarn to bulk properly, it must be untwisted to a greater extent than the theoretically required value to bring it to the zero

Figure 12.51(a). Air-textured yarn.

Figure 12.51(*b*).   Mechanically bulked yarn. (Courtesy, Dr. Gordon Wray)

twist state. This is required to prevent the filaments from reverting to the helical configuration instead of the required snarled and looped entangled state. They suggest that approximately 20-25% extra twist is required to produce bulking. Some typical values of processing variables suggested by Sen and Wray (25) for a nylon 66, 110-den (11-1 tex), 34-filament yarn are:

initial twist in supply yarn	15 tpi
overfeed	10%
feed-in speed	60 in./sec
false-twist spindle speed	1125 rev./sec
bulking zone length, L	1 in.

Some typical mechanically textured and air-textured yarns are shown in Figure 12.51; the former clearly resemble air-textured yarns in physical appearance. The physical bulk, tensile properties, and the percent instability* characteristics of a mechanically bulked yarn are essentially similar to those of a Taslan® yarn manufactured from the same parent yarn. The two most important factors that

---

*As defined by Wray (22), percent instability = $e_b - e_p$, where e_b = percentage extension of bulked yarn and e_p percentage extension of parent yarn, both measured at 3 gf/tex (0.33 g/d).

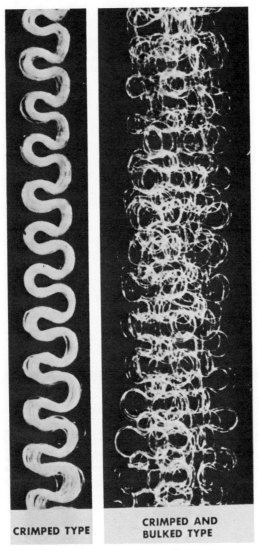

CRIMPED TYPE

CRIMPED AND BULKED TYPE

Figure 12.52. Photomicrographs showing 70 den. -34-fil. knit-de-knit Blue "C" nylon crinkle yarns made via the C-B-Tex process.

affect the bulked yarn characteristics are (1) the overfeed and (2) initial feed yarn twist, and, to a certain textent, (3) the linear density of individual filaments of the feed yarn. Exerting less influence are (1) the bulking zone length, (2) the percentage extra untwisting, and (3) total linear density of the yarn.

Some Additional Methods for Producing Bulked Yarns

Crinkle-Type Textured Yarns

These yarns can be produced by two basically different methods:

Knit-de-knit Method

In this method, the flat yarn is first knit, then heat set and unraveled to produce a crinkle-type structure as shown in Figure 12.52. There are various knitting machines available in the market that can be utilized to manufacture such yarns. The crimp frequency and shape can be varied by varying the needle gauge on the machine and the fabric structure (plain jersey, rib, double jersey, interlock, etc.). The fabrics produced from knit-de-knit crinkle yarns have a pronounced sparkling boucle texture, excellent stretch and recovery from stretch, and full hand. These yarns are torque free and therefore do not require any subsequent heat setting. Knit-de-knit yarns can also be produced by a continuous process called the C-B Tex Crimping process developed by the Textile Machine Works, Reading, Pa., USA. For details the reader is referred to reference 11.

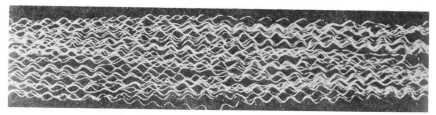

Figure 12.53. Miralon gear-crimped yarn.

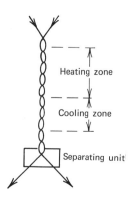

Figure 12.54. Principle of twist texturing.

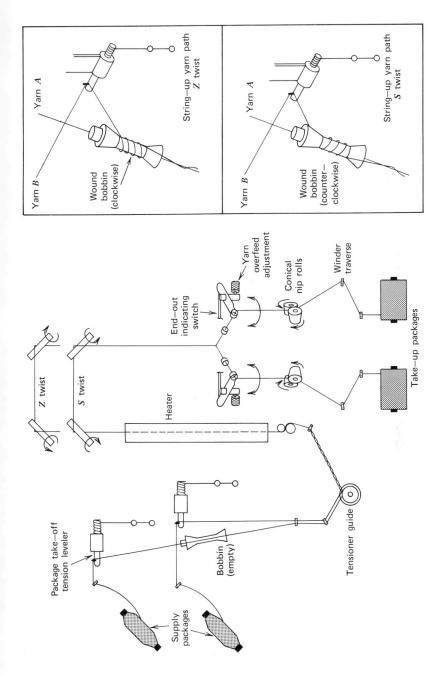

Figure 12.55. Diagram showing yarn path through the Turbo Duotwist machine. Insert shows thread-up bobbin winding detail. (Courtesy, Turbo Machine Co.)

467

Gear Crimping

Bulk can also be produced in a continuous-filament yarn when it is passed through closely meshed gears. The gear head is heated so that the crinkle produced in the yarn is permanent. The frequency and the amplitude of the crinkle or crimp can be varied over a wide range, depending on the type of gears used. One such gear crimping process called Miralon® has been developed by John Heathcoat and Sons, Ltd., England. A gear-crimped yarn has the appearance shown in Figure 12.53. These yarns are used in a variety of end-use applications, such as ladies' and children's knitted outerwear, sweaters, and ladies' blouses.

Twist-Textured Yarns

If two ends of yarn are twisted together around a common axis, rather than each yarn being twisted around its own axis, and the configuration is then heat set in the twisted state and finally untwisted, the yarn thus produced possess excellent bulk. The principle of twist-textured yarns is shown in Figure 12.54. This is a very simple and revolutionary concept of introducing texture into thermoplastic continuous-filament yarns. This is because this method enables the texturing of fine-denier yarns in the range of 15-70 denier at very high rates of

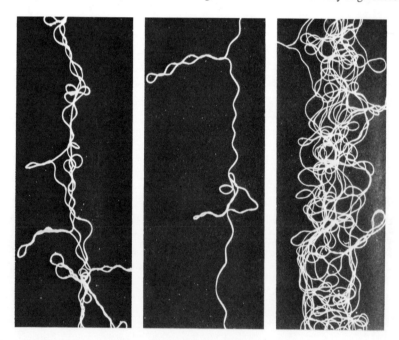

Figure 12.56. Nylon yarns textured on Turbo Duotwist.

production. Moreover, the yarns textured by this method have excellent dyeing uniformity, high cover, and extremely soft, smooth hand. Twist-textured yarns find uses in tricot fabrics, hosiery, and all types of knitted outerwear structures.

Twist-textured yarns are produced on the Turbo Duotwist Machine (Turbo Machine Co., Lansdale, Pa., USA) in addition to various others available commercially. A line diagram of the Turbo Duotwist Machine is shown in Figure 12.55. The twist in the yarn is first introduced and sustained throughout the operation because the yarn is wound around a bobbin (shown in the diagram). The twisted yarn is then passed through a heater (ranging in temperature from 110 to 240°C, depending on yarn denier and processing speed) and then separated before being wound onto separate packages. The yarns can be given either S or Z twist. Yarns produced by this method have the appearance shown in Figure 12.56. Typical yarn processing speeds used in twist texturing range up to 350 m/min, depending on the type of yarn and the heating efficiency.

REFERENCES

1. Finnie, T. A., papers presented at the Fall Technical Conference, November 9, 1973, Textured Yarns Association of America, Inc.
2. *Knitting Times,* November 1971.
3. Arthur, D. F. and A. F. Weller, "The Principles of Friction Twisting," *J. Text. Inst.,* 1960, **51**, T66.
4. Thwaites, J. J., *J. Text. Inst.,* 1970, **61**, T116.
5. El-Behery, H. M., *Clemson University Review of Industrial Management and Textile Science,* 1971, 13.
6. Arthur, D. F. and C. R. Jones, *J. Text. Inst.,* 1962, **53**, T217.
7. Morris, W. J. and A. S. Roberts, Bulk Stretch and Texture, Proc. Annual Conference of Text. Inst., 1966, **54**.
8. Denton, M. J. and W. J. Morris, *The Setting of Fibres and Fabrics,* edited by J. W. S. Hearle and L. W. C. Miles, Merrow Technical Library, 1971.
9. Jones, C. R. and J. Porter, *J. Text. Inst.,* 1965, **56**, T498.
10. Burnip, M. S. and J. W. S. Hearle, *Man-Made Textiles,* 1960, **37**, 54-57.
11. *Textured Yarn Technology,* vol. I, Monsanto Co., 1967.
12. Krause, H. W., Bulk Stretch and Texture, Proc. Annual Conference of Text. Inst., 1966, **54**, 147.
13. Burnip, M. S., J. W. S. Hearle, and G. R. Wray, *J. Text. Inst.,* 1961, **52**, P343.
14. Mebane, G. Allen, papers presented at the Fall Technical Conference, Nov. 9, 1973, Textured Yarn Association of America, Inc.
15. Mauretti, J., *Knitting Times,* 1974, **53**, 58.
16. Mattingley, D. A. E., *Textile Month,* 48, March 1974.
17. Klein, W. and A. Trummer, *Text Mfr.,* 16, January 1973.
18. Mey, W., Bulk Stretch and Texture, Proc. Annual Conference of the Textile Institute, 1966, 77.

19. Fitzgerald, W. E. and G. B. Hughey, *Am. Dyestuff Reporter,* 37, 1966.

20. Wray, G. R., Bulk Stretch and Texture, Proc. Annual Conference of the Text. Inst. 1966, 18.

21. Rozimarynowska, M. K. and J. Godek, Bulk, Stretch and Texture, Proc. Annual Conference of the Textile Institute, 1966, 29.

22. Wray, G. R., *J. Text. Inst.,* 1969, 60, T102.

23. E. I. du Pont Technical Bulletin X-154, October 1961.

24. Wray, G. R. and J. H. Entwistle, *J. Text. Inst.,* 1968, 59, T122.

25. Sen, H. and G. R. Wray, *J. Text. Inst.,* 1970, 61, T77.

26. Arthur, D. F., in *Modern Yarn Production from Man-Made Fibers,* G. R. Wray (Ed.), Columbine Press, London, 1960, 100.

27. Denton, M. J., *Textile Month,* 1969.

Appendix A

Recommended **SI Units** for Textiles

Property	SI Unit	Abbreviation	To convert to SI Units, multiply value in unit given by factor below:	
Length	millimetre	mm	inch	25.4
	metre	m	yard	0.914
	centimetre	cm	inch	2.54
Width	millimetre	mm	inch	25.4
	centimetre	cm	inch	2.54
Test or gauge length	millimetre	mm	inch	25.4
Thickness	millimetre	mm	inch	25.4
Linear density	tex	tex	Reference should be made to B.S. 947: 1970	
	millitex	mtex		
	decitex	dtex		
	kilotex	ktex		
Diameter	micrometre (micron)	μm	$\frac{1}{1,000}$ inch (mil)	25.4
	millimetre	mm	inch	25.4
Threads in cloth:				
length	number per centimetre	picks/cm	picks/inch	0.394
width	number per centimetre	ends/cm	ends/inch	0.394
Warp threads in loom	number per centimetre	ends/cm	ends/inch	0.394
Stitch length	millimetre	mm	inch	25.4
Courses per unit length	number per centimetre	courses/cm	courses/inch	0.394
Wales per unit length	number per centimetre	wales/cm	wales/inch	0.394
Cover factor (woven fabrics)	(threads per centimetre) $\sqrt{\text{tex}} \times 10^{-1}$	(threads/cm) $\sqrt{\text{tex}} \times 10^{-1}$	$\dfrac{\text{(threads/inch)}}{\sqrt{\text{cotton count}}}$	0.957
Mass per unit area	grammes per square metre	g/m^2	oz/yd^2	33.9
Twist	turns per metre	turns/m	turns/inch	39.4
	turns per centimetre	turns/cm	turns/inch	0.394

	(turns per centimetre) $\sqrt{\text{tex}}$	(turns/cm) $\sqrt{\text{tex}}$	$\dfrac{\text{(turns/inch)}}{\sqrt{\text{cotton count}}}$	9.57
Twist factor (or multiplier)			$\dfrac{\text{(turns/inch)}}{\sqrt{\text{worsted count}}}$	11.70
Breaking load	millinewton	mN	gf	9.81
	newton	N	kgf	9.81
Tearing strength	newton	N	lbf	4.45
Tenacity	millinewtons per tex	mN/tex	gf/den	88.3
Specific stress	millinewtons per tex	mN/tex	gf/den	88.3
Bursting pressure	kilonewtons per square metre	kN/m²	lbf/in²	6.89
Bending rigidity	millinewtons square millimetres	mN mm²	gf mm²	9.81

The Textile Institute

Index